The Antecedents of Man

W. E. LE GROS CLARK

THE ANTECEDENTS
OF MAN

An Introduction to the Evolution
of the Primates

THIRD EDITION

Chicago
QUADRANGLE BOOKS

First published 1959 by Edinburgh University Press.
Revised editions 1962, 1971.

Library of Congress Catalog Card Number: 73-162309
SBN 8129-0224-6

Preface to the First Edition

IN 1953 I had the honour of delivering the Munro Lectures at the University of Edinburgh under the title *The Palaeontology of the Primates and the Problem of Human Ancestry*. This book contains the substance of the lectures, somewhat differently arranged, and amplified by reference to additional discoveries which have accumulated during the last six years. The general presentation of the subject has also been adapted for use as an introductory textbook which is intended for students of whom some may have no special knowledge of comparative anatomy. Publication has been assisted by a generous grant from the Munro Lectureship Committee.

The treatment of my subject follows the main lines of an earlier book, *The Early Forerunners of Man*, published by Baillière, Tindall and Cox over twenty years ago, a book which has for a long time been out of print (partly because of the destruction of available stocks in the bombing raids of the last war), and which in any case is now long out of date. During the last quarter of a century there have been so many important accessions to the fossil record which bear on the lineage of mankind, as well as a considerable modification and clarification of conceptions regarding evolutionary processes which were once widely held, that it has been necessary to replace *The Early Forerunners of Man* by a book which not only incorporates the new factual data but also offers interpretations of the fossil evidence in accordance with palaeontological principles as they are now understood. I am glad to take the opportunity of thanking Messrs Baillière, Tindall and Cox for readily agreeing that I should make use of some of the illustrations in my previous book. Many of the illustrations in the present book, however, are original, and for these I am indebted to the artistic skill of Miss Christine Court.

Any author who composes a textbook which, though essentially introductory in nature, covers a wide field, necessarily owes much to the writings and researches of others. The present author feels heavily in debt to many authors past and present, and also to many of his colleagues, for sources of information which they have directly or indirectly provided. It is his sincere hope that this debt is adequately acknowledged either by specific allusions or by the references to treatises with comprehensive bibliographies.

The interpretation of fossil material (particularly when this is fragmentary) is often a matter for much discussion and argument, and some of the provisional interpretations of Primate evolution offered in this book will probably not find universal agreement. But as more fossil material accrues, uncertainties and disagreements have tended to become less prevalent. Even so, it may well be, of course, that new discoveries of more ample palaeontological material will require modifications of some of the views expressed by the present author. However, this is a prospect which is envisaged as a natural expectation by every student of the fossil record, and is likely to remain so for many years for the simple reason that well-preserved fossil remains are only by rare chance discovered. It needs to be recognized, indeed, that in the case of most groups of animals, and in spite of notable accessions which accumulate from year to year, the fossil record at present available is still very far from complete.

W. E. LE GROS CLARK, OXFORD 1959

Preface to the Third Edition

SINCE the second edition of this book appeared in 1962 many significant new observations and discoveries have contributed important new information bearing on the problem of human ancestry. There have also appeared a number of memoirs in which all or most of the fossil material available in different museums has been re-studied, leading to re-evaluations and re-assessments of the affinities of fossil Primates in relation to each other and to their modern successors. These careful and comprehensive studies have led not only to a clarification of Primate taxonomy and phylogeny, but also, in a number of cases, to nomenclatural changes. Clearly, therefore, a considerable degree of revision has been required for a new edition. It should be emphasized that, as its subtitle indicates, this book is primarily intended to give the reader a broad perspective within the framework of which he can add further details by more extensive and intensive studies. In spite of all the additional information that has accrued in recent years, and in spite of the ever growing volume of literature bearing on the main theme of the antecedents of man, the temptation to expand the book in its third edition has been avoided as far as possible. Nor has it seemed desirable to make reference to every one of the species and genera of modern and extinct Primates that from time to time have been enumerated; such a catalogue would defeat the introductory intentions of the book and transform it into an unwieldy compendium. It is also not feasible to include in the bibliography references to all the books, monographs and scientific papers dealing with the subject of what is now called Primatology that have proliferated in increasing abundance since the last edition appeared seven years ago. But it is hoped that such references as are listed will—each with its own bibliography—help those who wish to acquire a deeper and wider knowledge of the subject.

W. E. LE GROS CLARK, OXFORD 1969

vii

Contents

1. The Evolutionary Process and the Primates   1

2. A Preliminary Survey of the Primates in Space and Time   37

3. The Evidence of the Dentition   76

4. The Evidence of the Skull   126

5. The Evidence of the Limbs   170

6. The Evidence of the Brain   227

7. The Evidence of the Special Senses   265

8. The Evidence of the Digestive System   286

9. The Evidence of the Reproductive System   299

10. The Evolutionary Radiations of the Primates   315

Bibliography   361

Index   369

The Antecedents of Man

I

The Evolutionary Process
and the Primates

THE general conception that the more elaborately organized forms of life have evolved from simpler forms may be regarded as fairly established in the minds of biologists to-day, and indeed in the minds of most educated people. The evidence for such a thesis has accumulated in overwhelming detail since the time of Charles Darwin, and no alternative interpretation for this evidence has ever been offered which is in any way convincing or even plausible. It has been presented so fully by many well-qualified authorities, and is so readily available in current publications, that there is no need to recount it here except in relation to our main theme—the evolutionary history of the order of Primates. This order, one of the eighteen orders of living mammals which are now generally recognized, includes our own species, *Homo sapiens*, which according to our usual way of thinking represents the culminating peak of Primate evolution. In a sense such a point of view is justified, for all the evidence at present available indicates that *H. sapiens* was one of the last species, and certainly the most successful, of the whole Primate series to become differentiated. But it is not to be inferred therefrom that *H. sapiens* was, so to speak, the *objective* of Primate evolution, for there is no reason to suppose that the course of evolutionary change was a steady and continuous progression strictly confined in one direction—a predetermined direction leading to man. Any evolutionary sequence, considered in retrospect as it is displayed by the fossil record, might be arbitrarily construed as an inevitable trend directed by some unknown intrinsic agency, a conception to which the term ORTHOGENESIS has been applied. Such an appearance, however, is illusory and takes no account of the abundant ramifications which have characterized the evolution of every group of animals. Most of the ramifications were, in fact, rather short-lived (in terms of geological time) and became extinct. These were the unsuccessful lines of evolution, and it is only by ignoring them and limiting attention narrowly to the successful lines that an impression is gained of an orthogenetic trend.

Within the limits of the Primates the general scheme of evolutionary differentiation which ultimately led to the emergence of man himself has been discussed by many authorities. But the precise route by which the human family (that is to say, the family Hominidae of zoological nomenclature) came into existence, or what may have been the nature of its earlier progenitors, remains still a matter of conflicting opinions. It is agreed by biologists that the various living forms which are grouped together in the accepted scheme of classification as the order Primates represent divergent modifications of a common ancestral type, and that the latter primarily arose from a basal generalized stock which also provided the foundation for the development of other mammalian orders. From such an ancestral type there diverged in early geological times a number of separate groups each of which, while coming into the category of Primates by reason of a common general plan of organization, incorporated tendencies for evolution along its own subsidiary lines. These tendencies culminated in the development of distinctive patterns of structural characters superimposed, as it were, on the general plan. From one of these subdivisions, at some phase of geological time which still remains uncertain, the progenitors of *H. sapiens* arose, and in order to determine with some degree of probability the nature and origin of these progenitors it is first necessary to trace back the Primates to their evolutionary source, and then to mark out as far as the evidence permits the stages at which divergent trends of evolution became manifested within the group. The construction of genealogical trees purporting to demonstrate the evolutionary origin of *H. sapiens* is a common feature of popular scientific exposition and, because of the subject with which they deal, they naturally attract a considerable amount of interest. In fact, however, the origin of man forms only one item (though of course a supremely important one) in the more comprehensive problem of the development of the whole group of which he is a representative of a single family. We are here concerned with this major issue, recognizing that man's position in the scheme of things is likely to become more clear if it is viewed against the background of the evolution of the Primates as a whole.

It is mainly by reference to the data of comparative anatomy, blood reactions, and protein structure, and a study of the fossil record (PALAEONTOLOGY) that the nature of the genetic relationships between different subdivisions of the Primates has been determined, and that inferences can be drawn regarding their probable evolutionary history. It is true to say that the ultimate solution of these problems of PHYLOGENY (that is, problems relating to the study of the lineage and relationships of any natural group of animals) must rest almost entirely on palaeontology, for only when the geological record becomes sufficiently well documented by the discovery of the actual fossil remains of closely graded intermediate phases of development will it be possible to establish evolutionary sequences with any assurance. But even where the palaeontological sequence is still incomplete, sufficient data may have accumulated from the study of living forms and fossil remains from which it is possible to advance reasonable hypotheses on phylogenetic history. Thus, an incomplete fossil record may be adequate for demonstrating the main evolutionary trends which have characterized a major taxonomic group as a whole, although not yet adequate for establishing the precise line of descent, step by step, of one particular species or genus within the group. So far as the evolution of the Primates is concerned, it is still necessary to rely to a large extent on hypotheses based on an inadequate fossil record, particularly in discussions on the origin and differentiation of some of the earlier representatives of the order. But hypotheses are not to be disparaged because they are hypotheses; on the contrary, they are quite necessary preliminaries in the pursuit of any scientific enquiry, for only by the construction of hypotheses is the opportunity given of putting them to the test by further observation and discovery. 'The accrescence of interpretation and theory is as essential to the achievement of knowledge as is the accumulation of the raw materials from which these arise, and isolated observations of fact have little value until some progress has been made toward their integration' [Simpson, 146].

COMPARATIVE ANATOMY

One of the main fields of study from which evolutionary relationships have been inferred is that of comparative anatomy. But the evidence therefrom is, of course, indirect evidence, and is only open to verification or otherwise by the evidence of the fossil record when the latter accumulates in sufficient quantity. With the continued accession of discoveries of fossil remains it is perhaps not a little surprising that the inferences based on the indirect evidence of comparative anatomy are so commonly substantiated. Certainly this has been the case with the study of Primate phylogeny. Let us now examine the nature of this indirect evidence and the principles on which it is to be assessed.

Briefly, degrees of genetic affinity between existing (and also fossil) forms may be broadly determined by noting degrees of resemblance in anatomical details. Having defined a composite group of animals which, because they are constructed on a common general plan of anatomical design, are presumed to be closely related, it is possible by a sort of mental trianulation to postulate the approximate nature of the ancestral stock from which the component elements of the group may have originated in the evolutionary sense. Bearing in mind the variously modified characters which distinguish the component elements of such a natural group, and having a general conception of the direction of evolutionary development from the primitive or generalized to the advanced or specialized, we can hypothesize a morphological common denominator from which it is readily comprehensible that the diverse modifications in the existing group may have been derived. This method of reasoning is actually no more than exploratory, providing preliminary hypotheses of ancestor-descendant relationships which require to be tested by more rigorous enquiries, and it involves several important considerations.

In the first place, the question arises—how can we distinguish between anatomical features which are essentially primitive or generalized and those which are specialized or advanced? That is to say, in comparing homologous structures in different groups of animals, how far is it possible to infer that the type of structure found in one group represents (or closely approximates to) the prototype of which the other is derivative—that in the course of evolution the one represents the antecedent and the other the sequential? As a matter of practical experience, it has been found that this can usually be done (1) by reference to existing series of animals (*échelle des êtres*) which appear to form an ascending scale of organization leading from the more simply designed to the more elaborately designed forms, the assumption being that, in general, the former are those which have retained primitive features; (2) by reference to the early fossil representatives of the group under consideration, which appear to form the ancestral basis for its subsequent development; (3) by a detailed analysis of the anatomical structures concerned; and (4) by reference to their embryological development. Thus, for instance, PENTADACTYLY (that is, the possession of five digits on the extremities) is judged to be the primitive, ancestral condition among mammals, which has in some cases been replaced by specialization following the loss of one or more digits. For in living mammals of a relatively simple organization, and in the Reptilia (from which mammals were originally derived), pentadactyly is the rule; the fossil record has demonstrated that in the early precursors of those mammals which to-day have less than five digits pentadactyly was also a characteristic feature; and, lastly, a detailed study of mammals with less than five digits reveals in the adult vestigial or rudimentary remains of those which have been lost during the course of evolution or, in the embryo, transient traces of these ancestral structures.

In many cases the distinction between a primitive and an advanced trait is so obvious that it hardly requires discussion. The evolutionary history of the brain has clearly been marked by a progressive elaboration and increasing complexity of its higher functional levels (especially of the convoluted cerebral cortex), and in comparing two brains of a phylogenetic series there can rarely be any doubt which is the more primitive and which the more advanced. In some instances, however, there may be difficulty in distinguishing between the antecedent and sequential of two structural modificat ons, or only after protracted enquiry and prolonged discussion doei it become possible to affirm the direction of evolutionary chanse. This has been the case, for example, with certain features of th gdentition. The molar teeth in different mammals have a variabe number of enamel-covered tubercles, or cusps. At one time a theory was advanced that the primitive type of mammalian molar was furnished with a large number of cusps and that the less complicated molars of many existing forms are the result of a secondary simplification following on the suppression of some of the cusps. This was termed the multitubercular theory of molar evolution. Another theory, the tritubercular theory, involved the proposition that the primitive mammalian molar developed three main cusps and that progressive and divergent modifications proceeded rather by the addition of a varying number of new cusps. Without elaborating on this over-simplified reference to the two theories, it may be noted that the accumulation of evidence from comparative anatomy and palaeontology has finally established that the tritubercular theory is broadly correct. So far as the Primates are concerned, therefore, it may be assumed that molars with numerous cusps are the result of specialization (and this is so even if they occur in some geologically early forms), while tritubercular teeth represent a primitive and ancestral condition (even if they persist to the present day in some species, e.g. the small Primate, *Tarsius*.)

It is important to recognize that the progressive evolutionary modification of a morphological character is not necessarily associated with its increasing complexity. The reverse may occur, and in an antecedent phase of its evolution the character may actually appear to be more elaborately constructed than in a subsequent phase. Failure to recognize this has sometimes led to misleading conclusions. It is not always easy to decide finally (without a good fossil record) whether in some instances we are dealing with a truly primitive and ancestral feature, or with a secondary regression which only simulates one, but in such cases a close scrutiny will often resolve doubts. For example, the canine tooth in *H. sapiens* is a small, spatulate tooth which shows very little differentiation in comparison with the adjacent incisor teeth. It has actually been argued that this is more primitive than the condition in the anthropoid apes in which the canines are large, projecting, and pointed teeth. But, as we shall see, the real nature of the human canine as a tooth which has secondarily undergone some degree of reduction is betrayed by certain characters such as the length and stoutness of its root (which seem out of proportion to the functions it performs), its relative lateness in eruption, and the fact that the newly erupted tooth may occasionally be sharply pointed and somewhat projecting. The fossil evidence also makes it apparent that in the extinct hominids of earlier times the canine tooth was commonly more strongly developed than it is in modern man. It has been assumed by some authorities that the flattened nails characteristic of the Primates are more primitive than sharp claws just because the latter have a more elaborate and differentiated form. Here, again, the evidence of comparative anatomy and palaeontology, as well as a detailed analysis of their basic structure, shows clearly that nails have been derived secondarily from claws. In the brains of apes and monkeys there is a very distinctive folding of the cerebral cortex which is often called the *simian sulcus*. A superficial study might suggest that this is a specialized trait peculiar to apes and monkeys, and one that has been avoided in the evolutionary development of the human brain (which in this particular feature, therefore, would appear to be less specialized). In fact, however, traces of the sulcus are commonly to be found also in the human brain (though they are extremely variable—a characteristic feature of degenerative structures in general), and in the brains of some human races it may on occasion be almost as well developed as it is in the anthropoid apes. It can be inferred, therefore, that the simian sulcus is a primitive element of the brain of the higher Primates and that it has undergone a secondary retrogression in the modern human brain.

These examples have been quoted in illustration of the principles on which it may usually be determined which of two differently constructed homologous structures is more likely to be prototypic in respect of the other. In general, it is probably legitimate to assume that an anatomical character which in its simple construction has the appearance of being primitive and generalized is really so—provided it displays no details in its construction, development, or in the fossil record of evolutionary history, which point to a secondary simplification or retrogression.

It is of some importance to note that an extremely primitive condition in one anatomical system may not infrequently be associated with a high degree of specialization in another in the same species. In a sense this is to be expected, for the one serves as a sort of functional counterbalance to the other. But it has sometimes led to doubt being expressed whether the primitive features are not, after all, the result of secondary simplification since (it is argued) truly primitive characters are not to be expected in highly specialized animals. Reasoning of this kind, however, is not valid. *H. sapiens*, for example, has been able to preserve many remarkably generalized anatomical characters in virtue of the advanced development of his brain; indeed, this is to some extent true of the order Primates as a whole.

These considerations of the distinction between generalized and specialized, between primitive and advanced, are relevant to a principle which has been termed the *irreversibility of evolution*. Since evolution involves progressive morphological changes dependent on a succession of selective factors in the environment, and since the selective factors as well as the genetic basis of the morphological changes may be highly complex, the probabilities are all against a total reversal of evolution in the sense that an evolving group can, so to speak, retrace its steps and revert to a primitive organization precisely similar in all respects to that of the ancestral stock from which it was originally derived. Particularly is this the case if structural changes are related to (and responsible for) functional specializations which confine the species to a narrow and restricted environment. Most evolving groups tend to become more and more specialized in adaptation to one particular mode of life, and as they do so the possible variations which might be of use to them become progressively limited. Eventually, the only evolutionary changes open to them simply push them along the path which they have already pursued [30]. This process of directional evolution has been termed ORTHOSELECTION, and is not to be confused with the idea of orthogenesis which invokes inherent (but quite unknown) directional tendencies infused by some mysterious agency into the germ plasm of the evolving organism and impelling it along a predetermined course of evolution. As a general principle it is no doubt legitimate to discount the possibility of an evolutionary reversal if the morphological divergence from a more primitive condition can be assumed to be the cumulative effect of a long sequence of small mutational variations exposed to selective influences of a complex nature, or if there is reason to suppose that the genetic basis of the morphological characters concerned is so complex that an evolutionary reversal could occur only as the result of a multiplicity of mutational reversals. But the direction of the evolutionary development of single, individual characters having a relatively simple genetic basis can certainly be reversed, for mutational processes have actually been shown by genetic studies in the laboratory to be reversible. There are also many examples from the fossil record which show that individual morphological or metrical characters may undergo an evolutionary reversal; for example, the dimensions of a tooth may increase and subsequently undergo a secondary reduction, or a change in the limb proportions may show a trend first in one direction and subsequently in another. Instances of the loss of a character previously acquired are common features of evolution and may be termed *negative reversals*. On the other hand a *positive reversal*, that is to say the reacquirement of a complicated or composite morphological character in its exact form after it has been lost in the course of evolution, must surely be a rarity.

The possibilities of retrogression and negative reversals in evolution need to be thoroughly recognized, for some anthropologists have attempted to apply the general concept of the irreversibility of evolution rigidly to the smallest and simplest characters, and on this basis have then proceeded to argue that a particular extinct group could not be considered ancestral to some modern group because the former had developed some individual character which is not possessed by the latter. Thus, for example, it has been suggested that the fossil hominid for many years known by the generic name *Pithecanthropus*, but now regarded as a species of the genus *Homo*, i.e. *H. erectus*,* is debarred from consideration as the progenitor of the species *H. sapiens* because it was characterized by massive supra-orbital (brow) ridges which are absent in *H. sapiens* and were probably absent in the earlier evolutionary precursors of the Hominidae. But there is no theoretical or practical reason why, in the hominid sequence of evolution, the development and subsequent disappearance of prominent supra-orbital ridges should not be correlated with changing proportions of the jaw and braincase (see also p. 21).

While it is important to avoid such misapplications of the general concept of evolutionary irreversibility, it is equally important to recognize that the concept is valid for sustained and long continued phylogenetic trends directed and controlled by the selective action of the environment, for it leads to certain corollaries which have reference to phylogenetic studies. Thus it may be stated that the ancestral type from which a given natural group of animals may have arisen is likely to have been at least as primitive and generalized in its morphological organization as the most primitive member of that group. Conversely, no complex of anatomical structures in any species is likely to be more primitive in its total morphological pattern than the homologous structures of its evolutionary progenitors. For example, the most primitive type of brain among the existing monkeys is that of the marmoset, e.g. *Callitrix* (*Hapale*). It may be inferred, therefore, that in the ancestral stock from which the New World monkeys were derived the brain was probably at

---

* For convenience of reference, a generic name that has been changed in recent years will, where it seems expedient to do so, be placed in brackets following that now in use.

least as primitive. Again, an extinct group of lemurs which is more specialized in dentition and cranial characters than the generality of the modern monkeys is unlikely to have provided a basis for the evolutionary development of the latter. Thus, by reference to the pattern of organization of a number of anatomical systems, it is possible to construct a mental picture which bears some approximation to the general make-up of ancestral types, to differentiate those fossil types which probably represent aberrant or divergent offshoots of the main line of descent from those which may occupy a position on or close to the main line, and thus to gain some conception of the most likely scheme of evolution for any particular group of animals.

In the absence of an extensive fossil record, much of the evidence on which students have had to depend for a provisional reconstruction of the phylogenetic relationships of the Primates has been derived from the study of the comparative anatomy of existing (or Recent) types and it is therefore necessary to consider in somewhat more detail the nature and validity of this sort of evidence. As already noted, the unity of design shown in the basic anatomical structure of different types of animal (even when they appear superficially to be rather strongly contrasted) suggests a genetic relationship, and the closer the relationship the more numerous and far-reaching are the structural resemblances. The fact that most morphological characters are adaptive in function, relevant, that is to say, to some environmental need, is no objection to this thesis, for it is well established that species which are not very closely related may become sufficiently well adapted for the *same* requirements of the environment by making use of homologous characters of *different* design. The 'opportunism of evolution' was a phrase used by Simpson [147, 148] to express the fact that the evolutionary process by no means requires the development of a character in such a form that it is the absolute acme of perfection, and the only possible form for the function it is required to serve —there are functions which can be adequately served by making use of a structural modification in various forms. In other words, there may quite commonly be several possible solutions to an adaptational problem. The salient point of Darwin's exposition of natural selection is that advantage is taken of such heritable variations as by chance occur, and as a result of this process something is eventually built up which is 'good enough' for the species so far as its survival value is concerned. Thus different groups of animals may independently develop structures which are functionally equivalent but differ considerably in their basic pattern, and the chances are remote that the types which are not closely related will in this way acquire precisely similar patterns. This principle of opportunism may be illustrated by reference to the evolution of the cusp patterns of teeth. For example, the basic pattern of the first lower molar tooth in *Homo* is made up of a characteristic group of five main cusps with a complex of intervening grooves and associated crests, all disposed in quite a well-defined pattern, and they form an uneven grinding surface by which the food is triturated against the corresponding upper teeth. But it is not to be supposed that the molar would be any less efficient for performing its elementary grinding function if it were composed of an equivalent series of similar cusps grouped in a somewhat different pattern. Yet the basic pattern in human molars is identical with that of the large apes, so much so that in certain cases it may be difficult to distinguish the two. The identity of pattern is therefore presumed to be based on inheritance from a common ancestor, or at least (what amounts to much the same thing) the inheritance of a genetic mechanism conferring a common potentiality for developing the same pattern. In this particular example, the inference has been abundantly confirmed by collateral evidence derived from various sources, as well as by the evidence of the fossil record. By contrast, the cusp pattern of the molars of the Old World (cercopithecoid) monkeys, whose relationship to hominids and the anthropoid apes is much less close than either of these is to each other, is very different, and yet there is no reason to suppose that the dietary requirements in some of these monkeys differ from that of the anthropoid apes to the extent that such a different cusp pattern is necessitated. It is also the case that among the lower Primates the cusp pattern of the teeth may vary considerably between natural groups not immediately related to each other, even though their dietary habits appear to be quite similar.

While it may be broadly accepted that, as a general proposition, degrees of genetic relationship can be assessed by noting degrees of resemblance in anatomical details, it needs to be emphasized that morphological characters vary considerably in their significance for this assessment. Consequently it is of the utmost importance that particular attention should be given to those characters whose taxonomic relevance has been duly established by comparative anatomical and palaeontological studies. This *principle of taxonomic relevance* is rather liable to be overlooked, particularly in the uncritical attempts which are sometimes made to quantify degrees of relationship by statistical comparisons of isolated measurements. Each natural group of animals is defined by a certain pattern of morphological characters which its members possess in common and which have been found by detailed studies to be sufficiently distinctive and consistent to distinguish its members from those of other related groups. The possession of this common morphological pattern can be taken to indicate a community of origin (in the evolutionary sense) of all the members of the group. But as a sort of fluctuating veneer superimposed on (and sometimes partly obscuring) the common morphological pattern, there may be a number of characters, sometimes obviously adaptive, which not only vary widely within the group but overlap with similar variations in other groups. Such fluctuating characters may be of importance for distinguishing one species from another within the limits of a single family, but they may be of no value by themselves for distinguishing this family from related families. In other words, they are taxonomically relevant for assessing interspecific relationships, but taxonomically irrelevant so far as interfamilial relationships are concerned. The principle of taxonomic relevance may be illustrated rather crudely by reference to the Tasmanian wolf, *Thylacinus*. It would be easy to select a number of dimensional characters of the skull of this animal in which they correspond far more closely to an ordinary dog's skull than they do to (say) a cat's skull. But these particular characters have no taxonomic relevance for comparisons of this sort since other distinguishing, and much more fundamental, features show at once that the Tasmanian wolf belongs to a totally different group of mammals, the marsupials, and, in fact, the dog and cat are far more closely related to each other than either is to the Tasmanian wolf. As another example we may refer to the extinct hominid, *H. erectus*. The available evidence indicates that in this species the morphological features of the skull and jaws are very different from those of *H. sapiens*, while the limb skeleton (so far as it is known) is hardly distinguishable. Clearly, therefore, if the question arises whether a newly discovered type of ancient hominid is more closely related to *H. erectus* or *H. sapiens*, the morphological features of the skull and jaws are the relevant characters to which attention should be primarily directed.

Reference has been made above to the phrase 'total morphological pattern'. It is desirable to stress this concept of pattern in order to emphasize that, in using purely morphological criteria, the assessment of phylogenetic relationships must be based, not on the comparison of individual characters in isolation one by one, but on a consideration of the total pattern which they present in combination. By total morphological pattern is meant the integrated combination of unitary characters which together make up the complete functional design of a given anatomical structure. Many of the conflicting opinions expressed by comparative anatomists on the relationship of one group of Primates with another have been the result of the separate comparison of individual characters treated as though they were isolated abstractions. But it must be realized that individual components of a total pattern may undergo modifications as the result of a simple gene change without disrupting the pattern as a whole. Also, if isolated homologous features alone are compared in two groups, superficial resemblances may tempt conclusions regarding structural affinity which would certainly not be justified if these features are components of completely different basic patterns in each group.

It is clear from what has just been said of the principle of taxonomic relevance that the assessment of systematic affinities involves the selection, and appropriate weighting, of those characters which practical experience in systematics has demonstrated to be the most reliable. In the past, attempts have been made to divide anatomical characters into two categories—adaptive and non-adaptive, the former bearing an obvious and direct relation to some functional demand which itself is dependent on the particular circumstances of the environment. Adaptive characters, therefore, were regarded as of less taxonomic value in so far as it was supposed that they may be developed independently in unrelated forms which happen to be subjected to similar environmental forces. A non-adaptive character, on the other hand, was presumed to bear no direct relation to function and to be, therefore, an incidental feature in the structural composition of an animal which has no survival value for the latter. Such a character, it was supposed, might be expected to persist in the face of gross changes in the environment while other characters of less morphological stability would show wide modifications. However, it is now well recognized that it is quite impracticable to attempt to draw a sharp distinction between adaptive and non-adaptive characters. In the first place, the fact that it is not possible to attribute a functional significance to a character may merely mean that its functional significance is not yet determined. Secondly, it has been seriously questioned whether any character is non-adaptive in the sense that it has no relation to function at all; even if *by itself* it has no selective advantage in evolution it may be genetically correlated with some other phenotypic character which has.* For example, differences in the sutural patterns of the skull bones in different groups of animals appear by themselves to have no functional significance, but they may be the secondary result of changes in growth rate of the skull bones related to the development of cranial characters which are otherwise directly adaptive. In fact, a phenotypic character which by itself has no obviously direct selective advantage may much more easily undergo a rapid change by mutational variation than one which is directly adapted to a special environment, for any sudden disturbance of the second type of character might be presumed to place the animal at an immediate disadvantage. In other words, it might be expected that, at any rate in a group of animals which have become functionally specialized for a given environment, obviously adaptive characters would actually be more stable and therefore of more importance for assessing affinities. Apart from such considerations, the factor of functional specialization always needs to be taken into account in construing the morphological evidence for evolutionary pathways, and in trying to distinguish between evolutionary probabilities and evolutionary possibilities. For example, from an analysis of the mechanics of posture it appears more likely that the erect bipedal posture characteristic of the Hominidae was developed from the semi-erect posture of a small, lightly built anthropoid ape than from a heavily built large ape like the gorilla, for in the latter the centre of gravity of the body has been shifted forward to such an extent by the massive development of the shoulders and arms that raising the trunk to a fully erect position above fully extended thighs becomes mechanically more difficult.

Attempts have been made from time to time to institute statis-

---

* Non-adaptive characters may be established in small, isolated populations as the result of what has been termed random genetic drift, but the importance of this process for evolutionary development has been strongly controverted.

tical comparisons of metrical characters in different groups of animals, with the intention of expressing degrees of systematic relationship in quantitative terms. Many years ago this desirable objective was foreshadowed by T. H. Huxley in his oft-quoted statement 'Whatever system of organs be studied, the comparison of their modifications in the ape series leads to one and the same result—that the structural differences which separate man from the gorilla and chimpanzee are not so great as those which separate the gorilla from the lower apes' (by 'lower apes' Huxley was referring to the Catarrhine monkeys). This proposition was subsequently labelled by E. Haeckel the 'pithecometra thesis'. But though it is of course possible to express quantitatively individual measurements considered as isolated abstractions, or even combinations of measurements by the technique of multivariate analysis, it is not yet possible with present knowledge to give a strictly quantitative value to different morphological characters according to their relative taxonomic importance for the assessment of degrees of relationship. Thus, while it is possible to state that in certain of its dimensions, or in the degree of complexity of some of the cerebral convolutions, the anthropoid ape brain is morphologically closer to the brain of *H. sapiens* than it is to the brain of Catarrhine monkeys, it is not feasible to state in quantitative terms just how much more closely the ape brain *as a whole* resembles the human brain. Still less is it possible, of course, to state in numerical terms how much more closely in his anatomical structure as a whole man resembles the large anthropoid apes than the latter resemble the Catarrhine monkeys. The essays of amateur biometricians to apply unsuitable statistical methods to the elucidation of phylogenetic relationships have in some cases led to misleading statements because they have not realized the possible fallacies involved in such methods, and by so doing they have tended actually to discredit the value of appropriate statistical methods. One common fault in the past has been to employ only a few over-all measurements of a bone or a tooth and to treat these one by one as though they were independent and uncorrelated characters. Over-all measurements of complex structures, however, may be quite similar in two groups even though in their intrinsic composition of pattern and shape they are widely different, just as a sphere, a cube and a tetrahedron may have identical dimensions in breadth, length and height. So, for example, molar teeth may show identical over-all dimensions though their intrinsic composition is very different. Some uncritical writers, when comparing bones and teeth of different shapes, have even fallen into the trap of comparing a few isolated dimensions which because of *different* shape are not morphologically comparable and then, on the basis of this false comparison, have concluded that because these dimensions are similar the bones and teeth are actually of the *same* shape! Clearly, in comparing over-all measurements it is essential to ensure that the measurements refer to characters which are morphologically equivalent. The well-known aphorism 'Science is measurement' is strictly true, but it by no means follows that all comparative measurements are scientifically valid. While biometrical studies of immediately related forms belonging to the same restricted group (such as a species or sub-species) may be expected to give fair comparisons which approximate sufficiently closely in their morphological equivalence, the statistical comparison of different genera which show a greater disparity of shape and form needs to be carried out with a very critical appreciation of the technical difficulties involved. In particular, the *principle of morphological equivalence* is of paramount importance.

The fallacy involved in attempting to treat characters separately and independently, instead of in combination, was first discussed in some detail by Bronowski and Long [11]. They pointed out that a bone or a tooth is a unit and not a discrete assembly of independent measurements, and that to consider their measurements singly is likely to be both inconclusive and misleading. The right statistical method, they emphasized, must treat the set of variates as a single coherent matrix. This can be done by the technique of multivariate analysis, which is essentially a method (not possible with more elementary techniques) that can be used for comparing morphological *patterns*. In principle, the application of the technique is straightforward enough, but it requires care and discrimination, a sound knowledge of morphological principles, and also a considerable experience of statistical methods. The value of multivariate analysis was strikingly demonstrated when it was applied to a controversial issue which had arisen in regard to certain teeth of the South African fossil hominid, *Australopithecus*, for it served to demonstrate very positively their hominid character.

From what has already been said, it is clear that, in assessing degrees of phylogenetic affinity, it is always necessary to take into account the factors of parallelism and convergence in the evolutionary development of related or unrelated groups. These processes can lead to structural similarities which, taken by themselves, may be misleading. The term *convergence* is applied to the occasional tendency for distantly related types to simulate one another in general proportions or in the development of analogous adaptations in response to similar functional needs. But such similarities are superficial and easily distinguishable by a detailed comparative study of the animal as a whole. For example, the resemblance in general appearance, and even in a number of morphological features, of the Tasmanian wolf to a dog does not obscure the fact that in fundamental details of their anatomical construction they belong to quite different mammalian groups. On the other hand, the potentialities of *parallelism* seem often to have been much overestimated by some anatomists, for this phenomenon has sometimes been invoked in support of extreme claims for the independent evolution of groups which are almost certainly quite closely related. We can agree with G. G. Simpson that the whole basis of parallelism depends on an *initial* similarity of structure and the inheritance of a common potentiality for reproducing homologous mutations, and that, this being so, the initial similarity and the homology of mutations themselves imply an evolutionary relationship. Expressed in another way, it may be said that convergence increases resemblances (which are, however, no more than superficial), while parallelism does not so much increase resemblances as maintain and perpetuate (by development 'in parallel' so to speak) similarities which have already existed *ab initio* in the genetic make-up of related types. Thus, 'closeness of parallelism tends to be proportional to closeness of affinity'. It is important to recognize this principle, for it is frequently overlooked by students of Primate evolution. It may not always be possible to exclude finally the factor of convergence (as distinct from parallelism) in explanation of a similarity in certain individual features, but it is not permissible to dismiss a complicated pattern of resemblances (with a complicated genetic basis) as merely the expression of convergence without presenting convincing evidence in support of such an interpretation. On the basis of the extensive data of comparative anatomy and palaeontology which have now accumulated over many years, it is justifiable to assume that groups of animals which show a preponderance of structural resemblances in taxonomically relevant characters are genetically related to a corresponding degree, unless flagrant discrepancies exist in some basic features such as could only be explained by a long period of independent evolution.

It might be supposed that, since morphological characters have to be selected and weighted for their value in assessing systematic relationships, and since the selection and weighting are largely determined by the results of practical experience and are not easily susceptible to quantitative treatment, the methods ordinarily employed are likely to be too arbitrary to give reliable information. In fact, however, the validity of the methods and reasoning of comparative morphology have their justification in the confirmatory evidence derived from collateral sources. Particularly is this the case with the evidence of palaeontology, for the relationships which had been inferred from the comparative study of existing groups have so often been subsequently established by the discovery of fossil links which, so to speak, fill in the gaps between those groups which had been presumed to be closely related. As Watson has noted [163], in arriving at a natural classification 'individual animals . . . are allotted to their groups by structural similarities and differences, determined by observations directed by ordinary morphological reasoning. And the methods of morphology are shown to be valid because they enable us to make verifiable predictions.' So far as the Primates are concerned, the progress of palaeontological discovery has led to the verification of many of the 'predictions' regarding the relationships and evolutionary origins of the different groups of this order. Remarkable confirmation was also presented by Nuttall when, in 1904, he published his classic work on *Blood Immunity and Blood Relationships* [106].

Nuttall studied in many animal species the precipitin reaction of the blood, an immunity reaction which can be measured quantitatively. Briefly, a species A which is immunized against antigens in the blood of species B by repeated injections of small quantities of blood serum of the latter produces specific antibodies. Then, if the serum of A is mixed with the serum of B in a test-tube, a precipitate is formed. Moreover, a similar reaction is found to occur if the serum of A is mixed with the serum of a species known to be closely related to B, and the intensity of the reaction parallels the closeness of the relationship. On the other hand, no reaction is produced if the sera of quite unrelated species are mixed. The precipitin reaction is of considerable importance for the purpose of a natural classification because (1) it is determined by the bio-chemical constitution of protein in the blood—a most fundamental property of the whole organism, (2) it provides a means of testing the validity of conclusions based on comparative anatomy, (3) it is an objective test, and (4) the reaction can be expressed in quanti-tative terms. Lastly, there is good reason to suppose it may over-come one of the complicating factors that always needs to be considered by the systematist who relies only on morphological criteria—the factor of convergence. During more recent years, the specific configurations and reactions of blood serum proteins in general have been studied in greater detail and by more refined techniques. In a review on systematic serology, Boyden [9] has stated (on the basis of an extensive series of comparisons) that 'As far as serological convergence is concerned, we have as yet no proven case for protein antigens. As further work is undertaken they may appear but it is unlikely that they will be frequent.' With these considerations in mind, it is interesting to note that, at any rate so far as the relationships of the major taxonomic groups of mammals are concerned, serological studies, supplemented by the analysis of molecular structure of haemoglobins and of other pro-teins in various tissues of the body, have confirmed, in a most remarkable way, the inferences previously based on morphological criteria [35, 37, 46, 64].

These studies of what may be termed primate biochemistry have demonstrated a close relationship beween man and the large anthro-poid apes (particularly the gorilla and chimpanzee), a less close relationship with the Catarrhine monkeys, and no more than a dis-tant relationship with the lower Primates. Not only does it thus con-tribute the strongest validation of the accepted methods of morpho-logical reasoning, it also provides the final justification for associat-ing the Hominidae and the Pongidae in a common taxonomic group.

It might be supposed that a study of karyotypes, that is, of the chromosome complement of individual cells, would be of great assistance in the determination of taxonomic affinities, but, al-though it has certainly provided useful information in certain instances, the results so far have been rather disappointing. In-deed, in a recent review of chromosome cytology in the Primates [17] attention has been called to the great diversity in chromosome number and morphology among the major groups of the order.

This diversity in so basic a pattern of genetic elements will serve to emphasize the difficulties that face the taxonomist in his search for some fundamental morphological criteria on which to base his assessment of phylogenetic relationships.

EMBRYOLOGY

The great pioneer of comparative embryology, K. E. von Baer, published his classic studies on the development of animals in 1828, and called attention to the astonishing similarity of embryos of vertebrates which, when fully developed, are very different indeed. This was expressed in his formal proposition that during its development an animal species departs more and more from the structural organization of other species. Thus the early development of the human embryo (like that of mammalian embryos in general) involves a series of transformations during which one type of organization characteristic of lower vertebrates is exchanged for a more advanced type of organization. Such transformations are to be explained by reference to evolutionary history. The suggestion that, in a modified form, ONTOGENY (embryonic development of the individual) repeats PHYLOGENY (evolution of the type) has been termed recapitulation. But it must be emphasized that this does not mean (as it has sometimes been supposed to mean) that the successive stages in the embryonic development of an individual in any way closely represent the *mature* forms of successive stages of its evolutionary history. It means only that, broadly speaking, during ontogeny the developmental stages through which an animal passes reproduce the embryonic form of certain ancestral types [6]. Thus, it is well known that in the early mammalian embryo a foundation of gill arches is laid down in the pharyngeal region, precisely similar to that which, in fishes, finally leads to the establishment of functional gills. But the elements of these gill arches become completely reorganized so as to form, not the gills for which it seems certain they were originally intended in past evolutionary history, but quite different structures such as the framework of the larynx and its muscles, the facial musculature, the ossicles of the middle ear, and so forth. Many other examples of similar kinds of change or replacement could be enumerated, for example the sequence of events which leads to a partitioning of the heart, rearrangements in the patterns of many of the blood-vessels, and (in man and the anthropoid apes) the withdrawal of a projecting tail into the floor of the pelvis to form the rudimentary nodules of the coccyx. There is abundant evidence of this sort which accords with the general thesis that man is derived from a primitive and generalized vertebrate ancestry, but embryological data provide little direct information to indicate the nature of his more immediate progenitors. In certain respects, however, embryology does offer some contributions to this problem. For example, one of the distinctions between the Hominidae and the modern anthropoid apes which has often been cited for its taxonomic importance is that of body proportions, in particular the relative dimensions of the limbs. In association with the erect posture the lower limbs in man are relatively long, whereas in the apes their arboreal mode of life has led to a relative lengthening of the upper limbs. But there is found to be a much closer approximation between the limb proportions of man and apes in early developmental stages than in adult life, and this observation is clearly relevant to the problem of changing proportions of limbs during evolution. For there is no theoretical difficulty in the supposition that the relative lengths of the limbs in adult forms could be considerably modified by a single mutation of a gene determining their rate of growth during development. Indeed, experimental genetics have clearly demonstrated that such a mutational effect is by no means exceptional.

The fact that some mutations affect rates of development and growth not only explains how a simple gene change can ultimately produce quite large effects in the fully grown individual, it also provides the basis for paedomorphosis or neoteny, a process which involves the retention in the adult of morphological characters which were embryonic or juvenile features of ancestral types. This process of FOETALIZATION, as it has also been termed, is presumed to be the result of a retardation or cessation of growth and differentiation, and has been advanced to explain the evolutionary origin of a number of features in which some of the Primates appear to be less specialized (and to that extent actually more primitive) than the presumed ancestral stock from which they took origin. For example, in the relative disposition of certain elements of the base of the cranium, including the angle which the head makes with the trunk, the human skull approximates more closely to the foetal or juvenile skull of anthropoid apes than to the adult form. In other words, it appears that the human disposition during the course of evolution may have been achieved by arresting a sequence of development which proceeds further in the ape's skull. But it needs to be recognized that such a process of arrest or retardation is not the cause, but rather the concomitant, of evolutionary change, for the latter is primarily effected by the selective action of the environment on actively living and breeding populations as a whole. In some ways, the conception of paedomorphosis may even be illusory and for this reason has probably misled a few biologists into supposing that it explains more than it really can do. In the example of the cranial base just cited, for instance, the immediate and essential condition which leads to (or requires) a secondary retardation in the direction of growth is the greatly increased size of the brain of *H. sapiens*, together with the adoption of a fully erect posture; in other words, the operative factors are entirely different from those which obtain in the foetal ape. But the point is that paedomorphosis does occasion the opportunity and provide the mechanism for certain evolutionary changes which might otherwise appear to involve an unusually striking reversal in the direction of evolutionary development, and it certainly disposes of the arguments of those who reject certain interpretations of ancestor-descendant relationships on the grounds that in one individual feature or another the presumed ancestor is more specialized or more 'advanced' than the presumed descendant. Thus, so far as the Hominidae are concerned (and apart from other considerations already mentioned), there can be no justification for controverting the thesis that the extinct hominid species *H. erectus* is ancestral to *H. sapiens* simply because the former had developed massive brow ridges and the latter has not.

While in general it may be said that individual species belonging to different major groups diverge in their structural organization more and more during successive stages of their embryonic development, this must not be taken to mean that any feature which shows itself at an early stage must have a corresponding evolutionary antiquity. It is necessary to draw attention to this fallacy, for it has occasionally been assumed from von Baer's propositions (or, rather, from the rather crude derivation from these propositions which Haeckel elaborated as his Biogenetic Law) that since certain hominid characteristics (such as the disposition of the big toe) appear very early in human development, therefore the hominid sequence of evolution must have diverged from the anthropoid ape sequence at an extremely remote time. But there are many examples which make it clear that a phylogenetic novelty can sometimes display itself in the very initial stages of ontogeny. The embryo of a Manx cat can be distinguished from the embryo of an ordinary cat during the process of vertebral segmentation. But it is not to be inferred therefrom that the evolutionary independence of Manx cats and ordinary cats extends back to the initial stages of vertebrate evolution!

PALAEONTOLOGY

However extensive and compelling it may be, the evidence for phylogenetic relationships based on the comparative anatomy of existing types can only be indirect evidence. Direct evidence must depend on the actual demonstration from the fossil record of a succession of transitional stages representing the transformation in geological time of ancestral into descendant types, and the progressive diversification of the latter into their subsidiary groups. It is clear that if in the same geographical region numerous fossil remains are found in geological strata of different and successive ages, and if they exhibit a gradual series of structural changes linking up an ancient form X with a modern form Y, then here we have presumptive evidence of the evolutionary route by which Y may have been derived in the course of time from X. Unfortunately, palaeontological records even approaching such completeness are rare, and this is particularly the case with the Primates for, being mostly arboreal creatures whose evolution probably occurred in regions of tropical or sub-tropical forest, the circumstances have been by no means favourable for the preservation of their remains in a fossilized condition. Nevertheless, the accumulation of the fossil record has provided in some cases a direct demonstration of a succession of approximately intermediate types forming a temporal sequence, which thus provide concrete and objective evidence of the trends of evolution which had already been postulated on indirect evidence. It is one of the main tasks of palaeontology to discover whether a postulated sequence based on a graded morphological series actually does reflect a temporal sequence, and for this reason the time factor in palaeontological studies is of paramount importance.

The relative age of fossil remains is commonly determined on geological evidence, but it is of some importance to understand that, in the case of a scanty fossil record, the geological dating of the few specimens of a genus which have by a rare chance been found does not, of course, give the span of existence of the genus as a whole (it is known that a single mammalian genus may persist for ten million years or more). Failure to recognize this has been a not uncommon source of confusion in discussions on lines of evolution (particularly hominid evolution), when the suggestion that a particular fossil type may be 'ancestral' to a more advanced type is controverted on the grounds that the fossil remains of *the individuals so far discovered* are not very much more ancient than the supposed descendant type. But it has been well established that archaic types may persist for long periods in some parts of the world after they have given rise to more advanced types elsewhere. For example, individuals of the hominid species *H. erectus* may have persisted through the Middle Pleistocene period in the Far East to a time when the species *H. sapiens* was already in exeistnce in Europe. But, by itself, such evidence of contemporaneity in different parts of the world of *individual* representatives of a primitive and an advanced hominid would not preclude the possibility (now generally accepted as a probability) that the species *H. erectus* was ancestral to the species *H. sapiens*.

Considering the importance of palaeontological evidence for phylogenetic studies, it is unfortunate that it has serious limitations. Apart from the fact that it is only by a rare chance that animal remains come to be fossilized at all (and thus preserved for posterity), they are often fragmentary and distorted by pressure or fracture. Also, of course, the evidence is almost entirely restricted to skeletal structures, bones and teeth, and to such inferences as can be drawn from these about the soft parts (e.g., the size and convolutional pattern of the brain may be demonstrated by a cast of the inside of the skull, or certain details of the muscular system may be inferred from the impressions made by muscular attachments to bones.) Even when fossil remains are discovered which are fairly complete, difficulties of interpretation may arise unless palaeontological principles are well understood, particularly in the study of the earlier representatives of an evolutionary line which have not developed to the full extent the distinguishing taxonomic characters of its terminal products. This source of misunderstanding may best be explained by reference to the accompanying diagram (fig. 1) which represents schematically the diverging lines

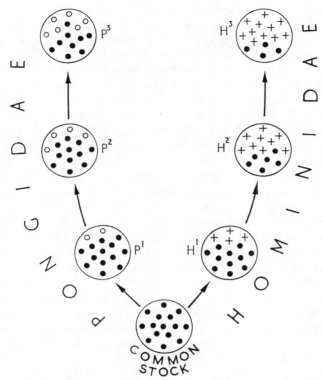

FIG. 1. Diagram representing the divergence of two evolutionary sequences, the Pongidae (anthropoid ape family) and the Hominidae (the family which includes Recent and extinct types of man). The two sequences inherit from a common ancestry *characters of common inheritance* (black circles). As the lines diverge each one acquires its own distinctive features, or *characters of independent acquisition*; those distinctive of the hominid sequence of evolution are represented by crosses and those of the pongid sequence by white circles. For further explanation, see text.

of evolution of the Hominidae (the zoological family of which *H. sapiens* is the only surviving species) and the Pongidae (the anthropoid ape family). In the generally accepted classification of the Primates these two families are grouped together in a common superfamily—Hominoidea, and this carries with it the implication that they had their origin in a common ancestral stock. In the diagram, the morphological characters of the common stock are indicated by black circles. Some of these ancestral characters are of course inherited and retained in common by both families—such characters may be termed *characters of common inheritance*. But when different lines of evolution segregate and branch out independently to form separate groups such as the Hominidae and Pongidae, each of them acquires its own special and peculiar pattern of morphological characters by which it comes to be distinguished from the other. Characters of this sort may be called *characters of independent acquisition*. In the diagram those which are distinctive of the Hominidae are indicated by crosses, and those of the Pongidae by white circles. Since the pongid sequence of evolution has been much more conservative than the progressive hominid sequence, its terminal products (the modern anthropoid apes) have preserved more of the original characters of the ancestral stock. As divergent evolution proceeds, characters of common inheritance will become progressively supplemented or replaced by characters of independent acquisition in each line. Conversely, if the lines are traced backwards in retrospect they will be found to approximate more and more closely to each other in the characters of common inheritance which they share. Thus, for example, in representatives of an early stage in the hominid sequence (H^1 in the diagram) it may well be found that characters of common inheritance predominate over characters of independent acquisition, the latter being as yet relatively few in number or only showing an incipient development. If, now, the remains of an individual corresponding to the stage H^1 is examined and the morphological characters are compared quite indiscriminately, if, that is to say, the characters are simply enumerated without giving to each one an appropriate weighting according to its evolutionary significance, the erroneous conclusion may be reached that because in the *sum* of its characters it shows a closer resemblance to apes than to modern man, therefore it is taxonomically a pongid. But this would be to ignore the highly important principle of taxonomic relevance in comparing morphological characters, to which reference has already been made. The decision as to the taxonomic status of a previously unknown fossil—whether it is a primitive representative of one or other of two divergent lines of evolution corresponding to two related families—must depend on a recognition of the fundamentally different trends which have distinguished the evolution of the two families and which are thus diagnostic of each of them as a natural taxonomic group. In this particular case the taxonomically relevant characters on which the diagnosis of *pongid* or *hominid* depends are the characters of independent acquisition which serve to distinguish the divergent trends in the two sequences, the nature and direction of the trends of each sequence being at once made evident by a consideration of the final stages actually reached by its terminal products. The stage H^1 of hominid evolution may be exemplified by the fossil genus *Australopithecus*, which is a particularly apt illustration for our present purpose. In the early discussions following the discovery of this primitive hominid, some anatomists, basing their judgement exclusively on the characters of common inheritance (such as the relatively small brain case and large jaws), were led to suppose that it really was an ape in the taxonomic sense. But more careful studies soon made it clear that many of the characters of independent acquisition distinctive of the Hominidae (particularly in the dentition and pelvis), but none of those distinctive of the modern Pongidae, had already been developed and superimposed on the characters of common inheritance. In other words, in those characters in which *Australopithecus* had undergone modification away from the ancestral stock of the Hominoidea, the direction of change was in that of the hominid sequence and divergent from that of the pongid sequence.* As we shall see, the need to distinguish between the two types of characters is of the greatest importance for assessing the taxonomic status of early representatives of any of the different groups of Primates. It is also a matter of considerable importance to recognize that the progressive modification of different systems of the body during the evolutionary development of any one group may (and frequently does) proceed at differential rates. This principle is well recognized by palaeontologists, but appears occasionally to have misled those (sometimes called neontologists) whose studies are limited to the comparative anatomy of existing types. From a quantitative study of the evolution of the Equidae, Simpson has postulated the two following theorems concerning the rates of evolution, (1) the rate of evolution of any character or combination

---

* This discussion, of course, applies as well to separate anatomical elements as it does to the organization of the whole animal. For example, in the earliest stages of evolution of the hominid pelvis (i.e. the pre-australopithecine stage), in the sum of its morphological or metrical characters it may well have shown closer agreement with the pelvis of modern apes than with that of modern man. But its correct interpretation would depend on the recognition of the incipient development of those features distinctive of the Hominidae, as opposed to those distinctive of the modern apes.

of characters may change markedly at any time in phyletic evolution, even though the direction of evolution remains the same, and (2) the rates of evolution of two or more characters within a single phylum may change independently. Such differential rates of somatic evolution may lead to structural contrasts which give an appearance of incongruity, and they are liable to be regarded as 'disharmonies' because they do not conform with the sort of correlation which studies confined to living species may lead one unconsciously to expect. The true affinities of such fossil forms may thus be overlooked and misinterpreted. Differential rates of somatic evolution must obviously be taken into account when selecting characters for their taxonomic relevance in the assessment of the phylogenetic status of fossil types.

One further point in the interpretation of palaeontological evidence in terms of evolutionary sequences demands emphasis, and here we borrow freely from the illuminating hypothetical example presented by Simpson [148]. As he has pointed out, a really complete sequence is never preserved or discovered, and such fossil material as does become available usually consists of a relatively small sample of the sequence scattered more or less at random in space and time. Now, consider the application of Simpson's diagram in fig. 2A, representing a series of fossils in which a general modification of structure is correlated with geological succession. As a direct practical illustration we have modified this diagram so as to make the successive groups correspond to some of the known fossils which have reference, directly or indirectly, to the evolution of the Hominidae. For example, the groups might be (1) some of the more generalized Pliocene apes, (2) the Early Pleistocene genus *Australopithecus*, (3) the middle Pleistocene species *H. erectus* and (4) *H. sapiens*. In their over-all structural organization these do appear to form a consecutive and gradational series. But each group has its own distinctive characters some of which have led those who assume that evolution is governed by a completely rigid principle of irreversibility to deny the possibility of any direct or indirect ancestor-descendant relationship between one group and another. They have therefore been driven to the interpretation shown in fig. 2B of an independent evolution of each group from some unknown (and unformulated) common ancestor. The more rational interpretation is that shown in fig. 2C, which represents the groups as lying on, or close to, the line of an actual phylo-

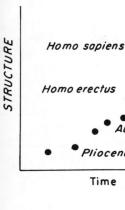

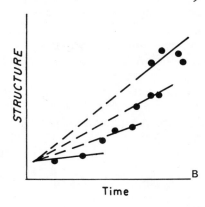

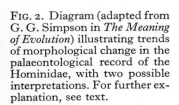

FIG. 2. Diagram (adapted from G. G. Simpson in *The Meaning of Evolution*) illustrating trends of morphological change in the palaeontological record of the Hominidae, with two possible interpretations. For further explanation, see text.

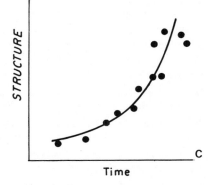

genetic sequence, in spite of the fact that the line shows a changing direction. That this is the more probable of the two interpretations is made evident by the following considerations—(1) it is based on a morphological trend which is demonstrably present in the series, (2) it does not have to rely on assumed principles of irreversibility and orthogenesis (which have in any case now been disproved or discredited), (3) it does not have to postulate a whole series of unknown intermediate types which are presumed (without any direct evidence at all) to have occurred in an independent ancestry of *H. sapiens* and of each of the other species or genera, (4) it does not make an inordinate (and seemingly impossible) demand on the potentialities of convergence and parallelism, and (5) it accords remarkably well with the sort of evolutionary sequence which had previously been hypothesized on the basis of the indirect evidence of comparative anatomy. Naturally, it has to be recognized that any inference based on a sparse fossil record is no more than a provisional inference, but as the record becomes more fully documented by the accession of more material, inferences drawn from it become less and less provisional and more assured. In the example cited, it is unlikely that any further discoveries will prove the interpretation shown in fig. 2c to be very wide of the mark, if only because the successive phases of hominid evolution, on purely theoretical grounds, could hardly have been very different from the fossil groups already known. It does not follow, of course, that these groups were *direct* ancestors, but almost certainly they were sufficiently closely related to the direct ancestors to provide an objective demonstration of the successive phases of hominid evolution.

TERMINOLOGY

In some discussions on phylogenesis, especially the phylogenesis of *H. sapiens*, confusion has often been introduced by the adoption of uncertain or alternative terminologies (and sometimes even by the author's invention of his own special terminology to reflect his own unorthodox views). But it is essential that the nomenclature employed in such discussions should be well recognized and understood by biologists in general, or at any rate defined as clearly as possible, in order to avoid profitless controversy. Obviously, from the evolutionary point of view, it is impossible to give a precise definition of a major natural group in terms of morphological characters which have only been acquired in their complete form in the terminal phases of its evolution. For example, the Primates include all those types which have existed since the immediate precursors of the Order as a whole became segregated from the basal mammalian stock which also gave rise to other Orders. Thus the earliest Primates would have shown none of the characteristic features of the fully differentiated groups such as a large brain, reduction of the jaws, and so forth. It is not feasible, therefore, to define a major group by the characters are which found in its *existing* representatives—the end products of long lines of evolution. In the pre-Darwinian days of the older naturalists of last century definitions of this kind were usual; for instance, Cuvier defined his equivalent of the horse family (Equidae) by characters such as a single complete digit on each foot and the complicated enamel pattern of the cheek teeth. But—so far as is practically convenient—the aim of modern taxonomy is to reflect phylogenetic relationships, and to this end major groups are defined in dynamic rather than static terms. Thus, the family Equidae comprises the whole evolutionary sequence of the horses from the small *Hyracotherium* (*Eohippus*) of Eocene times. Similarly, among the Primates the familial terms Hominidae, Pongidae, Cercopithecidae, etc., refer to the whole evolutionary sequence of each group from the time at which it became segregated from the evolutionary sequences comprising immediately related families. It is particularly important to note this in reference to the Hominidae, for much misunderstanding has been engendered by confusing this familial term (or its adjectival form 'hominid') with the colloquial terms 'man' and 'human'. It is now generally agreed that these latter terms should be restricted to representatives of the *later* phases of hominid evolution sufficiently advanced to have acquired the ability not only to use, but also to fabricate, tools and weapons. As a broad generalization there is much in favour of such a convention, for the ability to design and manufacture implements to serve a diversity of purposes, even if they are of quite simple construction, involves powers of imagination and abstraction far beyond anything found in the rest of the animal world. There is reason to suppose that this achievement was associated with the rapid expansion of the brain which culminated in the emergence of the *H. sapiens* in Pleistocene times. But the evolutionary sequence comprising the Hominidae must certainly have become segregated as an independent line of evolution long before such a development took place (most probably in Miocene or Early Pliocene times), and it is convenient to refer to this earlier phase as 'the pre-human phase of hominid evolution'. It is not uncommon to read a paper in which (say) the skull of a problematical fossil Primate is compared with a series of European skulls, with the result that a statement is made that in this or that character the fossil type resembles apes rather than 'man'. All that the author is entitled to say, of course, is that, in those particular characters, it resembles apes rather than the modern European variety of *H. sapiens*, for the term 'man' must presumably be taken to include not only all the racial varieties of to-day, but also a whole range of extinct and primitive types of Palaeolithic man (such as Neanderthal man and *H. erectus*) which were themselves in certain features also more like apes than the modern European. Unfortunately, the fallacy of extrapolating from single species or genera to larger taxonomic groups is not uncommon in the literature of primatology. Except when their meaning is unequivocal, there is a strong case for eliminating altogether the colloquial terms 'man' and 'human' in scientific discussions dealing with the evolution of our own species, if only to exclude as far as possible the emotional prejudices which so often unconsciously obtrude themselves—even in the scientific mind.

Uncertainties of nomenclature may lead one authority to deny that the Hominidae are derived from anthropoid apes because by the term 'anthropoid ape' he means an arboreal animal showing certain adaptive specializations such as are found in *modern* apes, and which would presumably be absent in any form ancestral to the Hominidae. This conception of anthropoid apes is based on a common text-book definition and, so far as it goes, the argument is valid. Another authority, who accepts the anthropoid ape ancestry of modern man, obviously uses the term 'anthropoid ape' with a much wider connotation. In so far as the term may be taken to refer to a group of Primates which occupy approximately the same evolutionary status as the modern apes in regard to brain development, general proportions of the skull and jaws etc., but in which certain specializations (such as the greatly elongated arms and the atrophy of the thumb) need not be essential characters, this second view is also correct. As we shall see, there is certainly evidence that the Hominidae have been derived from an ancestral stock whose general morphological characters would legitimately allow the application of the descriptive term 'anthropoid ape', even though it would not have presented some of the divergent modifications shown in the living apes.* Used in this way, the term has a wider application than it is accorded by some systematists who appear to confine their attention mainly to living mammals. In other words, recognition is taken of phylogenetic considerations in defining a natural group such as the anthropoid apes. Like the term 'man', it would no doubt make for clarity of thought and objectivity of approach if the term 'ape' were avoided in scientific discussions on hominid evolution, and the zoological term Pongidae (or pongid) used in reference to the familial category. It also

---

* It is odd that some of those who theorize on human evolution to-day still seem to ignore the warning of Darwin in 1871—'we must not fall into the error of supposing that the early progenitors of the whole Simian stock, including Man, were identical with, or even closely resembled, any existing ape or monkey'

seems that colloquial terms tend unconsciously to inject an erroneous idea of individualization and personification into discussions, a tendency which would be avoided by employing only the
terms proper to zoological taxonomy. In this connection, it is constantly to be borne in mind that *populations*, and not *individuals*,
are the units of evolution.

Yet another example of nomenclatural misunderstanding is
found in the argument that the whole stock of the Old World monkeys is excluded from any share in human ancestry. In so far as
the whole group of Old World or Catarrhine monkeys is techni-
cally defined by neontologists as being characterized by certain
specializations (such as bilophodont molars and strongly developed
ischial callosities)—that is to say, in so far as these features are
accepted as essential components in the make-up of an animal
which we can call an Old World Monkey—it is probably true to
say that such a type did not occupy any place in the ancestry of
the Hominoidea. But if an animal were discovered which in
general characters (e.g. dental formula, cranial features, limb structure, etc.) conformed closely with the known Old World monkeys,
but which lacked the few aberrant modifications which characterize
the latter (or only showed them to an incipient degree), it would
almost certainly be called a Catarrhine monkey, and the definition
of the Catarrhine monkeys would presumably be extended to
accommodate it.

It is clear from this brief discussion that current zoological
classifications often require to be treated as somewhat arbitrary
attempts to separate the known types into well-defined groups for
the convenience of description and cataloguing. On the other hand,
modern taxonomists also aim at classifications which, so far as is
practicable, reflect phylogenetic relationships, but in many cases
a compromise between these two objectives has to be effected in
order to give reasonable satisfaction to both objectives. It is clear,
moreover, that in order to reflect the realities of nature, definitions
based on anatomical characters should be made as general as possible, and (pending the accession of a reasonably substantial fossil
record) the larger the group the more elastic does its definition
require to be. 'Classifications serve an apparently dual purpose:
first, to systematize and reflect (not to express fully, for that is
impossible) what we can infer about the natural, that is, the genetic
and evolutionary relationships of the organisms concerned; and

second, to provide natural taxa in terms of which we can discuss the various groups clearly' [151].

THE ESTIMATION OF GEOLOGICAL ANTIQUITY

We have noted that the elucidation of lines of evolutionary development depends partly on the indirect evidence of the comparative anatomy of living forms, particularly those which permit the construction of a graded morphological series exemplifying a 'trend' through which the more primitive types appear to be linked with the more advanced types. But, as we have also emphasized, the obvious implications of such a series can only be confirmed by demonstrating a corresponding temporal sequence for the fossil record. Further, in order to establish this essential factor of time we need to know the geological age of fossil types recovered from various strata. The methods of estimating geological age are varied, and for their details reference should be made to special works concerned with the whole subject of GEOCHRONOLOGY [111, 180].

Unfortunately, the absolute antiquity of a fossil in terms of years can be determined only rarely with the methods at present available. One of these methods depends on the estimation in organic material of the relative quantities of radioactive carbon (C^{14}) and C^{12} of which the former is an isotope [84]. Radioactive carbon is assimilated into the substance of living organisms from the atmosphere (directly or indirectly), and at death it undergoes a gradual disintegration at a known rate. The proportion of radioactive carbon thus progressively diminishes in dead organic material in proportion to its antiquity, and it is estimated that after a period of 5,730 years its ratio will have dropped to half the value found in living material. With modern techniques, the accuracy of this method is limited to estimations of a period of not much more than 50,000 years—which does not take us very far back in the story of hominid evolution. Another method depends on the fact that the element uranium disintegrates at a known rate, the end-products of the breakdown being helium and uranium-lead. This process is extremely slow—one million grams of uranium produce $\frac{1}{7,600}$ grams of uranium-lead in one year, but, even so, in the case of ancient rocks of certain types it may be possible by estimating the proportion of uranium to uranium-lead to obtain a useful indication of absolute age. For example, an analysis of basaltic lavas of

Lower Miocene date has shown this geological level to be about thirty-two million years old. More recently a third method of estimating absolute antiquity has been developed that bridges the gap between the forward limit of the uranium-lead method and the backward limit of the radioactive carbon method [28, 29]. This new technique is the potassium-argon method, a chemical analysis which depends on the fact that all natural potassium contains 0·01 per cent of a radioactive isotope which in the course of gradual decay disintegrates to form calcium and argon, and which has a half-life of $1·3 \times 10^9$ years. A striking example of the application of this technique is the demonstration that the earliest fossil remains of undoubted hominids probably have an antiquity of at least two or three million years, and perhaps even more (see p. 49).

Indirect methods of estimating geological antiquity depend mainly on comparing the relative thickness of sedimentary deposits, the assumption being that, in general, the thicker the deposit the longer must have been the time taken in its accumulation. Clearly, estimates based on evidence of this sort can only be approximate and are subject to a number of possible fallacies; in particular, the rapidity of sedimentation must vary with the changing climatic conditions. But, in any case, the *relative* age of a fossil may be determined by purely geological data—that is to say, it may be inferred from the stratigraphical level of the deposit in which the fossils are indigenous. In the case of skeletal remains embedded in sedimentary formations which have been deposited layer by layer under the influence of agents such as running water, or in stalagmitic formations in caves, obviously the more ancient fossils will be found in the deeper layers. During the latter half of the last of the geological periods into which the Tertiary epoch is subdivided, the Pleistocene period,* there were rhythmic oscillations of climate of considerable magnitude, resulting in a succession of glacial and interglacial phases which altogether extended over rather more than half a million years. Evidence of these successive glaciations can be detected by the study of the characteristic geological deposits left by melting ice and so forth, and also by reference to the fossil remains of arctic or subarctic types of animals and plants which inhabited the fringes of the glaciated regions. By the determination of the rhythmical succession of the glacial and interglacial phases during the Pleistocene period, geologists have been able to provide

* The Tertiary epoch is the last of the three main subdivisions of geological time.

a relative time scale which has been of the greatest importance in interpreting the fossil record of the last stages of hominid evolution. In this connection, mention should also be made of certain chemical tests for relative antiquity. The fluorine content of fossil bones and teeth increases with the geological age, for by a process of ionic interchange this element is slowly taken up from the soil and becomes fixed in bony tissue in the form of a very stable compound, fluorapatite. The amount of fluorine in a fossil bone thus increases with time and gives an indication of the duration over which it has lain in position in a particular deposit. However, the amount of fluorine taken up also depends on the concentration of fluorine in the soil, and it does not therefore permit a comparison of the relative antiquity of fossilized bones derived from different deposits in which the fluorine content may vary widely. A further method, the radiometric method, depends on the observation that fossil bones and teeth often show some degree of radioactivity, a result of the gradual absorption of uranium from percolating ground waters, and the degree of radioactivity of a fossil depends partly on its geological age. But it also depends on the permeability of the deposit in which the fossil lies, as well as on the uranium content of the percolating waters in past ages.

By making use of all these methods, applying one to check estimates based on another, it has been possible to arrive at a general agreement on the durations of the successive intervals of geological time in the Tertiary epoch which saw the emergence and progressive evolution of the various subdivisions of the order Primates. The estimated durations are as follows, the successive periods being listed in their time sequence and stratigraphical level from below upwards.

Pleistocene	3 million years	
Pliocene	15 ,,	,,
Miocene	20 ,,	,,
Oligocene	10 ,,	,,
Eocene	20 ,,	,,
Palaeocene	15 ,,	,,

Thus, the total length of time occupied by the evolution of the Primates has extended over 80 million years, and of this time the genus *Homo* has been in existence for hardly more than half a million years.

2

A Preliminary Survey of the Primates in Space and Time

THE origin of mammals from a reptilian ancestry has for long been well established. Palaeontological studies have demonstrated the existence at the end of the Palaeozoic epoch (the first of the three major subdivisions of geological time) of a large number of mammal-like, or therapsid, reptiles which provide a remarkable series of structural transitions between true reptiles and true mammals. During the Permian age—which marks the last phase of the Palaeozoic epoch—certain small Therapsida flourished in South Africa which very probably included the ancestral group from which all mammals were initially derived. It was not until the middle of the Mesozoic or Secondary epoch, however, that undoubted mammals appeared and left their traces in the geological record. During the central phase of this epoch—the Jurassic age—mammals were relatively abundant and widespread, though it is now clear that most of these early types were, so to speak, abortive experiments and became extinct [145].

Of the various large groups of Jurassic mammals, one—the Pantotheria—almost certainly represents the stock which ultimately gave rise to the placental mammals and the marsupials.* Other great groups of early mammals, such as the Multituberculata, Triconodonta and Symmetrodonta, had already disappeared by the early phases of the Tertiary epoch and probably had no part in the evolution of later groups of mammals. At any rate, it is certain that at the very base of the mammalian stem, and quite soon after its emergence from the ancestral therapsid stock, there separated out two great evolutionary radiations which persist to-day as the Marsupialia (or pouched mammals), and the Placentalia.

---

* The egg-laying mammals of Australasia, the Monotremata, probably had an independent therapsid ancestry. Simpson [145] goes so far as to suggest that 'although the lower jaw bone does articulate directly with the skull and they do give milk, the monotremes might be classified as therapsid reptiles rather than as mammals'.

The latter are distinguished (*inter alia*) by the fact that their young are nourished in the uterus by a highly elaborated (allantoic) placenta and are born in a relatively mature condition. In the marsupials, on the other hand, only an elementary form of placenta is developed during the intra-uterine life of the embryo, the young are born after a very brief gestation still in an embryonic stage, and they continue their development in the marsupial pouch. The three subclasses of living mammals, Monotremata, Marsupialia and Placentalia represent broadly three levels of evolutionary development which have been termed prototherian, metatherian and eutherian. There is good reason to suppose that the placental mammals (which represent the eutherian stage) have in their phylogenetic history passed through prototherian and metatherian stages, but this does not necessarily mean that they were derived successively from monotremes and marsupials, which are surviving representatives of these stages. Rather, these two subclasses are to be regarded as side-branches of mammalian evolution which, while preserving primitive characters to a much greater extent than the placental mammals, have also undergone aberrant specializations peculiar to themselves.

By the beginning of the Tertiary epoch, the placental mammals had become differentiated into a number of separate orders, most of which have persisted to the present day. In fig. 3 is shown a diagram, based on the pioneer work of many students of palaeontology, which indicates the progressive differentiation of some of these main groups of mammals during the last phase of the Mesozoic epoch and the early part of the Tertiary epoch. We are concerned in this thesis entirely with the order of Primates.

It was probably in the latter part of the Cretaceous period that the ancestral or basal Primate stock began to undergo segregation from the common eutherian stock which also gave rise to the other placental orders, and thereafter it entered upon an independent evolutionary history which (in terms of geological time) rapidly culminated in the differentiation of a number of subsidiary groups, some well-defined, others (particularly some of the earlier groups which did not survive for very long) rather ill-defined. The earliest Primates must obviously have been extremely primitive in structure, differing hardly perceptibly from the earliest representatives of the other orders. They were characterized mainly by the fact that they incorporated a pattern of morphological features which en-

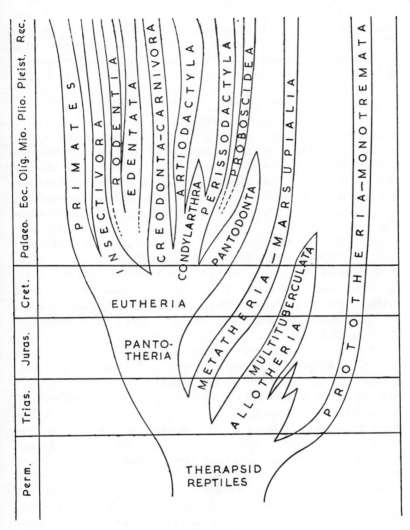

FIG. 3. Schema showing the progressive differentiation of the Orders of Mammals during the Mesozoic and Tertiary epochs. The geological subdivisions of these epochs are shown on the left.

dowed them with potentialities for evolutionary development along the lines diagnostic of the order as a whole, these potentialities at first manifesting themselves to no more than an incipient degree. With their continued realization, however, the early Primates acquired the complex of structural features which form the basis for the definition by which they are distinguished from other orders.

CLASSIFICATION OF THE PRIMATES

The orders of zoological taxonomy are subdivided into subsidiary groups such as families, genera and species, each of which is characterized by a certain combination, or pattern, of distinctive anatomical features. We have already noted in the previous chapter that this grouping is intended to reflect phylogenetic relationships, at any rate so far as this is compatible with the need for practical convenience which is required in any taxonomic system. But, since the components of any one taxonomic category have come into existence as the result of a gradual evolutionary diversification from a common ancestral stock, it is peculiarly difficult in many cases to give it a clear-cut, hard and fast, morphological definition. To some extent, therefore, definitions of taxonomic categories are arbitrary, and they need to be elastic and labile enough to allow for possible revision as phylogenetic sequences become more certainly elucidated by palaeontological studies.

The basic unit of zoological classification is the SPECIES, and, as a rough approximation, a species may be said to consist of a homogeneous community in which the individuals, or groups of individuals, closely resemble each other in the genetic foundation of their anatomical structure (except for relatively minor variations such as those of moderate differences in size, proportions and coloration), and which are usually capable of interbreeding freely and producing fully fertile offspring. On the other hand, different species are usually not capable of interbreeding with each other (or in natural circumstances do not do so), and if they do interbreed their offsprings are commonly infertile. The individual mammalian species which exist to-day are the result of a diversification in comparatively recent geological time and, in the evolutionary sense, the taxonomic category of GENUS may be taken to include all those closely related species which have become differentiated from the same immediate ancestral stock. The proximity of the relationship of the component species of each genus is indicated by the common morphological pattern which underlies their specific diversity, and it is on the basis of this common morphological pattern that the genus itself is defined. In evolutionary and palaeontological studies the genus is actually a more convenient taxonomic unit than the species, for generic differences are more abrupt and distinct than specific differences, and the latter may not be clearly reflected in the skeletal structure (which is usually the only material available for palaeontological study). Moreover, one genus can often be distinguished from another by single elements of the skeleton or dentition, and in the case of many of the earlier fossil Primates it is rare to find more than a few such fragments.

While the lower taxonomic categories, such as genera and species, may be defined in more or less static terms (at least in so far as they concern divergent rather than successional types), it is impracticable to draw up any comprehensive definition of larger categories, such as subfamilies, families, suborders and orders, except on the basis of evolutionary trends. As we have already seen, these trends may be inferred indirectly from a consideration of the various end-products of evolution in each group, or more directly from the study of palaeontological sequences.

The order Primates consists to-day of what seems to be a rather heterogeneous collection of types. In fact, it is not easy to give a very clear-cut definition of the order as a whole, for its various members represent so many different levels of evolutionary development and there is no single distinguishing feature which characterizes them all. Further, while many other mammalian orders can be defined by conspicuous specializations of a positive kind which readily mark them off from one another, the Primates as a whole have preserved rather a generalized anatomy and, if anything, are to be mainly distinguished from other orders by a negative feature—their lack of specialization. This lack of somatic specialization has been associated with an increased efficiency of the controlling mechanisms of the brain, and for this reason it has had the advantage of permitting a high degree of functional plasticity. Thus, it has been said that one of the outstanding features of Primate evolution has been, not so much progressive *adaptation* (as in other mammalian orders), but progressive *adaptability*.

Broadly speaking, the order Primates can be defined as a natural group of mammals distinguished by the following prevailing evolutionary trends:

1. The preservation of a generalized structure of the limbs with a primitive pentadactyly, and the retention of certain elements of the limb skeleton (such as the clavicle) which tend to be reduced or to disappear in some groups of mammals.

2. An enhancement of the free mobility of the digits, especially the thumb and big toe (which are used for grasping purposes).

3. The replacement of sharp compressed claws by flattened nails, associated with the development of highly sensitive tactile pads on the digits.

4. The progressive abbreviation of the snout or muzzle.

5. The elaboration and perfection of the visual apparatus with the development to varying degrees of binocular vision.

6. Reduction of the apparatus of smell.

7. The loss of certain elements of the primitive mammalian dentition, and the preservation of a simple cusp pattern of the molar teeth.

8. Progressive expansion and elaboration of the brain, affecting predominantly the cerebral cortex and its dependencies.

9. Progressive and increasingly efficient development of those gestational processes concerned with the nourishment of the foetus before birth.

10. Prolongation of post-natal life periods.

The circumstances of these main evolutionary trends (and of others) will be elaborated and amplified in later chapters. It may be noted here that not all Primates (even those that exist to-day) have developed all the characters which are usually cited as diagnostic of the order, or have completed their development to the same degree. And, of course, still less was this the case with the earliest Primates at the beginning of the Tertiary epoch. It may also be noted that the evolutionary trends associated with the relative lack of structural and functional specialization are a natural consequence of an arboreal habitat, a mode of life which among other things demands or encourages prehensile functions of the limbs, a high degree of visual acuity, and the accurate control and co-ordination of muscular activity by a well-developed brain.

The taxonomic subdivision of the main categories of the Primates into component groups has been a common subject for controversy, and many diverse opinions have been expressed on their probable inter-relationships. This diversity of opinion is partly due to the incompleteness of the evidence. Yet all workers in this field should agree to use the same classification, even if they regard it as no more than provisional, for only by so doing is it possible to enter on mutual discussions with a common understanding. In the present work we make use of Simpson's classification of the Primates [147] not because it is the only possible classification, nor on the assumption that it represents the last word on phylogenetic relationships (which no system of classification can do without a reasonably complete documentation from the fossil record). We use it because (1) it is based on recognized authority and a long practical experience of taxonomic methods, (2) it has the merit of comparative simplicity, (3) it appears to reflect, with at least some degree of approximation, such phylogenetic relationships as can be inferred from the evidence at hand (and as far as this can be done in any system of classification without seriously affecting its easy use as a sort of reference key), and (4) it has been provisionally accepted and recognized by most other authoritative workers in the same field. The main outline of this classification is seen below.

Order PRIMATES
    Suborder PROSIMII
        Infraorder LEMURIFORMES
            Superfamily *Tupaioidea**
                    Lemuroidea
                    Daubentonioidea
        Infraorder LORISIFORMES
        Infraorder TARSIIFORMES
            Superfamily *Tarsioidea*
    Suborder ANTHROPOIDEA
            Superfamily *Ceboidea*
                    Cercopithecoidea
                    Hominoidea

In the main part of this book we shall deal analytically with the various anatomical systems of the body and trace the evolutionary modifications which they have undergone in the main subdivisions of the Primates. This method of approach has its dangers, for it tends to concentrate attention on separate organs and structures

---

* It should be noted here that the inclusion of the Tupaioidea (tree-shrews) in the order Primates has been seriously questioned by some authorities. For a discussion on this point see p. 317.

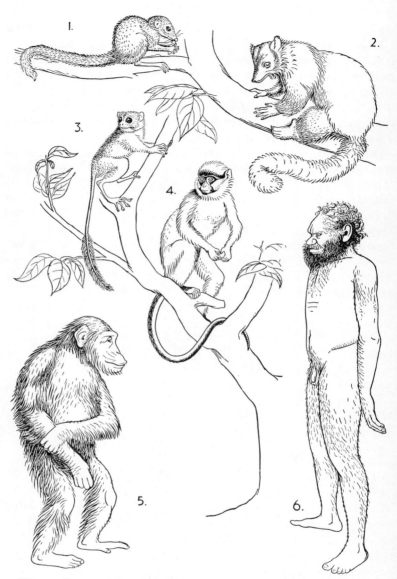

Fig. 4. Some of the living members of the Order Primates, representing a gradational series which, broadly speaking, links man anatomically with some of the most primitive of placental mammals. (1) Tree-shrew; (2) Lemur; (3) Tarsier; (4) Cercopitheque monkey; (5) Chimpanzee; (6) Australian Aboriginal.

44

as though they were isolated abstractions instead of integral parts of a living animal. It is well to emphasize, therefore, that although it is a common mode of speech to refer to the evolution of (say) a tooth or a part of the brain, evolution as a process concerns the whole individual or, more properly speaking, populations of individuals. The study of the evolutionary changes affecting a single organ is merely part of that type of analytical procedure which is a necessary preliminary to the formulation and solution of scientific problems in general. But it is a procedure which always needs to be followed by an attempt at resynthesis.

In the present chapter a general survey of the different groups of living and fossil Primates is given, the main purpose of which is to render intelligible the references which are made to them in the detailed discussion of Primate anatomy which follows. As a matter of convenience, we shall deal with them in four main categories, the suborder Anthropoidea, the superfamily Tarsioidea, the lemurs—that is to say the lemuriforms and lorisiforms exclusive of the tree-shrews, and lastly the tree-shrews (Tupaioidea). No attempt is made to give a complete list of all the known genera comprising these main groups.

There is one striking feature of the order Primates wherein it contrasts rather strongly with other mammalian orders; it is possible to arrange its *existing* members in a graded series which (quite apart from the evidence of the fossil record) suggests an actual scale or trend of evolutionary development within the limits of the order. For example, such a series may be constructed by comparing, in sequence, a tree-shrew, lemur, tarsier, monkey, anthropoid ape, and *Homo sapiens* (fig. 4). It is not to be inferred, of course, that man has actually been derived from an anthropoid ape of the specialized type which exists to-day, or that a monkey ever developed from a modern sort of lemur. The various members of the living Primates are all the terminal products of so many divergent lines of evolution; they are not in any way to be taken to represent a *linear* sequence of evolution. What a series of this kind does indicate is that, even without reference to palaeontological evidence, there are connecting links *of an approximate kind* between *H. sapiens* and small mammals of quite a primitive type; it also suggests the general trends of evolutionary change which have characterized the order Primates as a whole, and in a very indirect way the sort of transitional types which presumably had their

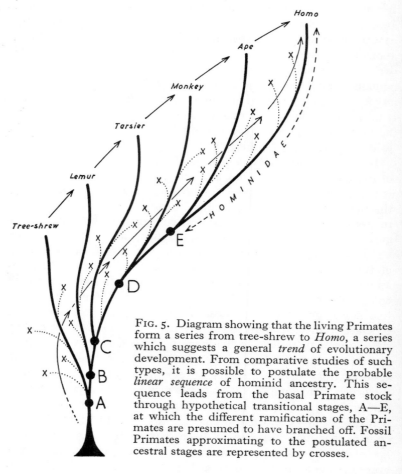

FIG. 5. Diagram showing that the living Primates form a series from tree-shrew to *Homo*, a series which suggests a general *trend* of evolutionary development. From comparative studies of such types, it is possible to postulate the probable *linear sequence* of hominid ancestry. This sequence leads from the basal Primate stock through hypothetical transitional stages, A—E, at which the different ramifications of the Primates are presumed to have branched off. Fossil Primates approximating to the postulated ancestral stages are represented by crosses.

place in a linear series in the actual evolutionary sequence of the existing Primates. In the diagram in fig. 5 the position of these postulated transitional types is indicated by the letters A to E. The chances of discovering the fossilized remains of such *direct* ancestral groups in early Primate phylogeny must be reckoned very slender indeed, if only for the reason that they were small arboreal, forest-living creatures—that is to say, they inhabited territories very far from conducive to the preservation of remains in a fossilized form. But a number of extinct types approximating to the postulated ancestral groups have in some cases already come to light. The relative positions of such approximating types are indicated quite diagrammatically by crosses in fig. 5, and the line of arrows is intended to make clear that, by reference to these types, it is at least possible to demonstrate the main trends of Primate evolution with far more assurance than by reference only to the comparative anatomy of living species. Or, to put it another way, they demonstrate a trend which approximates much more closely to the postulated direct line of descent (A to E) even though it by no means coincides with the latter. It cannot be too strongly emphasized that, while in the different fields of mammalian palaeontology the fossil record so far available can often establish *trends* of evolution, only if the record is sufficiently well documented by closely-graded and overlapping types can it be expected to demonstrate an actual linear sequence in terms of successive ancestor-descendant genera. So far as the Primates are concerned, the documentation of the fossil record in most cases still remains fragmentary.*

THE ANTHROPOIDEA

This suborder includes *Homo sapiens* as well as extinct types of hominid, the anthropoid apes and the monkeys. As the subordinal name implies, the members of the group are distinguished by their man-like appearance, which catches the attention of even the most unsophisticated eye. The human resemblance of the apes and monkeys on analysis resolves itself into a few outstanding characters such as the relatively voluminous and rounded brain-case; the flatness of the face; the position of the eyes, which look directly forwards and so appear rather close-set; the shrunken and fixed appearance of the external ear; the alertness and versatility of the facial expression, conditioned partly by the differentiation and complexity of the facial muscles; the mobility of the lips, and especially the upper lip which is not cleft and bound down to the gums as in most lower mammals; the absence of a naked moist surface (rhinarium) round the nostrils; the possession of a real

---

* T. H. Huxley (Lectures on Evolution, 1876), in reference to intermediate fossil types, suggested that 'it is convenient to distinguish these intermediate forms which do not represent the actual passage from one group to another, as *intercalary* types, from these *linear* types which, more or less approximately, indicate the nature of the steps by which the transition from one group to another was effected'. It is worth noting how very carefully Huxley was accustomed to express himself in these evolutionary problems.

'hand' employed for grasping purposes, combined with the functional independence of the thumb; and the presence of flattened nails on the digits of the hand and foot.

At the head of this suborder is placed the family Hominidae, a reasonable position in view of the fact that it represents in many ways the culmination of the main evolutionary trends which have characterized the Primates as a whole. Two genera of the Hominidae are now generally recognized, of which *Homo* is the only survivor and *H. sapiens* its only surviving species. The distinctive feature of *Homo* is the large brain which ultimately reaches the dimensions of modern man. One extinct type of *Homo*, Neanderthal man of late Pleistocene time, was commonly regarded as meriting specific rank, but this distinction does not now find general acceptance. The species *H. erectus* is known from fossil remains of Middle Pleistocene date found in Java and China, Africa and Europe. It is characterized by the relatively small size of the brain, the retreating forehead, and massive jaws and teeth. The second hominid genus is *Australopithecus*, known from abundant fossil remains found in Early and Middle Pleistocene deposits in South and East Africa. This genus is now regarded as representative of a subfamily of the Hominidae, the Australopithecinae, and some authorities have recognized other genera within the subfamily, e.g. *Plesianthropus* and *Paranthropus*. However, we shall here take the more conservative view that these different types are no more than so many different species or varieties of the one genus, *Australopithecus*. Two main types are now commonly recognized, the gracile and robust types of *Australopithecus*, which, by some, are taken to represent two separate species of the genus. When first discovered, there was considerable controversy on the taxonomic status of the genus—whether it was an exceedingly primitive hominid or a rather advanced type of anthropoid ape, and the atmosphere of discussion became rather confused by extreme claims put forward too hastily from two sides. It is now generally agreed that *Australopithecus* is properly to be classed as a genus of the Hominidae, for although the size of the brain had not advanced beyond the maximum so far recorded for anthropoid apes, the morphological details of the dentition, and the structural adaptations in the pelvis and other parts of the skeleton for an erect posture and gait, make it clear that the genus had already advanced a considerable distance along the evolutionary trends distinctive of the

Hominidae (and quite opposite to the trends which were followed by the anthropoid ape family).*

Even the extinct types of hominid so far known fill in the structural gaps between *H. sapiens* and the anthropoid apes in a most remarkable way, and they also make it clear (1) that the later stages of evolution which led to the emergence of the genus *Homo* and the species *H. sapiens* proceeded very rapidly (mainly in the latter half of the Pleistocene period) and (2) that some of the early hominids had a very wide geographical distribution in the Old World.

The anthropoid apes comprise the family Pongidae and to-day are represented by four genera, *Gorilla*, *Pan* (chimpanzee), *Pongo* (orang-utan), and *Hylobates* (gibbon).† Of these the first three are sometimes referred to as the large anthropoid apes, and are grouped as a subfamily, the Ponginae. The gibbons are placed in the subfamily Hylobatinae (or, by some authorities, in a separate family Hylobatidae).

The living anthropoid apes approximate structurally more closely to the Hominidae than do the monkeys in many features such as the relative size and configuration of the brain, details of the skull, skeleton and dentition, in the tendency to adopt a semi-erect or orthograde posture with which are correlated certain features such as a broadening of the thorax, the disposition of the abdominal viscera, the absence of a tail, and so forth. These resemblances have been noted and emphasized since the time of Darwin and Huxley, and they have been still further re-emphasized by discoveries of fossil hominids which show a much closer approximation to apes. It is for this reason that the Hominidae and Pongidae are taxonomically placed in a common superfamily, Hominoidea. Yet it is important to recognize that, since their divergence from a common ancestral stock, the Hominidae and Pongidae have followed very different and contrasting evolutionary trends. The anthropoid apes are pre-eminently arboreal specialists and, in association with their

---

* Since going to press new discoveries of the australopithecine group have been briefly described from deposits in the Omo Valley of Ethiopia. The antiquity of these deposits is estimated to range from two to four million years, suggesting that the genus *Australopithecus*, or its immediate antecedents, extended back to the terminal Pliocene (F. Clark Howell, 1969. 'Remains of Hominidae from Pliocene/Pleistocene Formations in the Lower Omo Basin, Ethiopia', *Nature*, *223*, p. 1234).

† One of the larger gibbons, the siamang, is commonly placed in a separate genus, *Symphalangus*.

FIG. 6. A gibbon (*Hylobates*). Note the inordinate length of the arms and the reduced thumb, an adaptation related to the highly specialized brachiating habits of these anthropoid apes.

habit of swinging among the branches with their arms (a mode of progression which is called *brachiation*), they have developed distinctive structural modifications such as a relative lengthening of the forelimb and a varying degree of atrophy of the thumb. The orang-utan and gibbon are the most exclusively arboreal and the most extreme brachiators, and in them, therefore, the accompanying structural specializations are most strongly marked (fig. 6). The *orang-utan* is an Asiatic ape, confined to-day to Borneo and Sumatra, but in Pleistocene times it extended its range as far north as China. It attains to a fairly large size, is slow and deliberate in its movements, and in the wild state only rarely descends to the ground from the trees.

The *gibbons* are found in the south-eastern parts of tropical Asia, extending over a wide area from Assam, Formosa, and the footslopes of the Himalayas, to Malaya, Java, Sumatra and Borneo.

They are slenderly built creatures of great agility, and in a number of features they provide a structural transition between the larger apes and the Old World monkeys, e.g. in the constant presence of well-developed ischial callosities ('sitting pads') and in the relative proportions of the brain. On the other hand, in their powers and range of vocalization and their aptitude for standing and walking in a bipedal manner, they offer points of particular interest in relation to the ultimate origin of the Hominidae. But it is important to note that erect bipedalism is not the usual mode of progression in gibbons except on their rare descents to the ground or along the larger branches of trees. Like the orang, they are exclusively arboreal in their habitat but very much more active in their movements. They are found in monogamous family groups of up to six individuals.

FIG. 7. A chimpanzee, illustrating the usual mode of progression on the ground—a modified type of secondary quadrupedalism in which the knuckles of the hand are used for support.

The *chimpanzee* is about the same size as the orang, but less obviously specialized for brachiation (fig. 7). While an active climber and almost constantly living in forests, it often comes to the ground and then walks quadrupedally and may occasionally elevate itself into a bipedal position. Chimpanzees inhabit the

forests of equatorial Africa, extending from Gambia in the west to
Lake Victoria in the east. Over a limited range south of the Congo
river is an interesting pygmy variety recognized by some authorities
as a distinct species, *Pan paniscus*.

The *gorilla* is the largest of the modern anthropoid apes, and like
the chimpanzee is found in equatorial Africa but in a more restricted
area. It is only slightly arboreal, spending much of its daily life
moving about in the undergrowth of the jungle. On the ground it
normally adopts a quadrupedal gait or, because of the inordinate
length of its powerful arms, what has been termed an obliquely
quadrupedal gait. Only rarely, particularly in attitudes of attack
or defence, does it rear itself momentarily on its hind limbs (fig.
8). The limb proportions of the gorilla (including the relatively
short thumb) indicate that it was primarily an arboreal brachiator
which, because of its great bulk, has secondarily taken to a terrestrial habitat and has become limited in its arboreal life to the large
lower branches of the trees. Both the chimpanzee and gorilla,
when assuming a quadrupedal gait, support their elongated forelimbs on the knuckles of the hand. In this connection, it is interesting to note that the term 'brachiator' was first used in 1859 by
Sir Richard Owen [112] in reference to the gibbon when he posed
the question 'Of the broad-breast-boned quadrumana, are the
knuckle-walkers or the brachiators, i.e. the long armed gibbons,
most nearly and essentially related to the human subject?' The
connotation of the term brachiation was later amplified and applied
by the late Arthur Keith to the mode of arboreal progression which
involves the habit of swinging from branch to branch by the arms
to the exclusion of movements of running, scampering, and leaping.
Brachiation in this sense, as Keith showed, is associated with functional and structural adaptations of a far-reaching type. Recently,
however, the use of the term has been extended by some writers to
include almost any action of arm-swinging, whether this is habitual
or not, so that even monkeys which are often terrestrial in their
mode of life may be called 'brachiators'. In order to avoid confusion it seems desirable to distinguish clearly between the term
brachiation in its original sense, and the incidental brachiation of
the quadrupedal monkeys. Gorillas live in small groups of a few
families whose numbers range from five or six up to thirty and, in
spite of stories which were at one time common, are not fierce or
aggressive by nature. On the contrary, in the wild they appear to be

rather docile creatures, restricting themselves to a vegetarian diet. The mountain gorilla (regarded as a distinct subspecies) is said to be more adapted to terrestrial life than the ordinary, or forest, gorilla, and the upper limbs are relatively shorter. The structural similarities between the gorilla and chimpanzee have led some authorities to doubt whether they are properly to be accorded separate generic rank.

Fossil types of anthropoid ape are now known in considerable numbers. One of the earliest is *Parapithecus*, represented by lower jaws recovered from Lower Oligocene deposits in Egypt.* But the pongid status of this genus is uncertain, and (as a sort of provisional device pending the discovery of more complete material) it is commonly assigned to a separate family of the Hominoidea, Parapithecidae. In this family is included another fossil Primate from Egypt, *Apidium*, known only from jaws and parts of the dentition. From the same Oligocene strata there has recently been described the frontal region of a skull which conforms in size to the *Parapithecus* mandible, but it may equally well belong to one or other of the smaller fossil Primates also known from these deposits. Also derived from Oligocene deposits in Egypt are several fossil apes of very primitive type. One of these is *Propliopithecus*, a genus which has been assumed, on the basis of its dental morphology, to be an early representative of the Hylobatinae and perhaps closely related to the ancestral stock from which the modern gibbon was evolved. However, another genus, *Aelopithecus*, is somewhat more specialized in the direction of Hylobatine evolution and for this reason is regarded by some authorities as more likely to be a forerunner of the gibbons of today. *Propliopithecus*, in the generalized characters of its dentition, may be ancestral to later hominoids and it has even been suggested that it was related to the basal stock whence the family Hominidae came to be derived. The oldest known, and almost complete, skull of a fossil ape found in the Fayum region of Egypt is of later Oligocene age, dating back to about 30 million years. It has been named *Aegyptopithecus* and may have been descended from the earliest Fayum ape

---

* Tiny fragments of Primate jaws from Upper Eocene deposits in Burma have been described under the generic names *Amphipithecus* and *Pondaungia*, and may possibly represent a still earlier stage in the evolutionary origin of the Pongidae. But the evidence they provide is too insecure to permit firm conclusions regarding their significance. It is perhaps more probable that they are relics of the basal stock of the Anthropoidea from which both the Hominoidea and the Cercopithecoidea took their origin.

so far known, *Oligopithecus*, which in its dentition shows resemblances not only to anthropoid apes but also to some of the Eocene prosimians as well as to monkeys [140]. The main interest of *Aegytopithecus* centres on the fact that, while the dentition is typically pongid, the skull and fragments of the limb bones so far known show certain primitive, or cercopithecoid, traits. Evidently this genus is to be regarded as marking a transition from the cercopithecoid level of primate evolution to the pongid level, and this interpretation receives support from the discovery of tail bones closely associated with the skull.

A great many remains of fossil apes have been found in Early Miocene deposits in East Africa which have an antiquity of about 18 million years. One of these is quite clearly an early type of gibbon but much less specialized for brachiation than the modern species. The dentition conforms to the hylobatine pattern, but the limb bones still show many resemblances to those of arboreal quadrupedal monkeys. Evidently it was an incipient brachiator. This fossil type was given the generic name *Limnopithecus* because its remains were found in lacustrine deposits, but there is now considerable doubt whether it can be properly separated from the previously known hylobatine genus *Pliopithecus*. We shall therefore, to avoid confusion, refer to the East African genus as *Pliopithecus* (*Limnopithecus*). *Pliopithecus* was common in the European region during Miocene times, possibly extending into the Pliocene. Apart from the generalized character of the limb bones, this genus probably still retained a tail. Following the Oligocene period, the most complete records of the large anthropoid apes are those of the Early Miocene species of the genus *Dryopithecus* from Kenya. These species were formerly grouped in a separate genus *Proconsul*, but it is now accepted by competent authorities that, at the most, they comprise only a subgenus of *Dryopithecus* [141].

The East African species of *Dryopithecus* (*Proconsul*) ranged in size from that of a pygmy chimpanzee to an ape of gorilloid dimensions, and it may well be that they include the ancestors of the modern chimpanzee and gorilla. They were remarkably generalized apes, with a limb structure showing no more than an early phase of brachiating adaptations and, like the Oligocene *Aegyptopithecus*, a number of features of the skull in which they approximated much more closely than modern apes to cercopithecoid monkeys. It seems likely, indeed, that they were actually

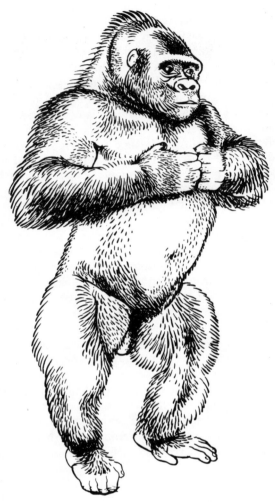

FIG. 8. A male gorilla, illustrating the imperfect type of erect posture which may be temporarily assumed. Note the great length of the upper limbs in proportion to the lower limbs. (Adapted from a photograph by the Zoological Society, Philadelphia.)

descendants of the Oligocene genus. During the Miocene and Pliocene periods *Dryopithecus* was also fairly widespread in Europe and Asia, extending as far east as China, but in these regions, apart from a single humerus and femur and two forelimb bones, the genus is known only from jaws and teeth. It is generally considered probable that the lineage of the Hominidae, as well as that of the modern pongids, may have stemmed from one or other of the groups of dryopithecine apes, and for this reason they are of considerable interest for students of human evolution. From the Siwalik hills of India other Pliocene types of ape have been recorded and assigned to distinct genera on grounds which, because of the fragmentary nature of the remains, do not seem to be always very sure, e.g. *Sivapithecus, Palaeosimia, Bramapithecus, Sugrivapithecus,* and *Ramapithecus.* Of these the last is of particular interest, for the small size of the teeth (particularly the canine) and the relatively simple construction of the molars have suggested the possibility that they were not far removed from the transitional phase which presumably marked the evolutionary derivation of the hominid type of dentition from that characteristic of most of the other Pliocene and Miocene apes. Indeed, it has been suggested that *Ramapithecus* may actually be a Pliocene representative of the family Hominidae.

Oreopithecus, a lower Pliocene genus whose remains have been found in Italy, is known from a number of specimens, and mainly because of certain dental characters it had until recently been grouped with the Cercopithecoidea. But the discovery of a skull, limb-bones, vertebrae and ribs has now made it clear that this genus is more closely allied to the anthropoid apes. It has even been claimed as an early type of hominid, but this interpretation is open to serious doubt. Because of its rather unusual combination of characters, it may for the present be assigned to a separate family (Oreopithecidae) of the Hominoidea. Close resemblances in dental morphology between *Oreopithecus* and the much earlier genus *Apidium* have been noted by Simons [136], suggestive of a phylogenetic relationship. From the Pleistocene of China and the Middle Pliocene of India have been discovered jaws of a very large ape that probably equalled in body size the mountain gorilla of today. It has been given the generic name *Gigantopithecus.* The flattened wear of the molar teeth combined with a reduction in size of the front teeth has suggested that it may have some affinity with the Hominidae, but such an interpretation has not found general acceptance. Evidently it was a true pongid of somewhat aberrant type, with certain specializations peculiar to itself.

The monkeys represent the lowest stratum of the Anthropoidea which exists to-day. Most of them are thoroughly arboreal, but many (especially the Old World monkeys) quite frequently come to the ground, and some are wholly terrestrial. Both in the branches and on the ground they commonly assume a horizontal or prono-grade posture associated with a quadrupedal gait. Unlike the Pongidae, the forelimbs are (with a few exceptions) not much longer than the hindlimbs, and in some of the Old World monkeys they are shorter. Thus, their limb proportions approach more closely the primitive mammalian condition in which the hind-limbs were probably slightly the longer. Most of them also have well-developed tails. The monkeys are divided into two clearly de-fined groups, Cercopithecoidea and Ceboidea. The former are also called Old World or Catarrhine monkeys, and the latter New World or Platyrrhine monkeys. The significance of the terms Catarrhine and Platyrrhine is related to differences in the dis-position of the nostrils, but this is a distinction which is not always very striking. Among other things, the two groups are contrasted in the characters of the dentition, certain features of the skull, and in the presence or absence of ischial callosities and cheek pouches. They are also strongly contrasted in their present distribution, and it is partly the circumstances of their geographical habitat which have led some authorities to suppose that the two groups probably evolved independently, perhaps from different prosimian ancestral stocks.*

The Cercopithecoidea comprise one family only, Cercopithe-cidae, which has a wide distribution over the Old World. The genus *Macaca* is perhaps the most familiar type (fig. 9). The macaque monkeys are robustly built animals of moderate size, usually with rather short tails (in one species practically absent). Their geo-graphical range extends from north Africa through southern and eastern Asia as far as Japan. *Cercocebus*, the mangabey monkey, is restricted to tropical Africa and is characterized by conspicuous white upper eyelids and a long tail. It is more thoroughly arboreal

---

* It should be noted that, in older schemes of classification, the category Catarrhini was taken to include the Hominidae and Pongidae as well as the Old World monkeys, and the latter were then frequently referred to as Cynomorpha to distinguish them from the anthropoid apes (Anthropomorpha).

FIG. 9. A macaque monkey (*Macaca mulatta*).

than *Macaca*. *Cercopithecus* (the guenon monkey) is the most abundant African genus and almost a hundred species and subspecies
have been described. It is similar to *Cercocebus*, but shows certain
differences in the dentition. Many of the species are brightly
coloured with rather unusual tints. One genus, *Erythrocebus*, distributed widely throughout equatorial Africa, is noted for the fact
that it is only slightly arboreal in its habits; the various species of
this genus are commonly found in open and sparsely wooded
country. The baboons (*Papio*) and mandrills (*Mandrillus*) inhabit
rocky open country, and the former seem to have abandoned
arboreal habits almost entirely (fig. 10). They are quadrupedal

animals of formidable size, with large projecting muzzles and (in the males) ferocious canines. The tail is rudimentary in the mandrills, and in the baboons relatively short. Baboons live in well-established communities; while predominantly vegetarian, they may on occasion attack and eat small mammals. *Cynopithecus* is a genus which is found in Celebes, and consists of a single species. So far as its external appearance and internal characters are con-

Wilson

FIG. 10. A baboon (*Papio*), a Catarrhine monkey adapted to terrestrial quadrupedalism. (By courtesy of the British Museum.)

cerned it appears to stand in close relationship to macaque monkeys. It is essentially arboreal in habit, with a much reduced tail.

All the cercopithecid genera hitherto mentioned are grouped as one subfamily, Cercopithecinae. In another subfamily, Colobinae, are collected the langurs (*Presbytis, Rhinopithecus*, etc.), the guereza monkeys (*Colobus*) and the proboscis monkey (*Nasalis*). They are distinguished as a whole by the absence of cheek pouches and the development of an elaborate and sacculated stomach in association with their strictly vegetarian diet. They are for the most part arboreal, feed largely on leaves and the young shoots of certain plants, and in captivity are much more delicate animals than the Cercopithecinae. The langurs are distributed over Malaysia, extending up to China and also into Kashmir and the southern slopes of the Himalayas.* *Nasalis* is confined to Borneo and is remarkable for the development of a conspicuous fleshy proboscis which is much larger in the male and gives the animal a most grotesque appearance. *Colobus* is the only African representative of the Colobinae. Over thirty species and subspecies have been described, and some of them possess a beautiful particoloured coat of long hair. In correlation with their specialized arboreal mode of life, the thumb is either very greatly reduced or may be practically absent altogether.

Fossil catarrhines have been found over a wide area in the Old World. Reference has been made (p. 53) to the Parapithecidae, a family of Oligocene Primates including two genera, *Parapithecus* and *Apidium*, and it was noted that, although they have been commonly regarded as exceedingly primitive anthropoid apes, this is no more than a provisional interpretation of their taxonomic status. They may have been early precursors of the cercopithecoid monkeys or, perhaps more likely, the family as a whole comprised a common ancestral stock from which both the monkeys and the apes took their origin. *Mesopithecus* is known from unusually complete fossil material of Lower Pliocene antiquity found mainly in Greece, but also in Czechoslovakia, South Russia and Persia. It is a fully differentiated cercopithecoid and is regarded (on its skull characters) as an extinct genus of the Colobinae. A number of jaws and teeth from the Lower Miocene of East Africa have also been assigned provisionally to this genus; their main interest is their demonstration that even at this early age the Cercopithecoidea had already undergone segregation to form a well-defined group. Closely similar to *Mesopithecus* is *Dolichopithecus* from the Pliocene of France. Another cercopithecoid, *Libypithecus*, is known from Middle Pliocene deposits in Egypt, and *Simopithecus* (very similar to the modern Gelada baboon of Abyssinia, and in some species reaching a massive size) from the Upper Pliocene and Lower Pleistocene of East Africa. Extinct Pleistocene baboons, *Dinopithecus*, *Gorgopithecus* and *Parapapio* also existed in South Africa. Remains attributable to the modern genus *Papio* have been found in Lower Pleistocene strata in North India, Algeria and Egypt, while *Macaca* seems to have been quite widely distributed during the Pliocene in Europe, India, China and North Africa. The Recent

---

* It was almost certainly a Himalayan species of langur which gained notoriety, from footprints in the snow, as the Abominable Snowman.

colobine genus, *Presbytis*, has also been reported from Upper Pliocene horizons in Europe.

From the palaeontological data it is thus evident that the Cercopithecoidea ranged over a considerable part of the Old World during the latter half of the Tertiary epoch. Especially noteworthy is the presence in past times of baboons in North India, and of macaque monkeys in North Europe even so late as the Pleistocene period.

The Platyrrhine monkeys, Ceboidea, which are to-day confined to South America, reaching as far north as southern Mexico, are grouped in two families, the Cebidae and the Callitrichidae* (= Hapalidae). The latter comprise the marmosets, of which several genera have been described (but the number of these genera varies considerably with the author who describes them). They include the smallest of living monkeys, being about the size of squirrels, and are characterized by having sharp curved claws on all the digits except the first toe or hallux (which bears a flattened nail) and by the loss of the third molar tooth. The latter is a constant feature except in the genus *Callimico*, which for this reason has sometimes been classified with the Cebidae.

The Cebidae are larger animals, though they do not reach the size of some of the Old World monkeys. They are thoroughly arboreal in habit, and some of them have developed prehensile or semi-prehensile tails by means of which they can hang and swing from the branches, or which they can even use as a 'third hand' for grasping objects such as food. A number of subfamilies have been recognized among the Cebidae, but it will be adequate for our present purpose to note a few of the better known genera. *Cebus* includes the small capuchin monkeys which are familiar as pets (fig. 11). The squirrel monkeys (*Saimiri*) are small creatures characterized among other features by an exceptional elongation of the occipital region of the skull and by a brain which, relative to the body weight, is unusually large. *Aotus* is a nocturnal genus, with rather large eyes and much reduced ears; colloquially these night monkeys are known as douroucoulis. The teetee monkeys (*Callicebus*) are found on the banks of the Amazon, and superficially are not unlike *Aotus* except that they are diurnal in habit, the eyes are

---

* Dr J. R. Napier informs me that this is the correct spelling, in place of the usual spelling Callithridae. In this case the generic term commonly in use, *Callithrix*, should more properly be *Callitrix*.

smaller, and the nostrils wider apart. *Pithecia* includes the saki monkeys which have long, but non-prehensile, tails. The spider monkeys (*Ateles*) represent the culmination of the tendency towards the kind of arboreal specialization which is manifested in varying degrees by the Cebidae in general (fig. 12). The tail is provided with a naked sensory surface on its under surface near the

Wilson

FIG. 11. A capuchin monkey (*Cebus*). (By courtesy of the British Museum.)

tip and is marvellously prehensile, and the thumb is either completely absent or extremely vestigial. In the relative elongation of the upper limbs and certain other features, the spider monkeys show some interesting resemblances to the gibbons, an expression of convergence related to somewhat similar arboreal activities. Lastly, the genus *Alouatta* contains the howler monkeys, the largest of the Platyrrhines. The skull of this monkey is unusually modified in association with the development of an enormously inflated hyoid bone. The latter functions as a resonating apparatus by means of which an exceptionally great volume of vocal sound can be emitted.

The geological record of the New World monkeys is very scanty. A portion of a skull and dentition of a small cebid from the Miocene of Patagonia has been given the generic name *Homunculus*. This genus, remains of which have also been found in deposits of a similar age in Colombia, is probably related to the howler monkeys. Two other genera from the Miocene of Colombia are known from much more complete material, *Cebupithecia* and *Neosaimiri*; they are closely similar to the modern *Pithecia* and *Saimiri*.

TARSIOIDEA

This group, the only superfamily of the infraorder Tarsiiformes, is represented by one living genus, *Tarsius*, the little tarsier which is found in Borneo, the Philippines, Celebes and Sumatra, and also in some of the smaller islands of the Malaysian Archipelago. This animal—which is the size of a small kitten—has for many years attracted the attention of comparative anatomists because it

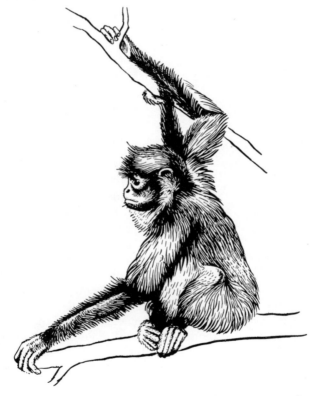

FIG. 12. A spider monkey (*Ateles ater*). Note the relatively long arms and the prehensile tail in this specialized arboreal type of Platyrrhine monkey.

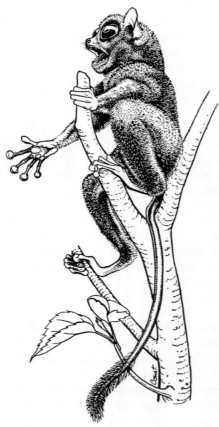

combines in its anatomical organization a number of remarkably primitive with unexpectedly advanced characters. The consensus of opinion today recognizes that *Tarsius* occupies a systematic position somewhere between the lemurs and the monkeys, but there is still no final agreement on its taxonomic status. By some authorities it has been separated from the lemurs altogether and accommodated in a central suborder of the Primates, but with the discovery of fossil types which in some features approximate much more closely to the lemurs it appears to be more reasonable to place it in the same suborder (*Prosimii*) as the latter, and at the same time to emphasize its distinctive characters by giving it an

Fig. 13. A male tarsier (*Tarsius spec-* independent infraordinal
trum) from a photograph by the author. status.

The modern tarsier is a crepuscular and entirely arboreal creature. It shows marked specializations in the relatively enormous size of the eyes, and in the peculiar modification of the hindlimbs for leaping among the branches (fig. 13). The ears are large, and the tail (which is not prehensile in the usual sense of the term) is long and naked except for a terminal covering of hair. In the structure of the nose and upper lip it resembles the New World monkeys and contrasts rather strongly with the lemurs.

In the early part of the Tertiary epoch, tarsioids were widely distributed over the world, and some twenty-five different genera have been reported from Palaeocene and Eocene deposits. Of these fossils more than half are derived from North America, and most of the remainder from Europe. A few remains of presumed tarsioids, *Hoanghonius* and *Lushius*, have also been found in China. The former is of Late Eocene, or possibly Early Oligocene, date and the latter of Middle Eocene. It is important to emphasize that some of the early genera are known only from quite small jaw fragments with very limited portions of the dentition; the tarsioid nature of these types has been inferred from the character of the teeth, but it remains to be determined by further discoveries whether this diagnosis is in all cases confirmed by the characters of the skull and limbs. Because of the fragmentary evidence the classification of the fossil tarsioids is peculiarly difficult, and (mainly for the purpose of convenience) most of them are usually grouped in two families, Anaptomorphidae and Omomyidae. The modern tarsier is given separate familial status in the Tarsiidae which also includes the extinct subfamily Necrolemurinae. Of the latter the Eocene genus *Pseudoloris* is remarkably similar to the modern *Tarsius*, so much so that its wide taxonomic separation tends to obscure a most interesting point, i.e. that *Tarsius* is a "living fossil" in the sense that it has persisted with little change for about fifty million years. Another extinct family of prosimians, which should probably be included in the Tarsioidea, is represented by the Carpolestidae, little creatures from the Palaeocene of North America with curiously specialized premolar teeth.

In very few fossil tarsioids, *Necrolemur*, *Tetonius*, *Nannopithex*, *Hemiacodon* and *Pseudoloris*,* the skull is known in whole or in part, and in some (*Necrolemur*, *Nannopithex*, *Hemiacodon* and *Teilhardina*) portions of the limb skeleton have been found which add corroborative evidence to that of the dentition. From these specimens it is clear that some of the characteristic specializations of the skull and hindlimb of the modern tarsier were already developed (but in an incipient and modified form) in early Tertiary times. The European genera *Necrolemur*, *Microchoerus*, *Nannopithex* and *Pseudoloris* are of special interest because in some structural features they seem to have approximated more closely to monkeys, and the group as a whole (Necrolemurinae†), or at any rate a closely re-

* It should be noted that some confusion has been introduced into the literature on fossil tarsioids by the misidentification of some of the remains of *Nannopithex* as those of *Pseudoloris* [137].

† Some authors use the term Microchoerinae for this subfamily.

lated group, may have provided the matrix from which the Anthropoidea took their evolutionary origin. Nothing whatever is known of the palaeontological record of the Tarsioidea after the Eocene, with the exception of a lower jaw and dentition from the Lower Oligocene of North America which has been assigned to a new genus, *Macrotarsius*, and perhaps the Chinese genus *Hoanghonius*.

LEMURS

As we have already indicated, the colloquial term lemurs is usually taken to include the lemuriforms (exclusive of the tree-shrews) and lorisiforms. The members of both these groups display their Primate status much less obtrusively than the monkeys, and in their appearance during life they seem to occupy an intermediate position between monkeys and lower mammals. It is for this reason that German naturalists refer to them as *Halbaffen*. In past years, indeed, the question had been debated whether they should not properly be excluded from the Primates altogether and accommodated in a separate mammalian order, *Lemures*. In later chapters the evidence for the Primate status of the lemurs will be reviewed in some detail, but we may note at this juncture that their evolutionary history has demonstrated a number of developmental trends which coincide closely with those of the Anthropoidea. In their external characters, the lemurs resemble monkeys in the functional adaptation of the extremities (manus and pes) for grasping purposes, in the freedom and mobility of the pollex and hallux, and in the presence of flattened nails on the digits. In all except one type of lemur (*Daubentonia*) only the second pedal digit is provided with a sharp claw. On the other hand, a strong contrast with the monkeys (and a corresponding approximation to non-Primate mammals) is provided by a variable elongation of the snout which in most cases projects forwards well beyond the level of the chin, the naked and moist skin of the muzzle (rhinarium), the median cleft in the upper lip where it is bound down to the underlying gum, the freely-moving and usually large ears, and the immobile expression of the face. The lemurs are almost all entirely arboreal creatures. Most of them are also nocturnal in their activities (with the exception of some of the Madagascar lemurs, e.g. *Lemur*, *Propithecus* and *Indri*), and the nocturnal habits are correlated with the rather large size of the eyes.

The Recent lemurs can be divided into two main groups,

Lemuriformes and Lorisiformes, which are distinguished by certain characters of the skull, nasal cavity, genital system, etc. The lemuriform lemurs* are confined to-day entirely to Madagascar and the neighbouring Comoro Islands, and are often referred to as the

FIG. 14. The ring-tailed lemur (*Lemur catta*).

Malagasy lemurs. They comprise two superfamilies—Lemuroidea and Daubentonioidea, of which the former is divided into two families, the Lemuridae and Indriidae. The family Lemuridae contains the genus *Lemur* which, though not so completely arboreal as most lemuriforms, is a very active climber and is provided with a long tail to serve a balancing function (fig. 14). Other genera

* This somewhat awkward tautology will perhaps be excused in the interests of clarity; it seems to be justified in order to avoid the confusion introduced by the various (and occasionally quite personal) nomenclatures employed by some primatologists.

include *Cheirogaleus* (the dwarf lemur) and *Microcebus* (the mouse lemur), both tiny creatures of attractive appearance which are entirely nocturnal and which go into a sort of hibernant sleep during the dry season, accommodating themselves in a nest of leaves and sustaining themselves by reserves of fat which they accumulate mainly in the base of the tail (fig. 15). The Indriidae, which include the largest of modern lemurs, are distinguished from the Lemuri-

FIG. 15. The mouse lemur (*Microcebus murinus*) from a photograph by the late F. W. Bond.

dae mainly by a numerical reduction of the dentition, as well as by a number of minor characters. They are thoroughly arboreal (though occasionally descending to the ground) and often leap from branch to branch in a more or less vertical position. Correlated with this mode of progression is the relative length of the hindlimbs.

The superfamily Daubentonioidea is represented by one genus only, *Daubentonia* (the aye-aye). This most peculiar and aberrant lemur is characterized by the enlargement and rodent-like appearance of its incisor teeth and the gross reduction of the rest of the dentition, by the retention of modified claws on all the digits except the hallux, by the curious attenuation of the third digit of the hand, and by unusual features of the brain. It is small wonder, then, that its taxonomic position has often given rise to controversy (the older naturalists actually placed it among the rodents). But while in many points of its anatomy it departs widely from other lemuriforms, its inclusion in this group of Primates seems well assured by a number of highly significant characters in the skull, limbs and viscera.

The lorisiform lemurs comprise a single family, Lorisidae, which is subdivided into two subfamilies, Lorisinae and Galaginae. The former include the genera *Loris*, *Nycticebus*, *Perodicticus*, and *Arctocebus*, of which the first two inhabit southern Asia and the others tropical Africa. They are all slow-moving creatures, climbing among the branches with a very deliberate crawling gait. The fore and hindlimbs are of almost equal length, the snout somewhat abbreviated, the eyes unusually large, and the tail very reduced or absent. The subfamily Galaginae contains one genus, *Galago*, and a subgenus *Euoticus*, both of which are characterized by the extreme modification of the elongated hindlimbs for jumping, and in this respect they parallel rather closely the tarsioid condition.

With the solitary exception of *Daubentonia*, in all the modern lemurs (both lemuriform and lorisiform) the lower incisor and canine teeth are curiously modified to form a dental comb with which the animals comb their fur, while the second pedal digit (which, as already noted, is furnished with a sharp claw) is used as a toilet digit for scratching purposes.

Fossil lemurs are known from early Tertiary deposits of Europe and North America. The Lemuriformes are represented by a widespread family, Adapidae, divided into the subfamilies Adapinae and Notharctinae. The former is confined to the Old World and includes two genera, *Adapis* and *Pronycticebus*. These have been found in Eocene deposits in France and are now known by well-preserved skulls with the dentition. Several species of *Adapis* have been described, of which the commoner are *Adapis magnus* and *Adapis parisiensis*. *Pronycticebus* is noteworthy because of certain primitive features of the skull and dentition. In Eocene times, lemurs similar to *Adapis* were evidently very widespread over the Euro-Asian region, for the fossil remains of one of these creatures has been described from deposits in China. From Lower Eocene (or perhaps Palaeocene) levels in France and England fragmentary mandibles of a primitive lemuriform, *Protoadapis*, have been described. It is generally agreed that (in spite of its name) this genus is not closely related to *Adapis*, but probably has affinities with the New World genus *Pelycodus*. The Notharctinae comprise the genera *Notharctus*, *Pelycodus*, and *Smilodectes* (= *Aphanolemur*)— all North American fossils. *Smilodectes* is known from several skulls with most of the dentition well preserved in the specimens. *Pelycodus* includes several species of which the earliest dates from the Palaeocene and the others from the early Eocene; they were exceedingly primitive Primates, and the whole group is probably to be regarded as including direct precursors of the more advanced genus *Notharctus*. The fossil remains of the latter include an almost complete skeleton, and it is therefore by far the best known of all the early Tertiary lemurs. It has been suggested that the Notharctinae may have provided the ancestral stock from which the New World monkeys were derived; such a proposition involves the assumption that the Cercopithecoidea and Ceboidea developed independently from different groups of prosimians, and that many of the resemblances between them are therefore the result of parallel evolution.

During Pleistocene times a number of remarkable extinct types of lemur made their appearance in Madagascar. One of these, *Megaladapis*, reached a gigantic size; indeed, it is probably the largest Primate known, for it almost certainly exceeded the gorilla in bulk. It belongs to the family Lemuridae. *Archaeolemur*, *Hadropithecus*, *Palaeopropithecus* and *Mesopropithecus*, which are also Pleistocene Malagasy lemurs of an aberrant type, are referable to the Indriidae (though the first two are sometimes allocated to a separate family, Archaeolemuridae). Both *Mesopropithecus* and *Archaeolemur* have attracted attention because of the monkey-like appearance of their skull characters.

Fossil lorisiforms are known from specimens of the skull, jaws and teeth found in Early Miocene deposits in East Africa. They have been referred to the genus *Progalago*, of which several distinct species have been recognized. In general, these creatures were rather similar to the modern *Galago*, but more primitive in some characters and in others showing an approximation to the Asiatic lorisiforms. The only fossil relic presumed to be an extinct representative of the Lorisinae consists of a single molar tooth from a

Pliocene level in India. This belonged to a fairly large animal (to which the generic name *Indraloris* has been given) and it was at one time assumed to represent the structural ancestor of all the existing Lorisiformes. However, apart from the obvious fact that a single tooth can hardly justify such a far-reaching conclusion, the assumption is contradicted by the discovery of the previous existence of lorisiforms in East Africa.

TREE-SHREWS

Tree-shrews are small arboreal creatures which are widely distributed in south-east Asia, extending over India, Burma, the Malay Peninsula, Sumatra, Java and Borneo [85]. The larger species tend to confine their activities to the forest undergrowth or to the lower branches of trees, while the smaller species occupy the heights of the forest canopy. They are characterized by their agility and quick movements, associated with which they have a particularly alert expression that seems to be accentuated by their relatively large eyes. They are in many ways very primitive and generalized, so much so that they have commonly been included (and still are by some zoologists) with other heterogeneous groups of primitive mammals in the order Insectivora, with a more particular relationship to an extinct branch of the latter called the Leptictidae. As we shall see, however, they show in their total morphological pattern such a remarkable number of resemblances to the lemurs, and particularly the Lemuriformes, that in spite of their primitive characters their affinities with the latter seem reasonably certain. We shall therefore continue to adopt the provisional classification which places them as one of the superfamilies (Tupaioidea) of the Lemuriformes. The living tree-shrews differ from other living Primates in the fact that all the digits of the manus and pes are furnished with sharp, curved claws, and the brain has a relatively simple construction. On the other hand, the digits possess a considerable degree of mobility for grasping purposes, and in the development of certain features of the brain the latter shows a much closer approximation to typical lower Primates than do any of the Insectivora.

The modern tree-shrews are divided into two main groups, Tupaiinae and Ptilocercinae. The former includes several genera, of which *Tupaia* is the most common. This genus has a close superficial resemblance to squirrels; indeed, the generic term is derived

FIG. 16. A young specimen of the Lesser Tree-shrew (*Tupaia minor*) from a photograph by E. Banks.

from the Malay word 'tupai' which means a squirrel. The resemblance is enhanced by the long, bushy tail, which is used as a balancer in arboreal activities (fig. 16). Although the Tupaiinae are predominantly insectivorous, feeding on cicadas, grasshoppers and so forth, they will eat fruit and seeds and also very small mammals. Thus, one species (*Tupaia glis*) has been observed to show great skill in attacking and killing mice, and to consume almost the entire animal. It is of particular interest, also, to note that it combs its fur (particularly the long fur of the tail) with its procumbent lower incisor teeth in exactly the same manner as a lemur [53]. While the Tupaiinae are diurnal in their habits, the only genus of the Ptilocercinae (*Ptilocercus*, or the pen-tailed tree-shrew) is nocturnal. Its colloquial name refers to the fact that the tail is naked except on the terminal part where the hair is disposed like the fringing barbs of a quill feather.

It is particularly unfortunate that no indubitable fossil tupaioids have so far been discovered. A skull from Oligocene deposits in Mongolia, given the generic name *Anagale*, was at one time assigned on what appeared to be good evidence to the superfamily Tupaioidea, but later studies failed to confirm this diagnosis.

During the Palaeocene and Eocene ages of Europe and North America there existed two families of extremely primitive Primates, Plesiadapidae and Paromomyidae, both of which, however, also developed specializations peculiar to themselves. They are mentioned here not with the implication that they are both more closely related to the Tupaioidea than to some other early groups of prosimians, but because in the degree of their primitiveness they do occupy a level somewhat comparable with that of the tree-shrews, though in certain respects they are even more primitive. Our knowledge of the Plesiadapidae rests largely on jaws and teeth, but portions of the skull, limb bones and vertebral column of one of the American species are known, and in recent years an almost complete skull and some limb bones have been discovered in Palaeocene deposits in France. A detailed study of these remains has led to the conclusion that the plesiadapids are definitely to be classed in the order Primates, and that they show evidence of relationship to the tupaioids on the one hand, and to the lemuroids (e.g. adapids and Madagascar lemurs) on the other. The family Plesiadapidae includes at least four different genera of which one, *Plesiadapis*, is of peculiar interest for the reason that it appears to be the only known genus of fossil Primates with a common distribution in the Old and New World. Closely related to the Plesiadapidae, as evidenced by similarities in the skull and dentition, is an Eocene family Microsyopidae which like the former, shows resemblances to the Tupaioidea. By some authorities the microsyopids are actually included in the Primates. Another family, the Apatemyidae (at one time included in the Plesiadapidae) has been commonly regarded as an aberrant and sterile ramification of the basal Primate stock which probably became extinct by the end of the Oligocene, but it is also possible (though less likely) that they were an aberrant group of insectivores which may have been an offshoot close to the insectivore-primate transition, but before the segregation of the Primates as a separate order. The plesiadapids, paromomyids and apatemyids (but particularly the last) are characterized by an enlargement of the first incisor teeth and a numerical reduction in the rest of the dentition. It has therefore been conjectured that they may have included (or were closely related to) the ancestral stock from which the aberrant lemur *Daubentonia* was initially derived. However, while this still remains an interesting possibility, there is at present no fossil evidence which would make it a real probability.

Although the matter will be referred to in later chapters, it is well at this juncture to emphasize that the classification of the early Primates of Palaeocene and Eocene times presents very considerable difficulties, mainly because of the fragmentary nature of their fossil remains. Indeed, it is partly for convenience of reference that some of them have been allocated to one or other of the main groups into which the Recent prosimians are divided; such a classification, of course, must be to a considerable extent arbitrary if only because the main groups had not at that early time undergone the degree of segregation and differentiation which they show to-day. The classification adopted here is to be regarded as no more than a provisional and temporary device, pending the discovery of more complete fossil material. But already, in recent years, there has been a tendency to separate off some of the fossil types into more independent major groups. For example, the genus *Necrolemur* has been allocated to a separate family Necrolemuridae [59], and the genus *Phenacolemur* (hitherto usually included in the Plesiadapidae) to a separate family Phenacolemuridae [149], the supposition being that both these groups were independent radiations from the basal Primate stock which became extinct, and which had no particular relationship to any of the existing groups of Primates. While the accumulation of a more abundant fossil record may well justify such a conclusion, and indeed may justify the recognition of a number of other families among the early Tertiary Primates, these subsidiary details of classification are not so immediately relevant to the main theme of this book. For the purpose of descriptive convenience, it seems more useful to employ some general classification which in the past has had the support of competent authorities (even though its tentative and over-simplified nature is well recognized), rather than to change the classification backwards and forwards every few years because of claims advanced for the taxonomic revision of one small group or another. In any case, however, not all of these suggested revisions seem to be based on evidence which is altogether convincing. For example there are reasonable arguments for retaining such early prosimians as *Necrolemur* in the broad adaptive radiation of tarsioids even though some recent authors have shown a tendency to transfer it to the lemuroids. Fluctuations in terminology which may be commended by some authors, but not by others, are a frequent source of confusion to students and are also a source of perplexity to the writer of an introductory text-book.

It is perhaps worth while stressing the obvious here. The evolutionary radiations of the very early Primates during the Palaeocene and Eocene continued over a period of about twenty-five million years. Those fossil remains which by good fortune have been discovered can represent the relics of no more than a scanty few of the numerous lines of diversification which must have developed during this period. Many of the lines became extinct, and the chances that any of the fossil remains so far found represent the actual ancestors of the modern groups of Primates are of course fantastically remote. But some of them may have been closely related to ancestral groups and may thus provide useful (if only approximate) indications of the main evolutionary trends of the latter.*

Finally it is to be noted that in this and the ensuing chapters we have not attempted to enumerate all the numerous genera and subfamilies of extinct Primates which have been described from time to time; rather, we shall confine attention to some of the main types that appear to be more particularly relevant to the evolutionary story of the Primates.

* Reference should be made here to the recent, and somewhat tentative, creation of a new family of Palaeocene Primates, Picrodontidae, and of a new subfamily, Purgatoriinae, of the Paromomyidae, but the evidence for the precise nature of their Primate affinities remains to be finally established. (F. S. Szalay, 1968. "The Picrodontidae, a Family of Early Primates", *Amer. Mus. Novit.*, No. 2329; L. Van Valen and R. E. Sloan, 1965. "The Earliest Primates", *Science, 150*, p. 743.)

3

The Evidence of the Dentition

In this and the ensuing chapters we shall deal with different anatomical systems one by one, with the intention of considering their relevance to problems of the interrelationships of the Primates. The present chapter is concerned with the evidence which the dentition contributes towards the solution of these problems, and is taken first because no anatomical features have yielded more fruitful evidence in enquiries of this sort than the morphological details of the teeth. Experience has proved that the inferences based on comparative odontology are in general remarkably consistent with all the collateral evidence indicating degrees of affinity. Isolated aberrations of tooth structure do occur here and there in related groups, of course, but they can usually be recognized as aberrations without any ambiguity. It may be accounted a happy circumstance, also, that, because of the durability of the enamel and dentine of which they are composed, in fossil material teeth are more commonly preserved than any other part of the skeleton; indeed, in many instances they are the only remains available for study.*

A major characteristic of the mammalian dentition, in contrast with that of the present-day submammalian vertebrates, is *heterodonty*. That is to say, the teeth are regionally differentiated in form so as to serve special functions. Thus, on either side of each jaw in a typical mammalian dentition there are relatively simple nibbling teeth or *incisors* in front, then a pointed and projecting *canine* which can be used for grasping purposes, and behind this a series of more complicated post-canine chewing teeth, the *premolars* and *molars*. Although a *homodont* dentition (in which all the teeth are of approximately uniform shape) is actually found in certain aberrant

---

* In emphasis of the morphological stability and individuality of the cusp patterns of teeth it is interesting to note that if an undeveloped tooth germ is explanted and grown in tissue culture, completely isolated from its normal environment, it still develops its characteristic cusp pattern, the cusps being almost normal in shape, number and arrangement, [33].

types of mammal (e.g. the porpoise), it is certain that in these cases it is a secondary condition, for the palaeontological evidence makes it clear that heterodonty was acquired even before the mammalian phase of evolution had been fully established. So far as they are known, no early mammals of Palaeocene (or even Mesozoic) times had a homodont dentition. The same was also true for the immediate precursors of mammals, the mammal-like reptiles. For, although in some of the earliest mammal-like reptiles such as the Permian therapsids all the teeth had a simple conical form, the first maxillary tooth was greatly enlarged to form an obvious canine separating the incisor series in front from the cheek teeth behind.

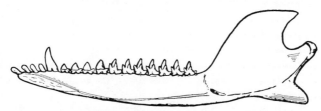

FIG. 17. Right half of the mandible of a Middle Jurassic Pantothere *Amphitherium*. After G. G. Simpson, *Quart. Rev. Biol.* 1935. Note the differentiation of the teeth into incisor, canine, premolar and molar series, and also the tricuspid character of the molars. (×3).

In the Triassic therapsids the establishment of a heterodont dentition was much more advanced, for in them the post-canine teeth had become complicated by the development of secondary cusps. Furthermore, in some of the pre-mammalian forms these teeth had already become clearly separated into distinct premolar and molar series.

In Mesozoic times, as we have already noted, there had become differentiated several orders of mammals, and of these early types the Multituberculata, Symmetrodonta and Triconodonta became extinct before the Tertiary epoch, probably without leaving any descendants [144]. On the other hand, the Pantotheria almost certainly provided the basis for the later development of marsupials and placental (eutherian) mammals, and in their day they were not only very abundant but also very varied. One genus, *Amphitherium*, which is actually the oldest pantothere so far known (Middle Jurassic), already had some of the basic features of the generalized type of dentition common to the primitive members of all the higher mammalian groups (fig. 17). One of the characteristics of this type of dentition is the fundamentally tricuspid nature of the molar teeth, an arrangement of cusps which proved particularly effective for the mechanical requirements of mastication. In the primitive reptilian type of dentition, the occlusal relationship between upper and lower teeth is one of simple alternation, that is to say, the conical crowns interlock with closure of the jaws. Such a dentition is designed for, and only allows, grasping functions. In the Symmetrodonta the crowns of the alternating molar teeth ac-

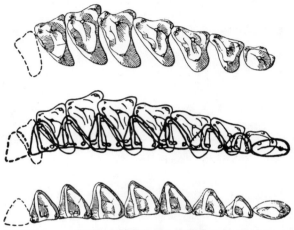

FIG. 18. Diagram illustrating the occlusal relationships of the cheek teeth in the Upper Jurassic Pantotheres, after G. G. Simpson, *Quart. Rev. Biol.* 1935. Above are shown the upper teeth, and below the lower. In the middle is shown the position of the interlocking cusps of the upper and lower teeth in occlusion. (×8)

quired a triangular form with sharp crests, so that in occlusion a shearing effect was added to permit cutting as well as grasping functions. On the other hand, some of the Multituberculata developed complicated flat-crowned molar teeth with many small cuspules, and in occlusion the upper and lower teeth came in direct opposition to form a highly efficient (but also highly specialized) triturating mechanism. As Simpson has pointed out [144, 145], an analysis of the pantothere type of occlusion makes it clear that in this early group the molar teeth were so designed as to combine in one mechanism all the advantages of alternation, shear and opposition (fig. 18). This was evidently one of the reasons for the evolutionary success of the Pantotheria (as compared with other

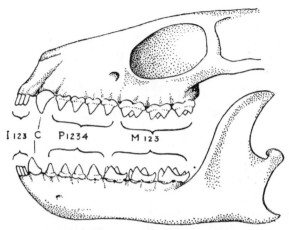

FIG. 19. Diagram showing the dentition of a hypothetical general-ized eutherian mammal, representing the type from which it is presumed that the dentition of the various groups of Primates was originally derived.

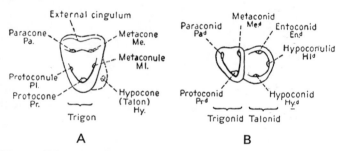

FIG. 20. Diagrams illustrating the basic cusp patterns of the left upper (A) and left lower (B) molar teeth in generalized primitive mammals. (After G. G. Simpson, *Biol. Rev.* 1937.)

Mesozoic mammals), a success ultimately manifested in the de-velopment of more advanced types which gave rise to the an-cestors of all the higher mammals, including the Primates. More-over, apart from the immediate functional advantages, the cusp pattern of their molar teeth provided a suitable morphological basis for the development by evolutionary modification of very varied forms of molar teeth adapted to diverse special needs, and it was evidently this pluripotentiality which permitted the deploy-ment of mammals over an exceptionally wide range of habitats.

Before discussing the main trends in the evolution of the dentition in Primates, it is convenient to refer to the basic characters of the eutherian dentition so far as these can be inferred from a study of primitive mammals still existing to-day, or of those extinct forms which are known to be of early geological date. It is widely accepted (and, indeed, may now be regarded as an established fact) that the primitive eutherian dental formula was $I \frac{123}{123} C \frac{1}{1} P \frac{1234}{1234} M \frac{123}{123}$, or more simply and in less detail, $\frac{3.1.4.3}{3.1.4.3}$. That is to say, the early placental mammals had on each side in the upper and lower jaws three incisors, one canine, four premolars and three molars (fig. 19). In most modern mammals the formula has been modified by some degree of reduction from the full complement of teeth, but the complete formula is found to be more and more common as modern types are traced back to their earlier geological history until, in the Lower Palaeocene, it was evidently the prevalent condition.

In the generalized form of the eutherian dentition the *incisors* are small teeth, cylindrical with rounded tips or slightly spatulate. In the upper jaw, they are implanted in the premaxillary element of the skull (see p. 128 and fig. 48). The *canine* is pointed and projecting, slightly recurved, and definitely larger than the incisors or premolars.* The upper canine is the first tooth implanted in the maxilla, and is thus situated immediately behind the premaxillary suture. The *premolars* are of simple construction, each with a single pointed, conical cusp somewhat compressed laterally, and the base of the crown thickened to form a ridge-like basal ring of enamel, the cingulum. This enamel ridge, which is commonly well marked on premolars and molars (and is also sometimes to be recognized at the base of the incisors and canine), is of considerable morphological importance, for the cingulum provides the foundation for the evolutionary upgrowth of certain additional cusps, leading in some types to an increasing complexity of the crown.

The upper *molars* of the generalized eutherian dentition are all tritubercular—that is to say, the crown of each tooth bears three

* We may note here that it is of some importance to recognize that the development of a relatively large canine occurred in the earliest stages of the evolution of mammals and is thus a very primitive character. It follows that the small canines found in some of the modern mammals (including certain Primates) are the result of a secondary reduction in size.

main cusps (fig. 20). Of these, one is medial, the *protocone*, and two are lateral, the *paracone* in front and the *metacone* behind. Each tooth is anchored to the jaw by three roots, two lateral and one medial, of which the last is the stoutest. The protocone is initially somewhat crescentic, and well-defined crests of enamel extend laterally from it along the anterior and posterior margins of the crown. On each of these crests a subsidiary cusp, or cuspule,* may be developed, the anterior being termed the *proto-* or *paraconule* and the posterior the *metaconule*. The inner side of the base of the crown is enclosed by a rounded rim of enamel, the internal cingulum, and on the outer side of the lateral cusps the crown may be extended to form a flattened shelf with a slightly raised marginal rim, the external cingulum. It is probable that in the earliest placental mammals this shelf was fairly extensive, forming a considerable proportion of the biting (occlusal) surface of the crown. If this was so, it became rapidly reduced in the early Primates, though it still remains quite well developed in some of the tupaioids. From the external cingulum little sharply-pointed cuspules may be developed at its anterior and posterior extremities, termed respectively the *parastyle* and *metastyle*, and also from its centre, the *mesostyle* (fig. 21F). These *styles* are a common feature of the upper molars of insectivores where they are sometimes developed to an exaggerated degree.

It will be observed that the three main cusps of the generalized type of upper molar tooth form the points of a triangle on the crown; this formation is called the *trigone*. It is from such a basic three-cusped or tritubercular pattern that the variously complicated upper molars of all Recent mammalian orders have been derived in the course of evolution. Such a conclusion derives from what has been termed the Theory of Trituberculy, a theory elaborated many

---

* The term *cusp* refers to a major elevation on the crown of a tooth, which is a broadly consistent feature of the dentition in related groups of mammals and which can readily be identified (i.e. homologized) in different types by reference to its position and relationships. A *cuspule*, on the other hand, is a secondary, and usually relatively small, elevation developed in relation to one of the main cusps or the ridges interconnecting them, and while it is a consistent feature within the limits of a species or genus it may be absent or very variably developed in closely related types. In contradistinction to these morphological entities are the apparently random crinklings of enamel which may vary considerably from one individual to another. It is of importance to recognize the existence of these individual variations (which are particularly common in the higher Primates), for in the study of isolated fossil teeth they have sometimes been assigned ataxonomic significance which is by no means justified.

years ago by the American palaeontologists Cope and Osborn. The alternative theory of multituberculy may be mentioned here for its historical interest. Adherents of this theory maintained that primitively the mammalian molar possessed a large number of small cusps—as seen, for instance, in the order Multituberculata of late Mesozoic and early Tertiary times—and that, by the elimination of varying numbers of these cusps, the more simple molar cusp patterns of Recent mammals have been derived. There are many objections to this theory, however, and even were it true in part for some groups of mammals, it is no longer disputed that it was the tritubercular type of molar which provided the basis for the evolutionary development of the different cusp patterns found in the Primates. In the more primitive Primates the upper molars approach very closely to the tritubercular form postulated by the original Cope-Osborn theory.

The crown of the lower molars of the primitive and generalized eutherian dentition consists of two parts, an anterior portion bearing three main cusps—the *trigonid*, and a posterior, heel-like, projection—the *talonid* (fig. 20). The trigonid is primitively raised well above the level of the talonid, and its cusps are disposed in a triangle, one laterally (the *protoconid*) and two medially (the *paraconid* in front and the *metaconid* behind). It will be noted that the three cusps of the trigonid of the lower molars appear to be arranged as a sort of mirror image of the trigone of the upper molars, and their names reflect this analogy. But the comparison is no more than superficial, and in occlusion the trigones as a whole alternate with the trigonids as a whole, producing a shearing action. The heel or talonid is hardly more than incipient in the mesozoic pantotheres (fig. 18), and in its primitive form, as seen for -example in some insectivores, it is narrower than the trigonid. Centrally it is hollowed out into a basin-like depression which is surrounded by a rather sharply defined raised rim. The latter is elevated laterally to form the *hypoconid*, and medially the *entoconid*. An additional cusp may be formed (particularly on the third molar) towards the middle of the posterior margin of the rim, the *hypoconulid*. In occlusion the protocones of the upper molars fit into the talonid basins of the corresponding lower molars, producing the action of a pestle in a mortar. The lower molars may have a narrow external cingulum, usually better defined anteriorly, and they are each implanted in the mandible by two roots, anterior and posterior. Accessory cuspules which occasionally develop from the basal cingulum are called *stylids*.

The term tuberculo-sectorial which has been applied to the type of lower molar just described is not very apt, and Simpson has suggested the term *tribosphenic* as descriptive of the generalized mammalian molar dentition as a whole [144]. This term has reference to the grinding functions of the protocone and talonid basin and to the wedge-like shearing action of the trigones and trigonids. It seems probable that the tribosphenic type of dentition first appeared in mammals of the Upper Cretaceous. According to Simpson, the molars of Palaeocene mammals 'are all tribosphenic or nearly so', though even at this early time in different groups they were beginning to show divergent modifications of the basic type.

GENERAL TENDENCIES OF DENTAL EVOLUTION IN THE PRIMATES

Having indicated the nature of the dentition in a generalized eutherian mammal such as might represent the ancestral stock from which the Primates have been ultimately derived, and before considering separately and in further detail the evolutionary modifications which have been manifested in the different groups of Primates, we may conveniently summarize in broad outline certain of the prevailing evolutionary tendencies which have characterized the dentition of the order as a whole.

The INCISORS in almost every known Primate have been reduced in number, for there are usually but two in each side of the upper and lower jaws. Indeed, the possession of two incisors is sometimes cited as one of the main diagnostic features of the order Primates, though it is probable that some of the representatives of the basal Primate stock in Palaeocene or Early Eocene times did have the full complement of three. The modern tree-shrews still retain three lower incisors, but as in Primates generally the upper teeth have been reduced to two. In the lemurs generally, there has been a common (but not exclusive) tendency for the upper incisors to become much reduced in size, and in some types they may vanish altogether; on the other hand, in all Recent genera of lemurs (except *Daubentonia*) the lower incisors are markedly specialized to form part of the dental comb to which reference was made in the previous chapter (p. 69), and have become so procumbent as to lie almost horizontal. That the evolutionary trends in the incisor series were far from uniform in the early ramifications of the Primates is very evident from the curious modifications which developed in several of the Palaeocene prosimians, for in some of these groups one of the lower incisors became grossly hypertrophied while in others they disappeared altogether. It seems that, of the whole dental series, the incisor teeth have shown the greatest variability in the lower Primates. Compared with the latter, the incisors in the higher Primates, i.e. the Anthropoidea, display a much greater constancy of form, and are characterized by their spatulate form and their relatively straight cutting edge.

The CANINES in many groups of Primates hypertrophy to form sharp and powerful dagger-like teeth, as, for example, the upper canines of *Lemur*, and the upper and lower canines in the larger monkeys and some of the anthropoid apes. On the other hand, they may be assimilated into the incisor series, becoming incisiform in appearance, as has occurred to some extent in *Homo sapiens* or more completely in the specialized lower dentition of modern lemurs.

The PREMOLARS in Primates show a tendency to a disappearance early in evolution of the first of the series, and in more advanced types of the second also. In the Indriidae among the lemurs, and in all catarrhine monkeys, anthropoid apes and hominids there are only two premolars (i.e. P_3 and P_4). In association with this tendency to reduction in total number, the third and fourth premolars (more particularly the latter) also tend to undergo a relative enlargement, and, as in many other mammalian orders, they may become progressively complicated by the upgrowth of cusps from the basal cingulum, thus assuming the appearance and characteristics of molars. This process is generally termed *molarization*, and in many of the lower Primates it has proceeded even further than in the higher Primates. Thus, while in the Anthropoidea the premolars are commonly bicuspid as the result of the addition of a second cusp to the medial side of the original main cusp, there is a tendency among some groups of lemurs for the last premolar to become quite distinctly molariform by the addition of two or more cusps. It is a general characteristic of the tarsioids and Anthropoidea that the premolars have almost entirely avoided this specialized feature of molarization, so that (particularly in the upper dentition) there is usually rather an abrupt contrast between the premolar and molar series. On the other hand, the tendency

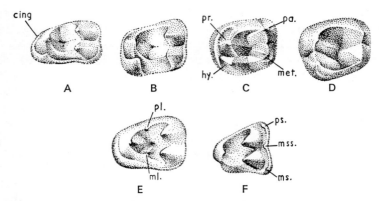

FIG. 21. Diagram illustrating the modifications of the simple tritu-bercular upper molar tooth which may occur in the Primates. In all cases a left upper molar is represented. (A) Primitive tritubercular tooth with internal cingulum, as seen in *Tarsius*; (B) Enlargement of the hypocone to form a quadritubercular tooth as seen in *Nyctice-bus*; (C) A more fully developed quadritubercular tooth as seen in *Dryopithecus*, with cusps of approximately equal size, i.e. protocone, paracone, metacone and hypocone; (D) An upper molar of modern man; (E) An upper molar of *Pseudoloris*, showing the development of a paraconule and metaconule; (F) An upper molar of *Tupaia*, showing the pronounced development of the external cingulum bearing three small cuspules, the parastyle, mesostyle, and metastyle.

towards molarization in the lemurs may proceed to the extent that the premolar and molar series appear to grade into each other without any obvious break. In certain aberrant groups of extinct prosimians (e.g. the Carpolestidae) the last lower premolar (P_4) became highly specialized to form a sharp longitudinal cutting edge.

The UPPER MOLARS in all the main groups of Primates show a progressive tendency to assume a quadritubercular form by the development of a fourth cusp postero-medially on the crown (fig. 21). It is of some taxonomic importance to recognize that this fourth cusp may be developed in more than one way. Usually it appears to spring up from the internal cingulum, and is then termed the *hypocone*. In the dentition of many primitive Primates the hypocone is either not developed, or is just evident as a slight thickening of the posterior part of the internal cingulum. In more advanced types it may become equal in its proportions to the main cusps of the original trigone. The fourth cusp appears to arise in some other types by a splitting or dichotomy of the protocone, and in this case it is termed a *pseudohypocone*. This type of quadritubercular molar is found in the Eocene lemurs of North America, the Notharctinae, and serves to distinguish this group rather sharply from the contemporary Eocene lemurs of the Old World, the Adapinae, in which the fourth cusp is a true hypocone. In its mode of origin, the Notharctine pseudohypocone shows some resemblance to the fundamental molar pattern of certain ungulates (e.g. Perissodactyla).

In the LOWER MOLARS a quadritubercular and quadrilateral crown is developed as an evolutionary tendency in the Primates in rather a different way (fig. 22). The paraconid undergoes a progressive reduction until it finally disappears altogether, while the protoconid comes to lie more directly lateral to the metaconid. At the same time the talonid broadens so as to equal or exceed the width of the trigonid, the two main cusps of the talonid (the entoconid and hypoconid) become larger, and the talonid basin fills up. The whole talonid may also become level with the trigonid so as to produce a more or less evenly flattened occlusal surface. Lastly, the hypoconulid tends to become more distinctly developed at the posterior margin of the crown rather towards the outer side.

It is interesting to note that, presumably in relation to their more uniform functions, the molar teeth in the different groups of Primates show far less deviation from the basic pattern than the other teeth often do. Thus, while the incisors, canines, and even the premolars, may in certain types show gross aberrations of form, the molars usually retain their fundamental Primate characteristics.

With this general conception of some of the main evolutionary trends of the dentition in Primates as a whole, attention may now be turned to a consideration in more detail of the changes which have occurred in the major subdivisions of the order, and the inferences which may be drawn therefrom regarding their phylogenetic differentiation and relationships.

TREE-SHREWS AND PLESIADAPIDS

We have noted that the most primitive Primates of which we have any knowledge are the Tupaioidea and Plesiadapidae. Although the precise degree of their phylogenetic relationship is still uncertain, it is convenient to consider these two groups together for the reason that, in spite of obvious differences, they do possess

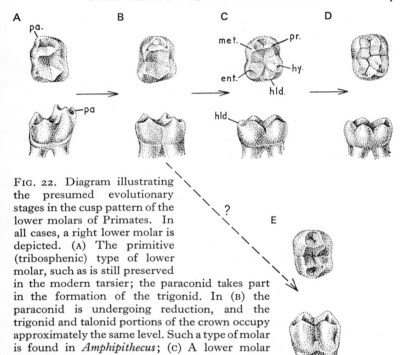

FIG. 22. Diagram illustrating the presumed evolutionary stages in the cusp pattern of the lower molars of Primates. In all cases, a right lower molar is depicted. (A) The primitive (tribosphenic) type of lower molar, such as is still preserved in the modern tarsier; the paraconid takes part in the formation of the trigonid. In (B) the paraconid is undergoing reduction, and the trigonid and talonid portions of the crown occupy approximately the same level. Such a type of molar is found in *Amphipithecus*; (C) A lower molar of *Parapithecus*, in which the paraconid has completely disappeared; the trigonid portion of the crown bears the metaconid and protoconid, while the talonid bears the hypoconid and entoconid, and a relatively well developed hypoconulid; (D) A lower molar of a Miocene ape, *Dryopithecus*, in which the five cusps are more or less equally developed and separated by a characteristic pattern of intervening grooves; (E) A lower molar of a cercopithecoid monkey showing the characteristic bilophodont pattern.

a number of morphological characters in common. Some years ago the plesiadapids were generally regarded as early tree-shrews showing certain aberrant traits, and since at that time the tree-shrews were placed among the insectivores the plesiadapids were similarly classified. Following on detailed anatomical studies, Simpson clarified the problem by demonstrating that the skeletal and dental characters of the plesiadapids approximate sufficiently closely to those of Eocene lemuroids to warrant their taxonomic inclusion in the Lemuroidea. At the same time, the resemblances which they also show to the tree-shrews have served to confirm conclusions already reached from other evidence, i.e. it is not possible to demarcate the tupaioids sharply from the other lower Primates.

In the modern tree-shrews the dental formula is $\dfrac{2.1.3.3}{3.1.3.3}$. That is to say, they have lost from the generalized placental dentition one upper incisor, and one upper and lower premolar (fig. 23). In the tupaiid genus *Urogale*, the third lower incisor has undergone a marked atrophy, suggesting an approximation to the dental formula characteristic of the Recent lemurs in which this tooth has finally disappeared. The upper incisors in most species of tree-shrew are relatively short and styliform, separated by a conspicuous gap in the mid-line to accommodate in occlusion the procumbent lower incisors (as in modern lemurs). In *Ptilocercus* the upper incisors are somewhat specialized and powerful teeth, the second incisor appearing almost premolariform by the development of a secondary cuspule at the base of its posterior margin. The procumbency of the lower incisors in the tree-shrews is very pronounced, approaching a horizontal position (figs. 24 and 48). In fact, as already noted, they comprise a dental comb which is specifically used for combing the fur in precisely the same manner as lemurs. Unlike the latter however, the lower canines have not become assimilated into this comb, but they are reduced in size and somewhat incisiform. In the Indian genus *Anathana*, indeed, they are incisiform and procumbent to such a degree as to suggest a phase of evolutionary development which surely must have preceded the more highly specialized dental comb characteristic of Recent lemuroids (fig. 24). The upper canines are also relatively small teeth in the modern tree-shrews; in *Ptilocercus* (and also as an occasional anomaly in some other genera) they have the unusual character of being double-rooted.

The first two upper and lower premolars (P_2 and P_3) are small and of simple construction, composed of a single pointed cusp. In almost all the modern genera the last premolar shows some degree of molarization, and in the upper dentition it may bear on its crown not only the three cusps of the trigone but also well-developed para-, meta- and mesostyles; it is implanted in the jaw by three roots. The lower last premolar is also distinctly molariform. In *Ptilocercus* the last premolars, although relatively large, are much less complicated, and to this extent are less specialized.

The upper molars show rather strong contrasts in the two sub-families Tupaiinae and Ptilocercinae. In the former their essenti-ally tritubercular character is complicated by the development of well-marked styles (of which the mesostyle is commonly bifid) and by a transverse system of V-shaped crests uniting the main cusps in zig-zag fashion (fig. 21F). The crests, as seen from the occlusal aspect of both upper and lower molars, form a double V, produc-ing a type of molar pattern called (from the Greek letter lambda) dilambdodont. In *Ptilocercus* the upper molars have a narrow ex-ternal cingulum with distinct paraconules on the first and second and small and inconspicuous blunt parastyles and metastyles (but no mesostyles), and as in the tupaioids generally the last upper molar is reduced in size (fig. 23A). The lower molars of the tree-shrews preserve the primitive tribosphenic pattern—that is, a raised tri-gonid surmounted by the paraconid, protoconid and metaconid, and a depressed, basin-like talonid bearing the entoconid and hypo-conid, and also (on the last molar) a distinct hypoconulid (fig. 23B).

Since the Plesiadapidae were for some time known mainly from jaw fragments, their affinities were determined to a large extent by dental morphology (figs. 25, 26 and 27). Here the cusp pattern of the molar teeth appears to be decisive, for according to Simpson [143] the resemblance to one of the genera of primitive lemurs of the family Adapidae, *Pelycodus*, 'is really amazing and extends to the apparently most insignificant details' (cf. fig. 30). The premolars are variable in number and conformation. There may be three (as in the Middle Palaeocene genus *Pronothodectes*), or the number may be reduced to two. The last premolars (P_4) show a pronounced degree of molarization, and to a lesser degree this may also be the case with P_3. We have already noted that molarization of the pre-molars has been a common tendency in the evolution of the lemur-oid dentition, and it is of interest to find that this tendency mani-fested itself so early among this group of prosimians. The gross enlargement of the anterior incisors, combined with absence of the lower canines and advanced reduction of the upper, is a plesiadapid specialization, but it parallels rather similar changes which occurred in other groups of early Primates and thus may be taken to represent a not uncommon tendency in the initial 'exploratory' phases of Primate evolution. The dentition of the Microsyopidae is very similar indeed to that of the Plesiadapidae. Resemblances to the Tupaioidea are also significant, as may be

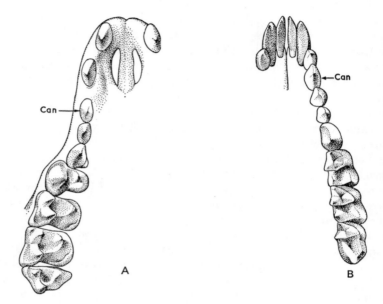

FIG. 23. Upper dentition (A) and the lower dentition (B) of the
pen-tailed tree-shrew, *Ptilocercus* (×4). In this and subsequent
illustrations the canine tooth is labelled.

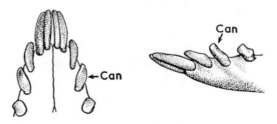

FIG. 24. Lower incisors, canine and front premolar of a tree-shrew,
Anathana wroughtoni seen from above and from the side. Note the
extreme procumbency of the incisors (which together form a dental
'comb'), and also the marked procumbency of the canine suggesting
an approach to the more extreme condition in the modern lemurs
(×3).

90

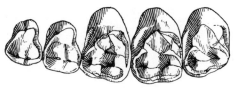

FIG. 25. Right upper teeth of *Plesiadapis anceps*, showing P³ and
P⁴, and the three molars (×4). (After G. G. Simpson, *Amer. Mus.
Novit.* 1936.)

seen by comparing *Plesiadapis* with *Ptilocercus* (figs. 27 and 48).
Such a comparison suggests that the specializations of the former
are the result of the extreme development of a condition present,
but less pronounced, in the latter, that is to say, the enlargement
of the vertical upper incisors, the procumbency and shape of the
lower incisors, and the greatly reduced canines.

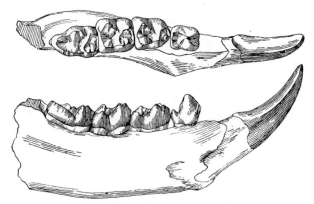

FIG. 26. Lower dentition of *Plesiadapis anceps* (×3). (After G. G.
Simpson, *Amer. Mus. Novit.* 1936.)

The Apatemyidae, which were at one time grouped with the
Plesiadapidae, are now regarded (on the basis of Jepsen's studies
62) as a separate and highly aberrant group which became extinct
by the end of the Oligocene. Their anterior lower incisors became
even more hypertrophied than those of the Plesiadapidae, the
massive roots extending back in the mandible as far as the level
of the third molar teeth (fig. 32A). Of the two lower premolars, the
last (P₄) dwindled to an insignificant size, while P₃ became highly
specialized to form a thin cutting blade. As already mentioned,
one of the reasons why both the plesiadapids and the apatemyids

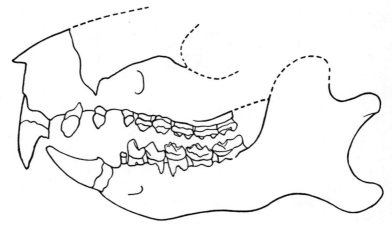

FIG. 27. Dentition of *Plesiadapis gidleyi,* showing also the left side of the mandible and part of the skull (×2). (After G. G. Simpson, *Biol. Rev.* 1937.)

have claimed attention is their supposed ancestral (or near-ancestral) relationship to the curious modern lemur, *Daubentonia.* Reference will be made to this question in the next section.

LEMURS

No system of the body demonstrates more emphatically than the dentition the aberrancy of all Recent groups of lemurs (figs. 28 and 53). In modern genera, with the single exception of the aye-aye (which shows an extreme specialization of quite another type), the incisor teeth and the lower canines are peculiarly modified. The upper incisors are shrunken in size and rudimentary, and in the genus *Lepilemur* they have disappeared altogether (though still retained in the milk dentition). Those on either side are separated by a comparatively wide interval, the gap serving to accommodate the procumbent lower incisors in occlusion as in the tupaiids. The lower incisors project forwards almost horizontally and (as already noted) with the incisiform canine make an exceedingly fine dental comb. The comb is used in the toilet of the fur by a forward thrust and scraping movement of the teeth with the mouth open. So far as is known the only other mammals which comb their fur in this

curious manner are the tree-shrews and possibly the aberrant
'flying lemur' *Cynocephalus* (= *Galeopithecus*), though most mammals clean their coats with a nibbling and biting movement of their
incisors. The extreme procumbency of the lower canines, together
with their complete relegation in position and form to the incisor
series, is common to all the diverse groups of modern lemurs, and
is apparently a fairly ancient character. At any rate, in the Early
Miocene genus *Progalago*, which lived in the tropical parts of
Africa about twenty million years ago, the upper incisors were
already much reduced in size and separated by a wide median gap,
and the lower incisors and canines were also procumbent. On the
other hand, in the Eocene lemurs of which the front teeth (or their
sockets) have been preserved, the lower incisors, though small, are
implanted more or less vertically, and both the upper and lower
canines are of the primitive mammalian type, i.e. sharply pointed,
conical teeth. The upper incisors in *Adapis* are of moderate size and
furnished with a chisel-like cutting margin, but in *Notharctus* they
are small and peg-like teeth (figs. 57B and C). In neither genus, however, are the upper incisors separated by a conspicuous median
gap. From the fossil evidence, then, it appears that the characteristic specialization of the front teeth seen in Recent lemurs must

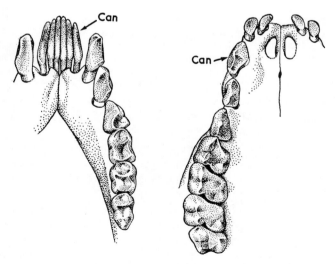

FIG. 28. Dentition of *Galago garnetti* (×2).

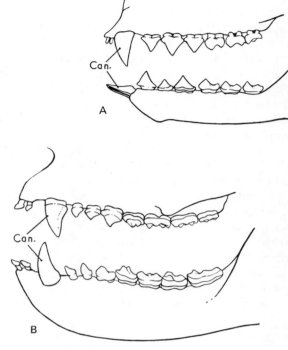

FIG. 29. Dentition of a modern lemur, *Lemur varius* (A) ($\times 1$)
and a fossil adapid, *Notharctus osborni* (B) ($\times 1\frac{1}{2}$).

have been acquired during the interval between the Upper Eocene
and the Lower Miocene, that is to say, during the intervening
Oligocene period. It is difficult to understand the requirements
which could have led to the development of such curious modifications, for there is no demonstrable reason to be found in the
texture of the furry coat in modern lemurs which might be supposed to demand the evolutionary construction of a fine dental
comb for keeping it in good condition (see p. 325)

In contrast with the incisiform lower canines, the upper canines
of modern lemurs are commonly large and dagger-like (fig. 29A).
This is particularly the case in some of the Madagascar lemurs,
and in the extinct giant lemur of Pleistocene times, *Megaladapis*,
they formed powerful tusks (fig. 59).

In almost all Recent genera of lemurs, the dental formula is $\frac{2.1.3.3}{2.1.3.3}$. That is to say, they have lost one incisor and one premolar from the primitive eutherian dentition. In most of the Indriidae the formula is still further reduced, by the loss of a second premolar and the lower canine, to $\frac{2.1.2.3^*}{2.0.2.3}$. The reduction of the incisors to two is clearly a very archaic character of the Lemuroidea in general for even the Eocene types (so far as they are at present known) had no more than two. On the other hand, in all these early types the full complement of four premolars was still retained. In both the genera *Adapis* and *Notharctus* the last premolars have undergone some degree of molarization; in the case of upper teeth they show a tricuspid pattern, so that there is no abrupt transition from the premolar to the molar series (fig. 31). This condition is not developed to any significant degree in the earlier and more primitive genus *Pelycodus*, nor in the Upper Eocene genus, *Pronycticebus*, and in both these fossil types the last premolars of the upper dentition have no more than two cusps, the main cusp or paracone, and a low internal basal cusp, the deuterocone. Many of the modern lemurs (e.g. *Lemur, Loris, Perodicticus*) have retained (or perhaps reverted to) this more generalized condition; in others (e.g. *Galago*), the tooth is molarized to the extent that it has developed a metacone and a rudimentary hypocone (fig. 28). It is evident, then, that while the process of molarization of the last premolar has been a common (and also a very ancient) trend in lemuroid evolution, it has not consistently manifested itself in this group of prosimians.

The anterior lower premolar in Recent lemurs is specialized to a varying degree by assuming the function of the true canine which—as recorded above—has been relegated to the incisor series. It has thus become caniniform, relatively large, pointed, and laterally compressed, and it shears against the upper canine in occlusion of the teeth (fig. 53). A somewhat similar change, but much less in degree, may even affect the anterior premolar of the upper series, which now becomes functionally related to the caniniform lower premolar in biting movements (e.g. in *Perodicticus*). In general, the posterior lower premolars of the lemurs are

---

* By some authorities the lateral front lower tooth in the Indriidae is identified as the canine, in which case the dental formula would be $\frac{2.1.2.3}{1.1.2.3}$.

fairly simple, unicuspid teeth, though the last of the series usually shows a small talonid. As in the upper dentition, the last lower premolar in *Adapis* and *Notharctus* is somewhat more molariform than it is in many Recent genera.

The upper molars in different groups of lemurs show every gradation from the typical tritubercular pattern to a quadritubercular form. In the genus *Lemur* there is a slight indication of a fourth cusp—the hypocone; in *Nycticebus* and *Loris* this may be quite conspicuous; in *Galago* it is large and the quadritubercular form of molar well developed. In the Eocene lemurs, a true (cingulum) hypocone is already present, though relatively small, in the molars of the Adapinae (*Adapis* and *Pronycticebus*). In the American genus *Notharctus*, the upper molars are also quadritubercular but, as in the Notharctinae generally, the fourth cusp is a pseudo-hypocone produced by fission of the protocone (and not as a secondary elevation from the internal cingulum). In the earlier forms of *Pelycodus* (primitive precursors of *Notharctus*) the upper molars are tritubercular (the pseudohypocone is not present in *P. ralstoni* or *P. trigonodus*, and is just beginning to appear in *P. frugivorus*), and the whole sequence of the fossil remains of this genus appears to demonstrate a progressive tendency towards the full development of the pseudohypocone as finally manifested in *Notharctus* itself (figs. 30 and 31). This unusual mode of development of the hypocone in the Notharctinae appears to distinguish the latter rather sharply from the Adapinae of the Old World, and it is thus commonly inferred that these two subfamilies of the Adapidae represent widely divergent lines of lemuroid evolution in the Eocene. On the basis of similar reasoning, it may be supposed that the Notharctinae were a somewhat aberrant line of lemurs, having no ancestral relationship to any of the modern lemurs (though the possibility has been entertained that some of them may have given rise to the New World monkeys).

The lower molars in the living genera of lemurs lack the para-conid, the protoconid lies directly opposite the metaconid, and the talonid basin tends to be rather broadened out (except in the last molar) and raised to the level of the trigonid. A more generalized type of lower molar is seen in *Notharctus* in which a small paraconid is present, at least in the first molar (fig. 31). In some species of *Pelycodus* the paraconid is well developed on all three molars (fig. 30) and the cusp pattern of the latter approximates very

closely to the primitive placental type. The gradual disappearance
of the paraconid is made very evident in the palaeontological
sequence of the Notharctinae. In all known forms of the Old World
genus *Adapis* the paraconid has already disappeared, but in the

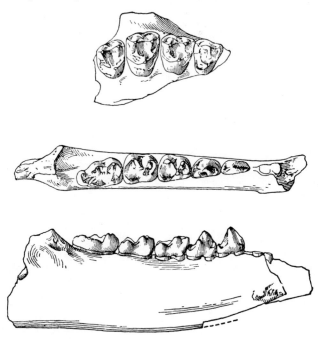

FIG. 30. Portions of the upper and lower dentition of *Pelycodus
trigonodus*. (After W. D. Matthew, *Bull. Amer. Mus. Nat. Hist.* 1915.)

more primitive genus *Pronycticebus* it is just apparent on the first
molar.

From the preceding account it will be evident that the outstanding characteristic of the lemuroid dentition is the extreme
specialization of the front teeth, and although this specialization
was not present in any of the known Eocene types, the fragmentary
remains of the early African genus *Progalago* indicate that it was
already far advanced by the Early Miocene. Moreover, since it is
found in all the widely diverse groups of modern lemurs, extending from the Far East to central Africa and Madagascar, and since
its unique and peculiar character makes it difficult to suppose that

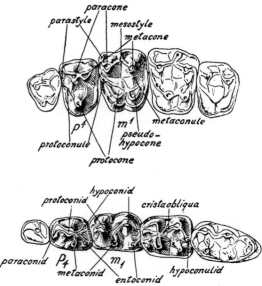

FIG. 31. Portions of the upper and lower dentition of *Notharctus* (×2) (partly diagrammatic). Note the pseudohypocone on the upper molars, which, unlike a true (cingulum) hypocone, appears to have arisen by a fission of the protocone. (After W. K. Gregory, *Mem. Amer. Mus. Nat. Hist.* 1920.)

it would have developed independently in different groups of lemurs, the inference follows that it was already present in the common ancestral stock which ultimately gave rise to these groups (or that, at any rate, these groups inherited a common potentiality to develop such a specialization). But in certain genera there are occasional curious aberrations which are of particular interest in so far as they demonstrate the possibilities of evolutionary reversal in individual characters of the dentition. Such aberrations are found among certain of the Pleistocene lemurs of Madagascar. In some of these the distinctive lemuroid pattern is retained; for example, the extinct giant lemur *Megaladapis* shows all the elements of the lemuroid trend of dental evolution (fig. 59), with procumbent lower incisors, incisiform canine, caniniform anterior premolar, and reduction of upper incisors (even to their complete disappearance in some species). As in the other large extinct lemurs of Madagascar, the molar teeth tend towards a bilophodont character somewhat similar to that of the cercopithecoid monkeys.

In *Archaeolemur*, the upper incisors are not only large, they are spatulate with a straight cutting edge, and are not separated by a median gap. It appears, then, that this Pleistocene type shows a reversal from what was presumably the ancestral condition of the upper incisors, i.e. gross reduction in size and wide separation in the mid-line. This secondary hypertrophy of the incisors is accompanied by an unusual modification of the premolar teeth, which are strongly compressed from side to side forming altogether a long antero-posterior cutting edge. On the other hand, the dental formula of *Archaeolemur* is $\frac{2.1.3.3}{2.0.3.3}$ (and thus approximates to that characteristic of the modern Indriidae by the loss of the lower canine), the lower incisors are moderately procumbent, and the anterior lower premolar is to some extent caniniform. All these characters betray the essential lemuriform (and particularly the indriid) affinities of this aberrant type.

Altogether the most puzzling item of lemuroid taxonomy is that presented by the aye-aye (*Daubentonia*), for the dentition of this creature shows no characters at all which can be regarded as distinctive of the lemurs as a whole. The front pairs of teeth in the upper and lower jaws (which are probably incisor teeth) are extraordinarily modified so as closely to resemble the incisor teeth of rodents. They are very large and appear to have continuous growth, being covered only on their anterior and lateral surfaces by enamel. The back teeth, by contrast, have undergone a very considerable degree of degeneration, and there remain, in a vestigial condition, one premolar and three molars in the upper jaw and three molars in the mandible. The dental formula is thus $\frac{1.0.1.3}{1.0.0.3}$.

Since *Daubentonia*, in its dentition, appears to have followed evolutionary trends very different from (and in some features diametrically opposite to) those which have characterized the lemurs as a whole, the proposition has been advanced that the genus has been derived from a different ancestry. Several authorities have pointed out the resemblances between the specialized features of the dentition of the modern aye-aye and that of the plesiadapids and apatemyids of early Tertiary times, with particular reference to the greatly enlarged and rodent-like incisors and the tendency towards a reduction of the cheek teeth. These resemblances are made evident by the comparison in fig. 32 of the skulls of Eocene

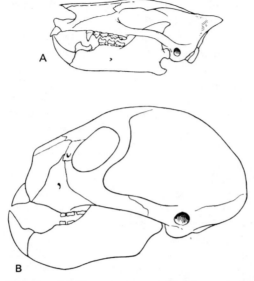

FIG. 32. (A) The skull and dentition of the apatemyid *Stehlinella* (the posterior part of the skull is restored) (×1), and (B) the skull and dentition of *Daubentonia* (×⅔). Certain resemblances in the dentition have suggested a phylogenetic relationship (but see text on this point).

apatemyid *Stehlinella* and *Daubentonia*. On the basis of such dental resemblances, it has even been proposed that *Daubentonia*, together with the extinct families Apatemyidae and Plesiadapidae, should be included in a separate subdivision of the Primates, Chiromyoidea. However, the general consensus of palaeontological opinion is against such a relationship, mainly for the reasons (1) that enlargement of the incisors was a rather common feature which developed independently in several groups of early Primates (including some of the Eocene tarsioids), and there is no compelling reason on the basis of this feature alone to infer a relationship of one group more than another with the modern aye-aye; (2) there is no palaeontological evidence which would indicate that the aye-aye was derived from a lineage separate from that of the other Recent lemuroids. The only fossil (or subfossil) record of the modern aye-aye refers to a Pleistocene species (*Daubentonia robustus*) which does not differ from the existing species except in its larger size. The lack of a palaeontological record does not, of course, eliminate the possibility that *Daubentonia* has been derived, independently of other lemurs, from a plesiadapid ancestry.* But, if account is taken of the several features of other anatomical systems wherein it shows affinities with the lemuriforms, it seems more probable that this odd creature represents a divergent and very aberrant branch of the main line of lemuriform evolution, even though this interpretation implies rather remarkable and abrupt deviations from the general course of evolution which has otherwise been characteristic of the lemuroid dentition (e.g. the directional change from a progressive reduction in the upper incisors to a gross hypertrophy). Incidentally, it has been remarked that the comparative lack of true rodents in Madagascar would obviously have provided ecological opportunities for the evolutionary development of a rodent-like creature of some sort, and it so happened that it was the lemuriform inhabitants of the island which exploited these opportunities. That they were able to do so is perhaps to be explained by the supposition that, even in later Tertiary times, the lemuriform stock still retained a residuum of the genetic potentialities manifested by many diverse groups of early Primates in the gross enlargement of the incisors.

TARSIOIDS

Taken as a whole, the dentition of the only existing tarsioid genus, *Tarsius*, is more primitive than that of any of the Recent lemurs, and, indeed, more so than that of many extinct prosimians (figs. 33, 34 and 35A). The dental formula of *Tarsius* is $\frac{2.1.3.3}{1.1.3.3}$. The incisors are implanted vertically in the upper jaw, and with no more than a very slight procumbency in the lower jaw. They are specialized to the extent that their number shows a reduction from the generalized placental formula, and also in the relative enlargement of the median upper incisors. The canines are conical, sharply pointed, and moderately projecting. The upper premolars have a simple form, the most anterior being very small and unicuspid; the two posterior are bicuspid, each having a prominent main cusp and a small internal basal cusp. There is no molarization of the last premolar, so that the premolar series makes an abrupt

* As already noted (p. 73), the fossil sequence of the Apatemyidae favours the interpretation that this was probably an aberrant side-line of early Primate evolution which did not survive the Oligocene.

contrast with the molar series. The lower premolars are correspondingly simple, the main cusp (topographically equivalent to the protoconid) being supplemented only by a very small and insignificant internal basal cusp. The upper molars display the simple tritubercular pattern, with the addition of a minute paraconule and a local thickening of the cingulum in the position where the hypocone develops in quadritubercular molars. The lower molars are likewise of a generalized type, with a complete trigonid bearing the three cusps protoconid, paraconid and metaconid, and a low, hollowed-out, talonid on the raised margins of which are the entoconid and hypoconid. There is also a small hypoconulid. The resemblance of the molar morphology to that of the pen-tailed tree-shrew is remarkably close (compare fig. 23 with 33 and 34).

This brief account leaves no doubt that the dentition of the modern tarsier is remarkably primitive. By contrast, many of the tarsioids of the Eocene, and even of the still more ancient Palaeocene period, showed quite marked specializations. One of the more generalized of fossil tarsioids is the European genus, *Pseudoloris*, of the Middle Eocene. Like its skull, the dentition of *Pseudoloris* shows similarities with that of *Tarsius*, but it is more primitive in the retention of four lower premolars (of which, however, the first is very small and vestigial), and more specialized in the enlargement of the most anterior tooth of the lower dentition. This latter tooth is possibly the canine, in which case the lower incisors have disappeared altogether. The upper and lower molars are hardly distinguishable from those of *Tarsius* except that the hypocone and the conules are rather more distinctly developed. Closely similar to *Pseudoloris* is *Nannopithex*, also of Middle Eocene date; apart from the dentition, considerable portions of the skull and skeleton of this genus are known. The Lower Eocene genus *Teilhardina* (from Belgian deposits) is known from incomplete mandible fragments which show that it had the dental formula 2.1.4.3; it is closely related to *Omomys* and *Hemiacodon* from the Middle Eocene of Wyoming. *Teilhardina*, *Omomys* (fig. 35B) and *Hemiacodon*, together with certain other genera of New World provenance, comprise the subfamily Omomyinae which is the only subfamily of fossil tarsioids common to the Old and New World. The subfamily Paromomyinae, of Palaeocene date, contains four New World genera in which the dentition resembles the Omomyinae in the primitive character of the molars and premolars, but they differ by

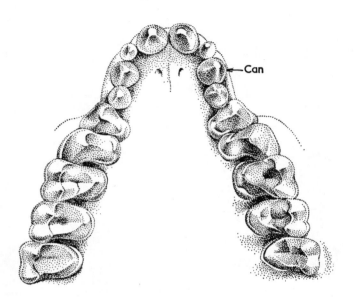

FIG. 33. Upper dentition of *Tarsius spectrum* (×5).

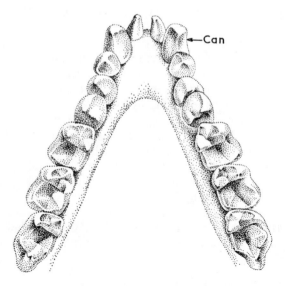

FIG. 34. Lower dentition of *Tarsius spectrum* (×5)

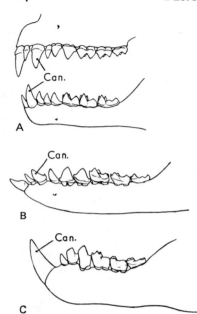

FIG. 35. Upper and lower dentition of *Tarsius spectrum* (A) and
the lower dentition of *Omomys* (B)
and *Tetonius* (C). (×2).

the development of a greatly enlarged lower front tooth (probably
an incisor).*

Apart from those groups which preserve a comparatively generalized type of dentition, the fossil tarsioids of the Old World and the
New World show rather contrasting tendencies in their dental
evolution. Two European genera, *Necrolemur* and *Microchoerus*,
commonly grouped in the subfamily Necrolemurinae, are characterized by progressive features of the molars (fig. 36). In the upper
dentition these teeth have developed an advanced quadritubercular
pattern, and those of the lower dentition have lost the paraconid
in M_2 and M_3 and have acquired a flattened occlusal surface by
the elevation of the talonid to the level of the trigonid. In these
features of the molar teeth, the Necrolemurinae appear to make a
significant approach to the Anthropoidea. In both *Necrolemur* and
Microchoerus the upper molars are complicated by accessory cuspules (paraconule, metaconule and styles), and in *Microchoerus* they

* Simpson [149] has suggested that the genera of the Paromomyinae should
be included with the aberrant Palaeocene genus *Phenacolemur* in a separate
family, Phenacolemuridae. The latter is taken to represent a sharply distinct line
of evolution, which became extinct without issue.

and the premolars are further specialized by numerous secondary wrinklings of the enamel. The dentition of *Necrolemur* is known in some detail; the formula is $\frac{2.1.3.3}{0.1.4.3}$. As in *Tarsius*, the tooth rows of the two sides converge anteriorly at a sharp angle. The upper

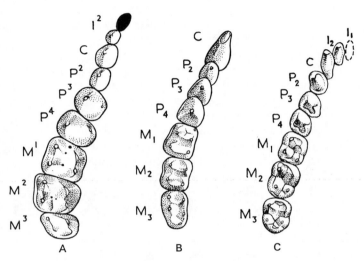

FIG. 36. The upper (A) and lower (B) dentition of *Necrolemur antiquus* (×3); note that the first lower premolar (which is diminutive) is not visible because it is obscured by P₂. The lower dentition of *Parapithecus* is also shown (C) (×2½).

incisors are small, with conical crowns. The upper canine is hardly larger and is slightly premolariform in general shape. There are three upper premolars, the most anterior small and of simple conical construction, and the two posterior larger and bicuspid. As commonly a feature of the tarsioid dentition, there is no pronounced molarization of the premolars. In the lower dentition of *Necrolemur* (and also in *Microchoerus* as far as can be determined from fragmentary specimens) the front teeth are highly specialized. The most anterior is a large, procumbent tooth, accepted by most authorities as the canine. If this diagnosis is correct, all the lower incisors have disappeared. Behind the canine are four premolars, of which the first is so diminutive as to be termed vestigial (in *Microchoerus* the first lower premolar has disappeared altogether).

Thus the Necrolemurinae contrast with the Adapinae (of approximately the same geological age) in the tendency for an early reduction in the premolar series. *Caenopithecus*, a European prosimian of Middle Eocene antiquity, is of uncertain status. According
to one interpretation (based mainly on certain details of the lower
molars), it has affinities with the Adapidae. On the other hand, the
construction of the upper molars, together with the abrupt contrast between the premolar and molar series, suggests tarsioid
affinities. The latter interpretation receives some support from incomplete remains of the skull which show a considerable abbreviation of the facial region (see p. 109). One other probable tarsioid
from the Middle Eocene of Europe may be mentioned here—*Periconodon*. This genus—known only from a maxillary fragment—is
distinguished by the formation of an additional cusp on the upper
molars, derived from the anterior part of the internal cingulum.
The main interest here is that a similar cusp pattern is found in
some of the Platyrrhine monkeys (e.g. *Saimiri*).

The New World tarsioids of the Eocene comprising the subfamily Anaptomorphinae differ from the Old World Necrolemurinae in the fact that, while preserving primitive molars of the
generalized tribosphenic type (similar to those still retained in the
modern tarsier), the last premolars tend towards a relative enlargement and lateral compression, particularly in *Uintanius* and
Absarokius. The more complete remains of *Tetonius* show that the
most anterior tooth of the lower dentition is greatly enlarged (fig.
35c). It is presumed to be a highly specialized canine similar to,
but even larger relatively than, that of *Necrolemur*. So far as it is
known, the dental formula of *Tetonius* appears to be $\frac{?\ 1.3.3}{0.1.3.3}$.

The Paromomyinae, which include some of the earliest of the
American tarsioids so far known and which date from the Middle
Palaeocene, had quite generalized molars and premolars. Of contemporary antiquity is another group distinguished as a separate
family of Primates, the Carpolestidae, whose relationships still
remain very obscure. Since they are known only from jaws and
teeth, their taxonomic allocation is particularly difficult. Their
distinctive feature is an extreme specialization of the posterior
premolars (fig. 37). In the lower dentition these teeth are much enlarged, and strongly compressed from side to side so that the crown
forms a sharp, shearing crest surmounted by a serrated edge. In

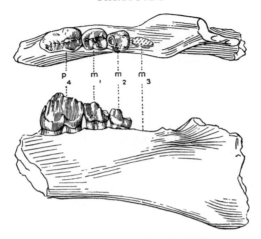

FIG. 37. Mandibular fragment and teeth of *Carpolestes nigridens*, showing the extreme specialization of the last premolar tooth (×4). (After G. G. Simpson, *Amer. Mus. Novit.* 1928.)

the upper dentition, both P³ and P⁴ are enlarged, and they are complicated by a series of secondary cuspules disposed in antero-posterior rows. Superficially these teeth show some resemblance to those of Mesozoic multituberculates, a resemblance still further enhanced by the presence of a greatly hypertrophied and pro-cumbent lower front tooth (as indicated by its socket). The infer-ence that the Carpolestidae are tarsioids depends on the evidence that the peculiar dental characters of the group are the culmination of a morphological trend which is apparent to a less pronounced degree in undoubted tarsioids. The skull and dental characters of *Tetonius*, for example, provide reasonably sound evidence of its tarsioid affinities, but it shows an incipient enlargement of the posterior premolars, and the last lower premolar is also laterally compressed (fig. 35c). Beginning with *Tetonius* it is possible to arrange a graded series of genera (i.e. *Uintanius—Absarokius—Elphidotarsius—Carpodaptes—Carpolestes*) which appears to de-monstrate a morphological sequence leading to the extreme special-ization of the premolar teeth already noted.* The functional significance of this specialization is uncertain, but it may have

* It is to be noted, however, that these genera do not form a temporal se-quence. For example, *Tetonius* is of Eocene antiquity, while *Carpolestes* extends back into the Upper Palaeocene.

served the purpose of crushing, or slicing into, the tough rind of fruit or the hard cuticle of insects.

In broad conspectus it is apparent that, while some of the most ancient tarsioids had a remarkably generalized dentition (and even in the only surviving representative, *Tarsius*, many of these generalized features are still preserved), during the evolutionary radiation of the group in Palaeocene and Eocene times there were manifested tendencies towards various specializations such as a reduction of the incisors, an enlargement and procumbency of the canines, the elaboration and hypertrophy of the last premolar (highly characteristic of some of the New World genera), and the complication of the crowns of the molars. Extreme structural modifications of so divergent a type are not uncommonly found in the earlier phases of the phylogenetic history of a taxonomic group, associated with the rapid development of adaptive radiations. The early tarsioids with a highly specialized dentition probably did not survive the Eocene, nor did they leave any successors, and so far as the palaeontological evidence goes there is a complete gap in the evolutionary history of the whole group between *Macrotarsius* of the Lower Oligocene and the present day. It is not unlikely, however, that a more generalized type such as *Pseudoloris* or *Nannopithex* may have been quite closely related to the ancestral stock from which the modern *Tarsius* was derived, for in many features besides the dentition they are closely similar.

So far as concerns the possible relationship of the Eocene tarsioids to the evolutionary development of the higher Primates, we may again refer to the fact that in the Necrolemurinae the conversion of tritubercular upper molars into quadritubercular teeth, together with the reduction or disappearance of the paraconid in the lower molars and the levelling of the trigonid and talonid moieties of the latter, is of particular interest, for these features appear to adumbrate the molar characters of the Anthropoidea. In the tarsioids generally, also, the contrast in structure between the premolar and molar series associated with the absence of any marked molarization of the former, and the tendency (in the European fossil types) for the last two premolars (P3 and P4) to approximate closely in size and shape, offer a further point of similarity with the dentition of the higher Primates.

The phylogenetic relationship between the early tarsioids and early lemuroids is by no means certain. Clearly at the commencement of the Tertiary epoch both groups were undergoing a rapid evolution during the same period and in the same geographical regions. It was evidently a time when the lately evolved basal Primates, comprising new adaptive radiations of great potentialities, were undergoing a remarkable diversification in response to all the new environmental possibilities opened to them. It is to be supposed, therefore, that besides the ancestral stocks which eventually gave rise to the modern lemurs and the only surviving genus of the tarsioids, subsidiary radiations of the early Primates must certainly have led to the development of numerous collateral types difficult of allocation to one or other of the main taxonomic categories. These taxonomic difficulties particularly obtrude themselves in the dentition. Generally speaking, it may be said that the early lemuriforms tended to develop specializations of the cheek teeth rather rapidly, as shown in the molarization of the last premolar teeth, and, in the lower molars, in the loss of the paraconid (which became merged in a ridge bounding the trigonid anteriorly). On the other hand, the Adapidae commonly retained all four premolars of the upper and lower dentition, and (unlike the Plesiadapidae and Apatemyidae) they tended to preserve a generalized type of incisor and canine. By contrast, in the early tarsioids the cusp pattern of the premolars and molars more often preserved the simple character of the generalized dentition (except for aberrant or progressive types). As already indicated, also, they tended to lose the first premolar more rapidly than their lemuroid contemporaries* and the last two premolars came to approximate rather closely to each other in size and cusp configuration. But while these contrasting evolutionary trends in the early lemuriforms and tarsiiforms certainly existed, they were not clear cut or consistent. It is for this reason, of course, that different opinions still exist regarding the taxonomic allocation of certain fossil prosimians. *Caenopithecus*, for example, has been accepted by some authorities as a tarsioid on the basis of its dental formula $\frac{?.1.3.3}{1.1.3.3}$, the morphological details of the upper dentition, the relative size of the orbit, and the reduction of the facial skeleton. But from the cusp pattern of the lower molars other authorities have assumed it to be an adapid

---

* In the Eocene tarsioid, *Anaptomorphus*, the second premolar of the lower dentition had also disappeared, so that this genus had acquired the dental formula characteristic of the Catarrhine Primates—2.1.2.3.

(in spite of the loss of the first premolar and the simple construction of those remaining). If *Caenopithecus* is to be classified provisionally in either of the two groups—and it is desirable to do so even if only as a temporary device for practical convenience—the balance of the morphological evidence seems slightly in favour of placing it among the tarsioids. In *Pronycticebus*, on the other hand, although the upper premolar-molar series suggests tarsiiform affinities, the balance of evidence from the dentition and the skull (see p. 143) is in favour of lemuriform affinities. But whether either of these extinct prosimians (and certain others also) are products of the same ramifications of Primates which gave rise to the modern tarsier or the modern lemurs, or whether they represent other ramifications which became extinct without leaving successors, can only be decided by the accumulation of more abundant fossil material.

So far as it is possible to conjecture the origin of the tarsioids on the basis of the dentition of the known genera, it is reasonable to infer that the common ancestral stock was characterized by the generalized placental dental formula, and by quite generalized features of all the component elements of the dentition. In other words it seems likely that the dentition of the ancestral stock would have shown no distinctive characters by which it could be labelled lemuroid. It has commonly been assumed that, in the evolutionary sense, tarsioids arose from lemuroids; if this were so, the separation of the one from the other must have occurred at a stage when the dentition was still exceedingly primitive and generalized.

MONKEYS

It is a fortunate circumstance for the student of the higher Primates that the dental morphology of the suborder Anthropoidea is far more uniform than it is in the Prosimii. Indeed, it may not always be easy to differentiate some of the genera of the higher Primates by reference to the dentition alone.

The Platyrrhine and Catarrhine monkeys differ significantly in their dental formula (figs. 38 and 39). In the former this is $\frac{2.1.3.3}{2.1.3.3}$, except only for most of the marmosets (Callitrichidae) in which the upper and lower last molars have been lost. The dental formula of these monkeys is thus $\frac{2.1.3.2}{2.1.3.2}$, a unique formula among

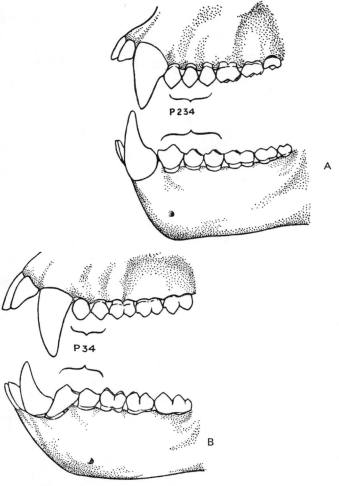

FIG. 38. Dentition of a Platyrrhine monkey, *Cebus* (×⁴⁄₃) (A) and a Catarrhine monkey, *Macaca* (×1) (B).

the Primates. But even in the Cebidae the last molars are commonly small and appear to be in process of regression, particularly in the upper dentition, and in the spider monkey (*Ateles*) they are actually absent in about 15 per cent of individuals.

Because of the loss of the last molars, the dentition of the marmosets must be regarded as specialized, particularly in view of the fact that the second molar may also be markedly reduced in size. On the other hand, the first upper molar preserves a primitive trituberculy, the hypocone being absent or extremely small.

In the Cebidae generally, the incisor teeth are spatulate with a straight or slightly oblique cutting edge. The upper premolars are

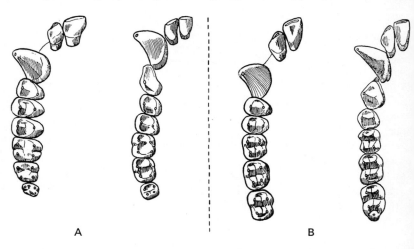

A   B

FIG. 39. Dentition of *Cebus* (A) (×⁴⁄₃), and *Macaca* (B) (×1). Note the atrophy of the last molar teeth in the Platyrrhine monkey, and the bilophodont character of all the molars in the Catarrhine.

bicuspid, the lateral cusp being the larger and more elevated. Of the lower premolars the most anterior (which is the largest of the three) is unicuspid and compressed from side to side, while the others have two cusps united by a rather conspicuous transverse crest. The upper molars are quadritubercular, but the hypocone is relatively reduced in the smaller species and the original tritubercular pattern of cusps is always quite evident. The lower molars all have four cusps of which the anterior pair are commonly united by an enamel ridge, and the trigonid and talonid portions of the crown are at the same level; the paraconid and hypoconulid have disappeared, but traces of these cusps are still to be seen in *Ateles* and *Alouatta*. The canine teeth in the Cebidae show a good deal of variation, being relatively more projecting and dagger-like in the larger species. If the canine is large in the upper dentition, it is usually separated from the lateral incisor by a conspicuous gap, the diastema, to accommodate the lower canine in occlusion of the teeth. In *Callicebus* the dentition is in some ways the most generalized of any of the Platyrrhines; in this genus the canines are but slightly projecting and have no diastema in relation to them, and the lower premolars retain a very simple form with the deuterocone represented by little more than a thickening of the internal cingulum.

In the Old World monkeys (Cercopithecoidea), the dental formula is $\frac{2.1.2.3}{2.1.2.3}$, as it also is in the anthropoid apes and the Hominidae. The upper central incisors are often quite broadly spatulate. The canines are usually sharp and projecting (particularly in male animals), and in some groups (e.g. the baboons) they form powerful weapons of offence (fig. 66). The premolars are typically bicuspid, but the anterior lower premolar is commonly enlarged and strongly compressed laterally to form an oblique cutting edge which shears against the upper canine. Such a type of premolar is generally termed *sectorial*. The molar teeth have four main cusps, the anterior and posterior pairs being each linked by strong transverse ridges. Together with a constriction (or *waisting*) of the crown between the anterior and posterior pairs, this produces a bilophodont type of tooth, which is to be regarded as a specialization distinctive of the whole group of Old World monkeys (fig. 43a). In the first and second lower molars the hypoconulid has disappeared; in the last molar it is commonly well developed to form a prominent heel-like projection.

The palaeontological record has so far yielded no certain evidence regarding the initial stages in the dental evolution of the ceboids and cercopithecoids. So far as the New World monkeys are concerned, the dentition of the extinct genera from the Miocene of Patagonia and Colombia differs in no fundamental characters from Recent genera, and earlier types are not yet known. The early history of the Cercopithecoidea is just as obscure. Two tiny fragments of mandible with a few of the cheek teeth, found in Lower Oligocene deposits in Egypt, have been regarded by some authorities as representing the initial phases in the evolutionary segregation of the Old World monkeys. The Primate nature of both these specimens (to which have been ascribed the generic names *Apidium* and *Moeripithecus*) has been called into question, and some authorities have supposed that they may be the remains of

small primitive ungulates called condylarths. But such an interpretation certainly does not apply to *Apidium* for, as Simons has
clearly demonstrated [136], the dentition of this genus closely
resembles that of *Oreopithecus*, whose Primate status is not in
doubt. *Apidium* has been regarded not only as a Primate, but as a
hominoid Primate, but like *Parapithecus* with which it is now included in a common family, Parapithecidae, it may also be interpreted as a generalized precursor of both the Cercopithecoidea and
the Hominoidea. The dentition of *Mesopithecus* from the Lower
Pliocene of Greece is quite similar to that of the modern Colobinae,
and fossil remains provisionally allocated to the same genus from
the Lower Miocene of East Africa appear to have similar dental
characters. It can at least be inferred from this meagre evidence
that the specialized bilophodonty distinctive of the Cercopithecoidea had developed at the beginning of the Miocene, and therefore that the group as a whole had by that time become segregated
from the Hominoidea. Minor variations in dental morphology have
been reported in later fossil cercopithecoids (e.g. *Simopithecus*,
Dinopithecus and *Parapapio*), but none of these shows any pronounced deviation from modern types.

ANTHROPOID APES

On first inspection the dentition of the Pongidae appears to
show no great difference from that of the Catarrhine monkeys.
They have the same dental formula, $\frac{2.1.2.3}{2.1.2.3}$, and they have certain
major features of dental morphology in common (figs. 40, 41 and
42). But they also display the following contrasts; the incisor teeth
in the Pongidae tend to be relatively broader (associated with a
characteristic widening of the premaxilla and the symphysial region
of the mandible), the bilophodont specialization is absent, all the
lower molars typically retain a hypoconulid, and the talonid of the
last molar is not extended into such a prominent heel. Except for
the hypertrophy of the incisors, it will be noted that in these
features the pongid is distinguished from the cercopithecoid dentition rather by the preservation of generalized characters. In more
detail, the central upper incisors have a wide chisel-shape with a
straight cutting edge, but the lateral upper incisors are relatively
much smaller with an oblique lateral margin. The lower incisors

have a narrow chisel-shape, and are all usually similar in shape and almost equal in size. It follows that while isolated central and lateral upper incisors are easily distinguished, this is often not the case with the lower incisors. The canines show a pronounced sexual dimorphism and in male animals may be long, sharp and dagger-like (figs. 68–70). In the upper dentition they are almost consistently separated from the lateral incisor by a wide diastema,* and in the mandible smaller gaps may in some individuals be found in relation to the lower canine (figs. 40A and 42). The

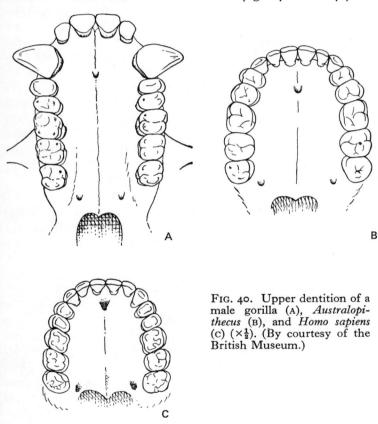

A

B

FIG. 40. Upper dentition of a male gorilla (A), *Australopithecus* (B), and *Homo sapiens* (C) ($\times\frac{1}{2}$). (By courtesy of the British Museum.)

C

---

* It should be carefully noted that a diastema may be absent until the canines have *fully* completed their eruption. Failure to recognize this may be responsible for some misleading statements on the frequency of the absence of a diastema in the pongids.

canines, it should be noted, are placed well behind the level of the incisor series. The premolar-molar series are aligned in straight antero-posterior rows terminating anteriorly with the canine, the rows of the two sides being more or less parallel, or even slightly convergent posteriorly. In closure of the jaws, the upper and lower canine teeth become interlocked. Obviously, in the position of occlusion this must limit the lateral excursion of the mandible and of the lower teeth (particularly the front teeth), and this restriction of movement is in part responsible for the differences in type of wear affecting the tooth cusps which distinguish the Pongidae from the Hominidae.

The upper premolars of the anthropoid apes are bicuspid, the lateral cusp being slightly the larger of the two. The lower premolars are rather strongly contrasted; while the posterior is bicuspid with cusps of approximately equal size and usually with a well-developed talonid, the anterior is predominantly unicuspid, the deuterocone, if present, being little more than a subsidiary cuspule situated on the inner slope of the main cusp. The sectorial character of the anterior lower premolar is its distinctive feature and appears to have been common among the known Miocene and Pliocene pongids as it is in those of today.

The cusp pattern of the molar teeth is relatively simple in the gibbon and gorilla, though there is of course an enormous difference in size in these two modern apes (figs. 43–45). In the orangutan, and to a much lesser extent in the chimpanzee, the main cusps are complicated by secondary wrinklings of the enamel. In absolute dimensions the molars of the chimpanzee approximate most closely to those of *Homo*; they also do so in the relative sizes of each of the three teeth, for while in the chimpanzee the decrease in dimensions from the first to the third molars accords more closely with the human molar series, in the gorilla and orang-utan, with the exception of the last upper molar, both dimensions of length and breadth commonly increase from M_1 to M_3.

Since the remains of the earliest known precursors of the anthropoid apes are almost entirely limited to jaws and teeth, the criteria of dental morphology are of particular importance for their diagnosis. Even so, however, these earlier remains are too fragmentary to permit of more than a provisional interpretation of their phylogenetic status. The lower teeth of *Amphipithecus* and *Pondaungia* (both from the Eocene of Burma) suggest affinities with the Hylobatinae, but they also present primitive features which possibly indicate a transitional phase linking a tarsioid level of evolution with the initial emergence of the suborder Anthropoidea as a whole. *Amphipithecus* had three lower premolars, but the diminutive root of the most anterior of these indicates that it was markedly reduced in size and perhaps almost vestigial. Of the relevant fossils next in geological time, *Parapithecus* and *Propliopithecus* (both known from jaws and teeth found in Oligocene deposits in the Fayum region of Egypt), the latter shows strong resemblances to undoubted hylobatines of later date, though the canine and the anterior premolar are less specialized. The dental formula of *Parapithecus* was at one time taken to be 2.1.2.3, but further discoveries have demonstrated that the correct formula is 2.1.3.3. The lower premolar teeth are of very simple construction, the deuterocone in the last premolar being represented by no more than a slight basal thickening of the cingulum. On the other hand the lower molars show an

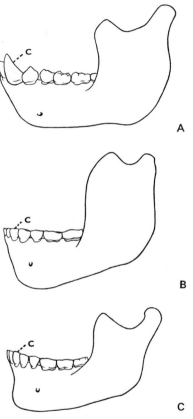

FIG. 41. Mandible and lower dentition of a male orang-utan (A), *Australopithecus* (B), and *Homo sapiens* (C) (×⅓).

advanced character in the total disappearance of the paraconid, and the trigonid and talonid portions of the crown occupy the same occlusal level (figs. 36c and 46). Moreover, the similarity of the cusp pattern of the lower molars in *Parapithecus*, *Propliopithecus* and some of the Early Miocene Hylobatinae is so striking that it

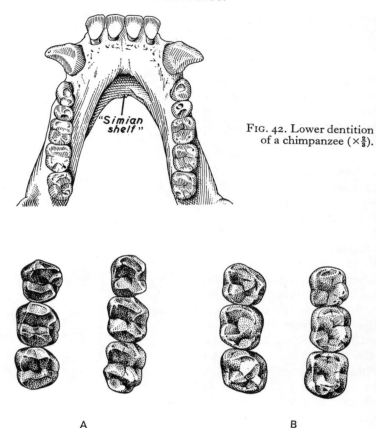

FIG. 42. Lower dentition of a chimpanzee ($\times\frac{2}{3}$).

"Simian shelf"

A   B

FIG. 43. Right upper and lower molar teeth of a Catarrhine monkey, *Presbytis* (A) ($\times 2$), and a gibbon, *Symphalangus* (B) ($\times\frac{4}{3}$).

would be difficult, by this criterion alone, to distinguish these types. No doubt the taxonomic position of *Parapithecus* can only be determined by the discovery of more complete remains of this extinct genus, and, while it is certainly an Oligocene Primate—and perhaps an exceedingly primitive representative of the Pongidae, some palaeontologists have suggested that the features of the dentition are too generalized to permit the latter of these assumptions. Of the other Oligocene Primates more recently brought to light in the Fayum, it has already been mentioned (p. 53) that in its dentition

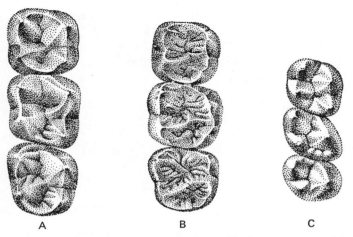

FIG. 44. Right upper molar teeth of a gorilla (A), orang-utan (B), and *Homo sapiens* (C) (×⁴⁄₃).

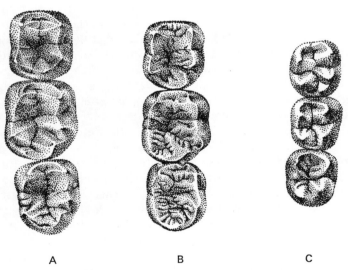

FIG. 45. Right lower molar teeth of a gorilla (A), orang-utan (B), and *Homo sapiens* (C) (×⁴⁄₃).

Aelopithecus shows more definite hylobatine affinities than *Proplio-pithecus*, and that the dentition of the earliest Fayum anthropoid ape so far known, *Oligopithecus*, possesses certain features resembling some of the Eocene prosimians and catarrhine monkeys

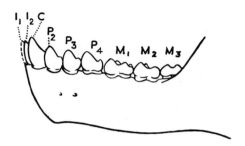

FIG. 46. Lower dentition of *Parapithecus*. (×3).

as well as the pongids. By contrast, the dentition of the somewhat later genus, *Aegyptopithecus*, is typically pongid in its morphological features.

The dentition of the apes which inhabited East Africa in Early Miocene times is known in considerable detail. Of these, *Plio-pithecus* (*Limnopithecus*) presents no essential difference from the modern gibbons, and this makes it clear that the hylobatine type of dentition had already become established at this early geological date. The larger Miocene apes of East Africa, belonging to the genus *Dryopithecus* (*Proconsul*), display considerable variation in size. Their dentition is typically pongid, but shows certain differences from that of the modern anthropoid apes; for example, the incisors are relatively smaller (the tooth rows tending to converge anteriorly), and on the upper molars the internal cingulum is strongly developed and often rather elaborately crinkled. *Plio-pithecus* from later Miocene and Early Pliocene deposits in Europe is similar to the East African representative of this genus in its dentition, though in some features somewhat less specialized. On the basis of dental morphology it has been suggested that this European genus was most likely the product of an offshoot from the *Parapithecus—Propliopithecus* group, this divergence presumably having occurred at a morphological stage somewhat earlier than that represented by the subgenus *Limnopithecus* [82].

The dentition of *Dryopithecus*, a genus known from Miocene

and Pliocene levels of Africa, Europe and Asia, approximates closely to that of the modern anthropoid apes except in the relatively small dimensions of the incisor teeth and certain other minor features. Of the extinct pongid types discovered in India, one is of special interest by reason of the relatively small size of the teeth, the simple construction of the molars, and certain features of the dentition in which it seems to approach the Hominidae. This genus is *Ramapithecus*. Its upper premolars and molars undoubtedly have a remarkably hominid appearance which is further enhanced by the parabolic contour of the upper dental arcade, the small size of the canine, the lack of a diastema, the relatively low degree of prognathism, and the flat wear of the molars. It has been surmised, from the dental evidence, that *Ramapithecus* may have been quite closely related to the ancestral stock of generalized anthropoid apes which gave rise to the Hominidae.*

The Lower Pliocene Primate, *Oreopithecus*, displays a number of interesting features in its dentition, particularly the bicuspid character of the front lower premolar. Mainly because of the latter feature, combined with the small size of the canine and the absence of a diastema, it has been proposed that this genus should actually be included in the Hominidae. That it is related to the Pongidae no longer remains in doubt, for this relationship is clearly demonstrated by the details of the skull and post-cranial skeleton (see pp. 163 and 215). However, the dentition is not typically pongid for it displays a number of unusual characters, particularly the peculiar conformation of the premolar cusps, the feeble development of the hypoconulid on M_1 and M_2, the pronounced backward extension of the talonid of M_3, and what appears to be a partial or incipient bilophodonty of the lower molars; such considerations suggest that *Oreopithecus* is a rather aberrant type of hominoid Primate.

HOMINIDAE

The structural affinities which justify the inclusion of the Hominidae and Pongidae in the common superfamily Hominoidea are reflected in the general pattern of the dentition as a whole.

---

* It should be noted that, apart from those mentioned, about a dozen different genera of extinct pongids have been named from fossil teeth found in Tertiary deposits. The relationship of these genera to each other and to Recent types is obscure, and in some cases it may be doubted whether the generic distinctions which have been made are really valid.

This applies as well to certain details of individual teeth as to main proportions; for example, it is by no means easy to distinguish the incisors and molars of *Homo* from those of some of the Tertiary apes. But there are also quite strong contrasts between the two families, which are known to extend back to at least the beginning of the Pleistocene period, and—conjecturally but almost certainly—well into the Pliocene.

The main characters in which the hominid type of dentition differs from the pongid type may be listed as follows (figs. 40B and C, and figs. 41b and c). (1) The dental arcade of the Hominidae has an evenly curved contour of parabolic or elliptical form, with no diastemic interval. (2) The incisors avoid the tendency to a gross hypertrophy such as is manifested in the Recent pongids. (3) The canines are relatively small, spatulate, and bluntly pointed; they wear down flat from the tip only, and at an early stage of attrition do not project beyond the occlusal level of the other teeth; they also show no pronounced sexual dimorphism. (4) The anterior lower premolar is a bicuspid, non-sectorial type; the two cusps are subequal in primitive hominids (e.g. *Australopithecus* and *H. erectus*), but in *H. sapiens* the lingual cusp has undergone a secondary reduction. (5) In the molar teeth the cusps tend to be more rounded and more closely compacted than in the Pongidae; even in the early stage of attrition, their occlusal aspect tends to become worn down to an even, flat surface. In modern *H. sapiens*, the hypoconulid is commonly much reduced or entirely absent. (6) The canine teeth erupt early, usually before the second molars, whereas in the Pongidae they erupt relatively later, after the second molars and sometimes even after the third molars. (7) In the deciduous dentition corresponding differences are found; for example, the lower milk canines are spatulate in contrast to the sharply pointed pongid teeth, and the first lower milk molars are multi-cuspid instead of being predominantly unicuspid and conical in form.

In the most primitive of the Pleistocene Hominidae, *Australopithecus*, all the typical features of the hominid dentition are well developed (figs. 40B, 41b and 71B). The teeth of this extinct genus, particularly those of the post-canine series, are relatively large, but they show considerable variation in their absolute dimensions. In *H. erectus* [165] the teeth are likewise massive in comparison with modern *H. sapiens*, and in some cases their dimensions come within the range of variation of the australopithecine teeth. The dentition of *H. erectus* also shows primitive characters in the canine teeth (in some individuals these are unusually pointed and projecting and may even be associated with a small diastema), and in the size of the last molar relative to the first and second molars. In their general proportions, the teeth of Neanderthal man, and also those of the Middle Pleistocene Heidelberg jaw, approach more closely those of *H. sapiens*, though they are relatively large by modern standards.

THE EVOLUTIONARY DEVELOPMENT OF THE DENTITION IN THE HOMINOIDEA

As we have seen, the problematical fragments of *Amphipithecus* and *Pondaungia* suggest that the evolutionary appearance of the earliest representatives of the Anthropoidea may have occurred in Eocene times. This inference would accord with the palaeontological evidence provided by the progressive Eocene tarsioids, such as the Necrolemurinae, which in certain features of the dentition appear to indicate an approximation to that of the higher Primates. By Oligocene times, it is evident from *Propliopithecus* that a dentition of a generalized hominoid type had already become established, while the somewhat enigmatic jaws of *Parapithecus* and *Oligopithecus* suggest that the Hominoidea may have been ultimately derived from a tarsioid ancestry. In the early Miocene, anthropoid apes with dental characters distinctive of the Pongidae had become diversified in considerable variety. We have also noted that some of the extinct Asiatic apes of later date (Pliocene) display dental characters which in some respects approximate to the dentition of the Hominidae. The question arises whether these resemblances indicate an actual phylogenetic relationship. The view has been expressed that the short blunt (brachydont) canine of the hominid dentition is a *primitive* feature, and that the conical, projecting canines of the Pongidae, associated with a diastema and a sectorial lower premolar, are *specialized* characters. It has thus been argued, presumably on the mistaken assumption that evolution can only proceed in straight lines (orthogenesis), that the hominid type of dentition could not have been preceded by a pongid type, and that the evolutionary divergence of the Hominidae from the Pongidae must have been very remote—even earlier than the Miocene. In

fact, however, there are good reasons for supposing that the so-
called primitive characters of hominid canine teeth are the result
of a secondary simplification, or retrogression, from more strongly
developed characters [118]. For example, the newly erupted and
unworn canine of *H. sapiens* (particularly the deciduous canine)
may project quite markedly beyond the level of the adjacent teeth
and may occasionally also be sharply pointed. Again, the permanent canine is provided with an unusually robust root, and the
latter is also longer than that of the immediately adjacent teeth.
Such features are difficult to explain on a purely functional basis,
for in modern man the canines have no special function to perform, but they do become intelligible if we suppose that they had
special functions in the past. Finally, the eruption of the permanent
canine in *H. sapiens* is still late relatively to that of the adjacent
teeth; it may come into place only after the two premolars, and
sometimes even after the second molar. There is also the evidence
of fossil hominids that the modern human canine has undergone
retrogressive modifications, for, as we have noted, in *H. erectus*
the canines (in some individuals at least) were relatively large
teeth associated with a definite diastema. That the predominantly
unicuspid and sectorial type of the anterior lower premolar,
characteristic of the Pongidae, is really a primitive and not a
specialized feature of the hominid dentition, and that the bicuspid
tooth characteristic of the Hominidae is secondarily derived, are
indicated by several considerations. For one thing, the sectorial
type of tooth is a functional correlate of a projecting and overlapping upper canine, since it is shaped in conformity with the
occlusal requirements of the latter. Again, in every known genus
of the earlier and more primitive anthropoid apes the front lower
premolar is a predominantly unicuspid tooth, and, indeed, this is
a characteristic of the primitive mammalian dentition as a whole.
Lastly, the first deciduous lower molar (i.e. the temporary precursor of the front lower premolar) offers very suggestive evidence,
for even in modern *Homo sapiens* it is often quite markedly compressed from side to side, with the anterolateral surface sloping
obliquely somewhat as it does in a typical sectorial tooth. This
sectorial-like feature of the human milk molar, which appears to
have been even more strongly marked in *H. erectus*, is not easy of
explanation unless it has reference to past evolutionary history.
    Taking into account this morphological evidence of the second-

ary simplification of the hominid canine and the associated changes in the first lower premolar, it seems clear that there is no theoretical reason why the hominid type of dentition should not have been derived from the more generalized pongid type found in some of the Miocene and Pliocene apes. Indeed, such indirect evidence as is available certainly supports this proposition. We know, from the abundant remains of *Australopithecus*, that the hominid dentition was fully established in all its essential characters by the Early Pleistocene [77]. It may be presumed, then, that transitional stages bridging the morphological gap between the pongid and hominid patterns of dentition occurred in Miocene or Pliocene times, perhaps more probably the latter in view of the fact that such generalized types as *Ramapithecus* date from the Pliocene.

4

The Evidence of
the Skull

THE distinguishing features of the skull of the Primates are in the main related to the fact that this mammalian Order is characterized by a progressive evolutionary tendency towards an enlargement of the brain, the development to a high degree of the apparatus of vision with a corresponding reduction in the apparatus of smell, the adoption to various degrees of an upright (orthograde) position of the trunk, and the replacement of the grasping functions of the teeth by the use of the forelimb for prehension rather than simply for support and progression. There is evidence, also, that in the initial stages of their development all these characters were directly associated with an arboreal mode of life.

In relation to the expansion of the brain the brain-case or neurocranium is usually voluminous and rounded, and this in many cases provides sufficient surface for the attachment of the temporal muscle (one of the masticatory muscles) and the neck muscles without the need for the development of conspicuous bony crests and flanges such as are commonly found in non-primate mammalian skulls. Moreover, the employment of the forelimb for prehensile functions (which in other mammalian Orders are served by the teeth and lips) is correlated with a reduction in the size of the jaws so that the masticatory muscles themselves tend to become relatively small. The reduction of the jaws is shown in a recession of the snout region of the skull, though this is no doubt also partly determined by the diminishing importance of the nasal cavities. In general, the tooth-bearing part of the skull is often of moderate size only.

The orbital cavities of the Primate skull are rather large, and they are surrounded (at least in all Recent types and in almost all the known fossil genera) by a bony ring which is completed by the articulation of the frontal with the zygomatic bone. The orbital aperture, also, is usually directed somewhat forwards. The zygomatic arch, which gives attachment to another of the masticatory muscles (the masseter), is relatively slender.

The foramen magnum on the base of the skull, instead of facing almost directly backwards as it does in completely pronograde mammals, becomes rotated to varying degrees so as to look more downwards. This displacement of the foramen magnum is in part associated with a change in the poise of the head following the

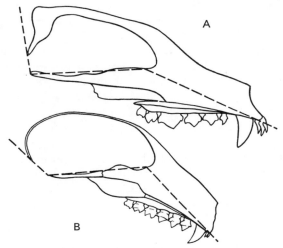

FIG. 47. Diagrammatic sagittal sections through the skull of a dog (A) and a lemur (B) to show how, in the latter, the facial region is bent downwards on the basicranial axis and the plane of the foramen magnum at the back of the skull is less vertical. Such changes in the angulation of the component elements of the cranial base are characteristic (to varying degrees) of the Order Primates as a whole.

development of an orthograde posture of the trunk; it is also correlated with a bending of the basal axis of the skull so that the face —in addition to a general reduction in size—comes to be placed more below the front part of the brain-case than directly in advance of it. This tendency towards a recession of the facial skeleton is illustrated diagrammatically in fig. 47, which shows a median longitudinal (sagittal) section through the skull of a lemur compared with that of a dog. But the relegation of the foramen magnum from its posterior position to the basal aspect of the skull (which is also indicated in fig. 47) is partly conditioned by, or associated with, the progressive development of the occipital lobes of the cerebrum which results from the expansion of the visual area of the cerebral cortex. Thus in most modern Primates the occipital region of the skull is rounded and prominent, projecting well back behind the level of the foramen magnum.

Lastly, it may be noted here that the tympanic cavity of the Primate skull (which contains the ossicles of the middle ear) is completely encased in bone, the osseous floor of the cavity being formed mainly by an expansion of the petrous bone (that is, the bony capsule which encases the cochlea and semicircular canals of the inner ear).

In considering these broad generalizations on the main features of the skull of the Primates, it is important to bear in mind that within the limits of the order there is quite a considerable variation in morphological details as well as in general proportions, particularly if fossil Primates are taken into account. We may best deal with these variations by discussing the skull characters of the different groups of Primates separately, and in giving attention first to the tree-shrews we shall take the opportunity of briefly describing the essential bony elements which constitute the mammalian skull and the manner in which they combine and articulate to form the complete structure.

TREE-SHREWS

The skull of the pen-tailed tree-shrew, *Ptilocercus*, is shown in lateral view in fig. 48. The snout region of the skull is formed mainly of two bony elements, the maxilla and the premaxilla. The latter contains the sockets of the incisor teeth and is separated from the maxilla by the premaxillary suture. The identification of this suture is of some importance for determining the dental formula, for the first tooth immediately behind it is normally the canine. The maxilla shows a large foramen just in front of the orbital cavity—the infraorbital foramen—which transmits the infraorbital nerve, an extension of the second division of the fifth cranial nerve (maxillary nerve). The roof of the snout region of the skull, including the roof of the front part of the nasal cavity, is provided by the narrow, elongated nasal bones. The roof and walls of the brain-case are formed mainly by the parietal bones, but also (to a much lesser extent) in front and behind by the frontal and occipital bones respectively. The squamous part of the temporal bone (squamosal) contributes to the lower part of the lateral wall. Much of the side wall of the cranium gives origin to the large temporal muscle, the upper boundary of this muscular area being demarcated by a roughened ridge, the temporal line. Posteriorly this line runs into confluence with a well-defined bony crest, the nuchal crest, which is disposed transversely and separates the

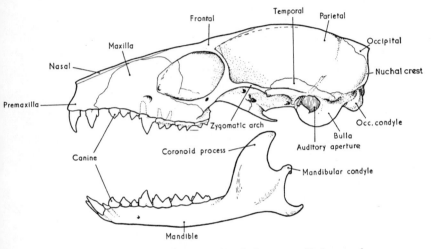

FIG. 48. Lateral view of the skull of the pen-tailed tree-shrew, *Ptilocercus* (×2). (*Proc. Zool. Soc.* 1934.)

parietal from the occipital region. The mid-point of the nuchal crest forms a small prominence called the occipital protuberance. The crest and the occipital area behind and below it provide attachment for the dorsal muscles of the neck. The margin of the orbital aperture forms a closed ring, completed posteriorly by the articulation of the orbital process of the frontal with the orbital process of the zygomatic bone. The latter articulates below and in front with the maxilla, and posteriorly is extended back as a slender temporal process which articulates with a zygomatic process of the temporal bone to complete the zygomatic arch. From this arch arises the masseter muscle; this and the temporal muscle are both masticatory in function and are inserted into the lower jaw. The auditory aperture is a relatively large, rounded opening placed close to the posterior extremity of the zygomatic arch; it leads into the tympanic cavity and is situated on the lateral aspect of a prominent, bubble-like swelling of bone, the auditory bulla. Behind this are visible in lateral view the occipital condyles; by these articular surfaces, which are oval, strongly convex antero-posteriorly and less so from side to side, the skull as a whole articulates with the first cervical (atlas) vertebra.

The lower jaw consists of a horizontal part, the body of the mandible (that is, the tooth-bearing part), and a vertical part, the ramus. Near the front of the body is a small mental foramen for a cutaneous nerve, the mental nerve. The ramus terminates above in the coronoid process which provides for the attachment of the

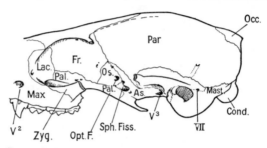

FIG. 49. Lateral view of the skull of *Ptilocercus*, with the zygomatic arch removed to display the component bony elements of the orbito-temporal region (×2). *As.*, Alisphenoid; *Cond.*, Occipital condyle; *Fr.*, Frontal bone; *Lac.*, Lacrimal bone; *Mast.*, Mastoid region of temporal bone; *Max.*, Maxilla; *Occ.* Occipital bone; *Opt. F.*, Optic foramen; *Os.*, Orbitosphenoid; *Pal.*, Palatine bone; *Par.*, Parietal bone; *Sph. Fiss.*, Sphenoidal fissure; *Zyg.*, Zygomatic bone; V^2, Infraorbital foramen for the maxillary nerve; V^3, Foramen ovale for the mandibular nerve; *VII*, Stylomastoid foramen for the facial nerve. (*Proc. Zool. Soc.* 1934.)

temporal muscle, and the articular condyle whereby the mandible articulates with the glenoid fossa on the base of the skull. The lateral surface of the ramus makes a broad area of attachment for the masseter muscle. The lower border of the mandible, at the angle of the jaw, is extended backwards into a hook-shaped angular process. The two sides of the mandible are joined together in the chin region at a joint called the symphysis menti; in some Primates this joint may be ossified to form a bony union or synostosis.

In fig. 49, the skull is shown with the zygomatic arch removed in order to display the construction of the inner wall of the orbital cavity by the sutural contact of several different bony elements. It is of some importance to take note of this osseous mosaic, for its pattern varies significantly in different groups of Primates. In the tree-shrews (and also, as we shall see, in the lemuriform lemurs)

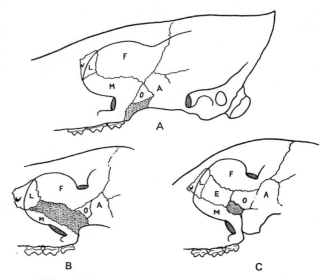

FIG. 50. Diagram showing the bony elements which may contribute to the formation of the orbito-temporal region of the skull. The palatine bone is indicated by stippling. (A) A common mammalian condition, in which there is a wide fronto-maxillary contact. (B) The condition in recent Lemuriformes (including the Tupaioidea) in which the orbital plate of the palatine extends forwards to articulate with the lacrimal. (C) The condition in the Lorisiformes, *Tarsius*, and the Anthropoidea, in which the ethmoid separates the frontal bone from the maxilla, and the palatine from the lacrimal. *A*, Alisphenoid; *F*, Frontal; *L*, Lacrimal; *O*, Orbitosphenoid.

the orbital wall is formed above by the orbital plate of the frontal bone, and this articulates below directly with the orbital process of the palatine bone. Anteriorly is the lacrimal bone, which is pierced immediately in front of the orbital margin by the lacrimal foramen; the latter transmits the naso-lacrimal duct by which the secretion of the lacrimal gland ('tears') is conveyed into the nasal cavity. The lacrimal bone extends beyond the orbital margin on the face to varying degrees in different Primates. Posteriorly the inner wall of the orbit is completed by two elements of the sphenoid bone—the orbitosphenoid and the alisphenoid. The floor of the orbit (not seen in fig. 49) is formed by the orbital surface of the maxilla. The orbitosphenoid is pierced by the optic foramen through which passes the optic nerve, and between the orbito- and alisphenoid is a large irregular foramen, the sphenoidal fissure, through which the first division of the fifth nerve (ophthalmic), the maxillary nerve, the nerves to the ocular muscles, and a number of bloodvessels gain exit from the intracranial cavity. In the higher Primates the maxillary nerve is accommodated in a separate foramen, the foramen rotundum. Near the posterior margin of the alisphenoid is a large opening, the foramen ovale, for the third division of the fifth nerve (mandibular nerve).

Three of the main variations in the mosaic construction of the orbital wall in mammals are shown in fig. 50. In many primitive mammals (for example in some genera of the order Insectivora) the orbital surface of the maxilla reaches up from the floor of the orbit to effect a wide contact with the frontal bone, while anteriorly the lacrimal, and posteriorly the orbitosphenoid, alisphenoid and palatine bones, contribute to the mosaic. Now, in animals in which the eyes are large, and particularly if they become rotated forwards to some degree (e.g. Carnivora and Primates), the maxilla is displaced downwards more completely into the orbital floor and thus comes to be separated from the frontal by a broad interval. This interval may become filled up in two different ways. In one case the orbital plate of the palatine bone extends forwards so as to reach, and make sutural contact with, the lacrimal (as in the tree-shrews and the lemuriform lemurs); in the other the ethmoid bone (the main part of which occupies the nasal cavity) becomes exposed in the orbit, separating the palatine from the lacrimal, and the frontal from the maxilla (as in the lorisiform lemurs and the higher Primates).* As we shall see, there are occasional exceptions to these broad generalizations; nevertheless, the variations just noted appear to have a taxonomic significance.

In fig. 51 is shown the skull of *Ptilocercus* from above. This view displays the nasal bones and shows the oblique inclination of the plane of the orbital aperture. The zygomatic arch is also seen to better effect; beneath it passes the tendon of the temporal muscle. The frontal is separated from the parietal bones by the coronal suture while, posteriorly, the lambdoid suture intervenes between the parietals and the occipital. The temporal lines or ridges marking the upper limit of the attachment of the temporal muscle (indicated in fig. 51 by an interrupted line) extend back in more

---

* A small portion of the ethmoid may become exposed in the orbit of other types of mammal, for example in some modern insectivores, e.g. *Erinaceus*; but, if so, it only forms a very minor part of the inner orbital wall.

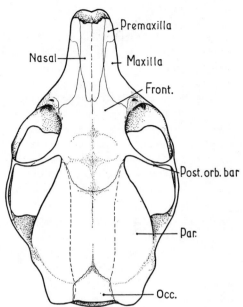

FIG. 51. Skull of *Ptilo-cercus* seen from above. (*Proc. Zool. Soc.* 1934.)

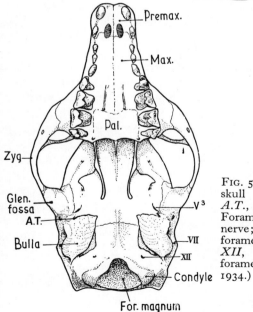

FIG. 52. Basal aspect of the skull of *Ptilocercus* (×2). *A.T.*, Auditory tube; *V³*, Foramen of mandibular nerve; *VII*, Stylomastoid foramen for facial nerve; *XII*, Hypoglossal nerve foramen. (*Proc. Zool. Soc.* 1934.)

133

or less parallel formation. In the genera of the *Tupaiinae* they coalesce posteriorly, and in some Primates in which the temporal
muscles are large and powerful, or in which the brain-case is small
relatively to the size of the jaws, the lines may fuse into a bony
sagittal crest extending along the mid-line of the cranial roof to a
varying extent.

The basal aspect of the skull (fig. 52) shows the bony palate.
This is formed by the palatine process of the maxilla in most of
its extent, anteriorly by the premaxilla, and posteriorly by the
palatine bone. Behind the latter are seen the posterior openings
of the nasal cavities (posterior nasal apertures). Through the
foramen magnum at the back of the skull the spinal cord becomes
continuous with the medulla oblongata of the brain, and accompanying it are the vertebral arteries; the foramen is flanked on
either side by the occipital condyles. At the posterior end of the
zygomatic arch is seen the glenoid fossa, for articulation with the
condyle of the mandible. Behind this, again, is the conspicuous
auditory bulla, the cavity of which contains the small ossicles of
the ear. It also encloses within its cavity a bony ring, the tympanic
ring, across which is stretched the tympanic membrane or ear-
drum, and which is kept in place by a membranous attachment
to the bulla wall, the annular membrane. The osseous bulla is
formed developmentally from a bony element which has a separate
centre of ossification, the entotympanic, but, though such an interpretation has been disputed, it seems probable that this is to be
regarded as an extension of the petrous bone which has secondarily
acquired a developmental independence.* Behind the bulla is the
mastoid region of the temporal bone—a region which is not distinctly demarcated in the tree-shrews, but which in some Primates
may undergo some structural differentiation. Of the foramina in
the base of the skull, note should be taken of the foramen ovale
and, immediately behind this under cover of the anterior margin
of the bulla, the opening for the auditory (Eustachian) tube. Between the condyle and the bulla is seen the hypoglossal aperture
for the twelfth cranial (hypoglossal) nerve, while lateral to the
bulla and close to the posterior margin of the external auditory
opening is the stylomastoid foramen which transmits the seventh
cranial (facial) nerve. One other foramen (not shown in fig. 52)
should be mentioned here; it is the foramen through which the

* On this point see R. Saban's paper on the tupaioid skull [126].

entocarotid artery (internal carotid artery of human anatomy) perforates the skull base. In the tree-shrews the foramen pierces the bulla close to its posterior margin, and through it the entocarotid artery enters a bony canal within the tympanic cavity. Here, after giving off a relatively large branch, the stapedial artery, which perforates the ear ossicle called the stapes, it continues as a small and relatively insignificant vessel, the 'arteria promontorii'. As with many other details of the tupaioid skull, this arrangement of the entocarotid foramen and artery is found in the skull of all the Lemuriformes. The small size of the continuation of the entocarotid artery is related to the fact that in these Primates (as in lower mammals generally) the brain receives most of its blood supply from other sources. In the higher Primates the entocarotid supply becomes of relatively greater importance, and in correlation with this the 'arteria promontorii' increases in relative size.

The foregoing description of the tupaioid skull has been mainly concerned with that of *Ptilocercus*. In the Tupaiinae (*Tupaia, Anathana, Urogale* etc.), the skull is very similar except that, in association with the more advanced development of the cerebral hemispheres, the brain-case is relatively more expanded. The plane of the orbital apertures is also directed more laterally, and in the larger species the snout region is extended more prominently.

LEMURS

Apart from the tree-shrews, the most primitive expression of the skull in modern types of Primate is to be found in the lemurs (fig. 53). In these, as in most lower mammals, the facial part of the skull tends to be relatively large in comparison with the brain-case. Thus they often have a conspicuous and elongated snout, which is most strongly developed in the Lemuriformes. On the other hand the well-developed brain is indicated by a relatively voluminous brain-case, the foramen magnum is also directed downwards as well as backwards, and the cranium lacks a conspicuous bony crest for the attachment of the temporal muscle. In some of the Lorisiformes (especially in the more diminutive species) the snout has been reduced to a degree which almost parallels the condition seen in *Tarsius*.

As we have noted in the previous chapter, in all the living lemurs with the exception of *Daubentonia* the upper incisor teeth have undergone a marked reduction in size and, associated with this, the premaxillary bones have dwindled to very small proportions. This atrophy has not progressed so far in the fossil Adapidae, and is not present at all in *Daubentonia* (indeed, in this aberrant creature the premaxilla is unusually big—see fig. 32B). The

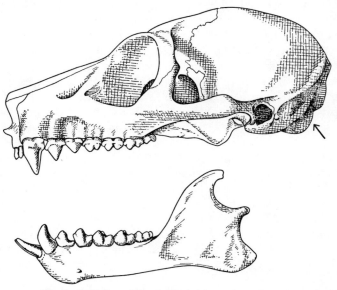

FIG. 53. Skull of *Galago garnetti*. In this and the following illustration, the direction of the axis of the foramen magnum is indicated by an arrow ($\times\frac{4}{3}$).

orbital aperture is relatively large, particularly in the more nocturnal species such as the galagos, pottos and lorises, and the plane of the aperture tends to be directed somewhat forwards. In the lorises it looks upwards to a marked degree, a disposition which is also present in the fossil *Adapis parisiensis* (fig. 57B). The lacrimal bone has usually no more than a moderate facial expansion in living lemurs, and in the lorisiform group the bone as a whole is considerably reduced in size. It articulates directly with the zygomatic bone in the Lemuriformes; in other lemurs it may be separated from this element by the maxilla (which thus takes a part in the formation of the orbital margin). In all Recent genera the lacrimal foramen is situated at, or immediately in advance of, the margin of the orbit.

The mosaic pattern of bony elements forming the orbital wall shows significant differences in the two main groups of lemurs. In the lorisiforms the orbital plate of the ethmoid is well exposed, intervening between the palatine bone posteriorly and the lacrimal anteriorly, while in the lemuriform lemurs the orbital process of the palatine extends forwards to articulate with the lacrimal as it does in the tupaioids.* A remarkable exception to this generalization is seen in *Daubentonia*, in which a more simple and seemingly more primitive mammalian condition is preserved, that is, a wide fronto-maxillary contact separates the palatine from the lacrimal. This disposition is also found in all the known Eocene Adapidae (at any rate in the European types), but it is clear, from *Progalago*, that the lorisiform type of mosaic had been fully developed by the Early Miocene. From this morphological evidence it may perhaps be inferred that the divergent lorisiform and lemuriform types were probably derived from a primitive and common type in which the fronto-maxillary contact was as yet undisturbed. If this is so (and the inference needs, of course, to be confirmed by more extensive palaeontological evidence), it seems likely that the two main Infra-orders of the lemurs underwent an evolutionary segregation before the Early Miocene.

The structure of the auditory region of the skull shows similar contrasts in the two main groups of lemurs, and it is convenient here to consider these differences in relation to those in the Primates generally. In certain primitive mammals the tympanic cavity, the roof of which is mainly formed from the petrous bone, has no ossified floor, while its lateral wall is formed by the tympanic membrane or ear drum which is supported in a fine bony ring, the tympanic ring, derived from an element called the os ectotympanicum (fig. 54A). In most mammals, however, the floor of the tympanic cavity becomes ossified to a variable extent, and this may involve different bony elements of the skull in different taxonomic groups. For example, in the modern insectivores, the basisphenoid and alisphenoid bones contribute a considerable proportion of the bulla wall. Thus the morphological details of the construction of the tympanic chamber have a considerable interest for the assessment of phylogenetic relationships [65]. In the

---

* In one of the lemuriform genera, *Microcebus*, the ethmoid becomes exposed in the orbital wall somewhat as in the lorisiforms, but unlike the latter the orbital process of the palatine is also extensive.

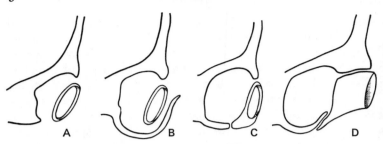

FIG. 54. Schemata illustrating the various relations of tympanic ring (os ectotympanicum) to the tympanic bulla. (A) Primitive mammalian condition, in which the ring is exposed and the floor of the tympanic cavity is unossified; (B) The lemuriform type (probably a characteristic feature of primitive Primates in general as well as in the Eocene leptictids), in which the ring is enclosed within an osseous bulla; (C) The lorisiform and Platyrrhine type, in which the ring is placed at the surface and contributes to the formation of the outer wall of the bulla; (D) The Catarrhine type (also seen in the modern *Tarsius*), in which the ring is produced outwards to form a tubular auditory meatus.

Primates (including the tree-shrews) the osseous floor is formed mainly from the petrous bone itself (or what appears to be its entotympanic extension), and in the lower members of the order this becomes distended to form the smooth rounded swelling on the skull base which is called the bulla (figs. 52 and 55).

In the Lemuriformes the bulla is relatively large and envelops the tympanic ring so that the latter becomes completely enclosed within it and attached to its inner aspect by the annular membrane (fig. 54B). This disposition, as we have already seen, also occurs in the tupaioid skull. Moreover, the placement of the entocarotid foramen, and the course of the entocarotid artery and its division in the tympanic cavity to form a large stapedial artery and a small *arteria promontorii*, are the same. In the Lorisiformes, on the contrary, the tympanic ring is exposed at the surface and, becoming somewhat expanded and applied to the outer margin of the tympanic process of the petrosal bone, contributes to the formation of the bulla wall (fig. 54C). Such a condition is also characteristic of the Platyrrhine monkeys. In *Loris* and *Nycticebus* the ectotympanic is further produced to form a very short tubular canal, the external auditory meatus; this extension is also present in the tarsioid skull, and is still more conspicuously developed in the Catarrhine monkeys and the Hominoidea (fig. 54D). The lorisi-

form lemurs are further distinguished from the lemuriforms by
the fact that the entocarotid artery divides into its two terminal
branches before it reaches the base of the skull, forming a large
arteria promontorii and a very small stapedial artery, that is, the
reverse of the lemuriform condition. Moreover, the *arteria promontorii* does not run a tortuous course through the tympanic
cavity, but takes a short cut by passing directly into the skull
through the foramen lacerum in front of the bulla. This foramen,
it should be noted, is not present in the Lemuriformes, for in the
latter it is obliterated by an overlapping of the base of the bulla

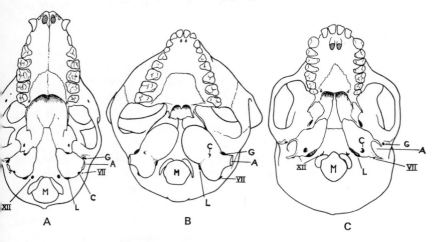

FIG. 55. Basal aspect of the skull of (A) *Lemur catta* (× ⅔); (B)
Tarsius spectrum (× ⁴⁄₃); (C) *Cebus fatuellus* (× ⅔).
A, Auditory aperture; *C*, Entocarotid foramen; *G*, Postglenoid
foramen; *L*, Jugular foramen; *M*, Foramen magnum; *VII*, Stylomastoid foramen; *XII*, Hypoglossal foramen.

wall. In the Lorisiformes the small stapedial artery pierces a
minute foramen on the posterior aspect of the bulla and runs the
same course as in the Lemuriformes.*

It is probably fair to say that in the past it has generally been
accepted that the intrabullar position of the tympanic ring in the
Lemuriformes is a divergent specialization, an aberrant development of a sort of 'fortuitous' nature which presumably had no

* In the sub-family Cheirogaleinae the arrangement of these arteries appears
to correspond to the lorisiform condition though in other features the auditory
region of this subfamily conforms essentially to the lemuriform type.

place in the ancestry of the other Primates in which the tympanic ring retains its position at the surface. But there is another interpretation which, on the evidence now available, seems to be more probably correct—that the intrabullar ectotympanic and its attached tympanic membrane is characteristically a primitive Primate feature. The evidence for such an assumption is derived from several sources. In the first place, this disposition (as we have seen) is consistently present in the most primitive of the living Primates, the tree-shrews, and is also stated to be characteristic of the Early Tertiary Leptictidae, a family sometimes referred to as Insectivora *sensu lato*. (Incidentally, this phrase '*sensu lato*', not infrequently used by students of basal Primates when comparing the latter with insectivores, illustrates the taxonomic dilemma confronting them.) It is worthy of note that a closely similar disposition is found in one of the most primitive and generalized of the existing marsupials, *Dasycercus*, though here the intrabullar ectotympanic is fixed and extended to the external auditory aperture by the ossification of the annular membrane (as in the fossil genera *Necrolemur* and *Adapis*, *vide infra*). Again, in the few early prosimian skulls of Eocene antiquity so far known, the ectotympanic is enclosed partly or completely within the bulla. Lastly, it is to be recognized that the simple annular form of the ectotympanic in the Lemuriformes, combined with the fact that it is more or less freely suspended, is itself a primitive character of embryonic type. The proposition that the modern Lemuriformes retain what was essentially the primitive Primate construction of the tympanic region of the skull is of some importance for the consideration of Primate phylogeny. For example, we know from *Progalago* that the lorisiform type of construction was already developed by the Early Miocene, but this would not preclude the possibility, or even the probability, that this type may have been evolved after the Eocene as a secondary modification of the lemuriform type previously developed in the Adapidae.

In the modern African lorisiforms, the mastoid region behind the bulla is characteristically distended by air cells. The tympanic bulla is also relatively large and spherical as in lemurs generally, but in the Asiatic genus *Nycticebus* it has an irregular 'deflated' appearance. It is interesting to note, therefore, that in this respect the African Miocene genus *Progalago* is similar to *Nycticebus*.

The mandible in the modern lemurs commonly shows a rather

slender body, the symphysial region slanting strongly forwards in conformity with the procumbency of the incisors and lower canine. The vertical ramus is relatively low, with well-marked coronoid and angular processes. The two halves of the mandible do not diverge from the symphysis to more than a slight degree. The symphysis menti is normally unossified in the adult, but synostosis did supervene in some of the extinct lemurs e.g. *Adapis* and *Archaeolemur*.

Attention may now be turned to the skull form of fossil lemuroids. The almost complete skull of *Plesiadapis*, from Palaeocene deposits in France [125], displays very primitive features in the small size of the brain-case, the large premaxillary bone, the massive zygomatic arches, and the absence of a post-orbital bar. On the other hand, as in Primates generally, the tympanic bulla is completely ossified and derived from the petrous bone. But (somewhat

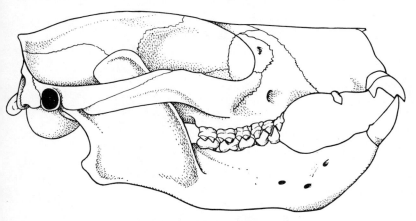

FIG. 56. The skull of the Palaeocene Primate, *Plesiadapis*, Natural size. Redrawn from an illustration (corrected for some distortion) kindly supplied by Dr Don Russell (× 1).

unexpectedly) the ectotympanic is extended outwards into a tubular meatus, presumably in secondary relation to the widely expanded zygomatic arch which must have accommodated an exceptionally large temporal muscle. In the Eocene genus *Microsyops* the premaxilla was reduced in size and, in this respect, the skull rather more closely resembles that of *Ptilocercus*. It has already been mentioned that the Eocene lemuroids show certain primitive features in the construction of the bony orbit, and in the relatively larger size of the premaxilla. Other primitive characters are largely correlated with the fact that at that early time in their geological history the brain had not as yet undergone any considerable expansion (fig. 57). Thus, the brain-case in the Adapidae is by no means voluminous, and in order to provide sufficient area for the attachment of the temporal and neck muscles, strong sagittal and

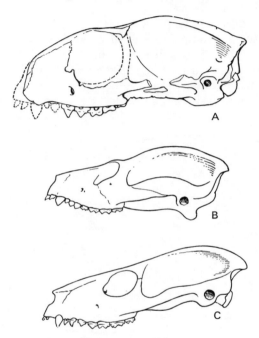

FIG. 57. Lateral view of the skull in certain fossil Adapidae. (A) *Pronycticebus gaudryi* (×1); (B) *Adapis parisiensis* (×$\frac{2}{3}$); (C) *Notharctus osborni* (×$\frac{2}{3}$). Note the relatively small brain-case, the muscular crests, the unreduced premaxillary region, and in B and C the small orbits.

nuchal crests are developed. The pronounced post-orbital constriction of the skull, as viewed from above, and the small size of the frontal bones, are also outward manifestations of the relatively poor cerebral development, and the foramen magnum faces almost directly backwards as it commonly does in non-Primate mammals of a pronograde habitus. In the American Eocene genus *Smilodectes* the front part of the brain-case is relatively more expanded and the sagittal crest has partly disappeared as the result of the separation anteriorly of the temporal ridges. It has been suggested that this fossil lemuroid may represent a transitional phase linking the early Notharctinae with the Platyrrhini, but strong claims for its much closer relationship to *Notharctus* have been advanced [32]. In some Eocene types the orbital cavity was smaller than in modern lemurs, suggesting that the visual apparatus was not so highly elaborated, or, perhaps, that these extinct creatures had not adopted nocturnal habits. The lacrimal bone in both *Adapis* and *Notharctus* is much reduced in size, while the lacrimal foramen is at, or immediately within, the orbital margin. Since a reduction of the lacrimal bone is generally regarded as an advanced character, it is curious to find it in Eocene lemurs (though there is less reduction in *Pronycticebus*). It has already been noted that in all the known adapid skulls the tympanic region is constructed precisely on the lemuriform plan, except that the annular membrane may be ossified.

The skull of the Upper Eocene genus *Pronycticebus* calls for some special attention because it presents certain features which appear to be unusual for an adapid. The earlier description of the skull suggested that the orbit was not encircled by a complete bony ring, but subsequent studies of the only skull available have made it clear that the apparent deficiency in the ring is the result of its erosion during the course of fossilization. The globular shape of the brain-case, the much abbreviated snout region, the relatively large size of the orbits, the forward extension of the lacrimal foramen, and the general shape of the palate—in all these characters the skull contrasts with that of *Adapis* and *Notharctus* and shows some approximation to the tarsioids. As we have already noted in the previous chapter, certain features of the dentition are also suggestive of tarsioid affinities. On the other hand, the constructional details of the base of the skull comprise such a close replica of those characteristic of the Lemuriformes as to justify the inclusion of *Pronycticebus* in this infraorder. Its phylogenetic relationships can, of course, only be finally established by the discovery of more complete remains, but its curious mixture of taxonomic characters is perhaps to be explained on the assumption that it represents an independent radiation of the early Prosimii from a time when the tarsioids had not yet become completely segregated as a taxonomic group from the lemuroids. *Pronycticebus* certainly emphasizes the difficulty of distinguishing closely between tarsioids and lemurs in the initial stages of their evolutionary differentiation, and supports the system of classification which places the Tarsiiformes in a common suborder with the Lemuriformes and Lorisiformes.

The skull of the Miocene genus *Progalago* is typically lorisiform in proportions and morphological details, and thus accords with the suggestion that *Progalago* may actually represent a direct ancestral antecedent of the modern Galaginae. The astonishing ramifications of subsequent lemuroid evolution in Madagascar in Pleistocene times gave rise to aberrant forms in which the skull shows wide variations. It is not practicable to give consideration to all these diverse types, but mention may be made of two extremes, *Archaeolemur* and *Megaladapis*. In the former (fig. 58) the general proportions of the skull are remarkably simian, the facial skeleton being much abbreviated, the brain-case relatively large and globular, and the orbital apertures facing almost directly forwards. On the other hand, there is a pronounced postorbital constriction, the nuchal crests are strongly developed, and the tympanic region is constructed on the lemuriform plan. The skull of

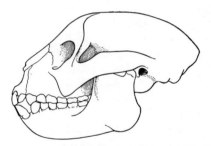

FIG. 58. Skull of *Archaeolemur majori* ($\times\frac{1}{3}$).

Megaladapis is very different in proportions, and in general appearance it quite closely resembles that of large ungulates (fig. 59). The brain-case is relatively small and furnished with strong sagittal and nuchal crests, the facial skeleton is massive, the foramen magnum and occipital condyles are directed posteriorly, the zygomatic arch is unusually broad, and the orbital apertures are small and surrounded by a raised bony rim. The tympanic bulla is continuous with a long meatus that appears to be constructed mainly from the squamous element of the temporal bone to form a so-called 'false external auditory meatus', but the ectotympanic appears to occupy the lemuriform position inside (though, instead of lying free, it is fused with the bulla wall). The entocarotid

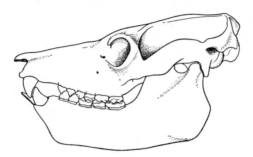

FIG. 59. Skull of *Megaladapis grandidieri* ($\times \frac{2}{9}$).

foramen is placed at the postero-internal angle of the bulla, and its small calibre indicates that, as in the lemuriforms generally, the brain was vascularized mainly by the vertebral and other arteries.

The evidence provided by the morphology of the skull for deciphering the evolutionary history of the lemurs has several points of special interest. For one thing, it adds considerable substance to the thesis that their relationship to the tree-shrews is a reasonable proposition. The original classification of the Tupaioidea in the order Insectivora (still favoured by a few systematists) finds little support in the skull structure, for the latter shows a number of strong contrasts in the two groups. In the Insectivora, for example, the brain-case is relatively small (except for that part which accommodates the olfactory bulbs), the orbital cavity is not encircled by an osseous ring, and the bony elements forming the orbital wall usually preserve what is assumed to be a more primitive mammalian pattern; the tympanic cavity also has an archaic construction, for its floor is incompletely ossified and its wall is largely formed from an expansion of the basisphenoid element of the cranial base. On the other hand, in the tree-shrews (as we have seen) the brain-case is relatively expanded, the orbital apertures are completely ringed with bone, and the orbital wall and tympanic cavity (with the associated foramina) are disposed precisely as they are in the Lemuriformes. Altogether, these structural details (and others) constitute a similarity in total morphological pattern which is difficult of explanation except on the basis of a natural affinity. Perhaps the most persuasive evidence of this kind is demonstrated by a direct comparison of the skull of a tree-shrew with that of a small lemur of comparable size, *Microcebus* (see fig. 60). In an unusually detailed study of the skull, Saban [126] concludes 'L'examen complet de l'ensemble des caractères de la tête osseuse établit ainsi les affinités certaines des Tupaioidea avec les Lemuridae.'

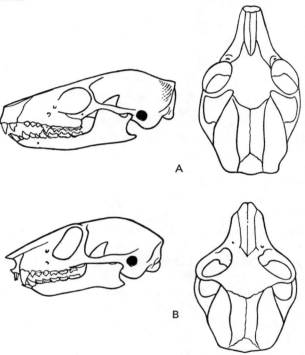

FIG. 60. Lateral and dorsal views of the skull of (A) *Ptilocercus* ($\times \frac{3}{2}$) and (B) *Microcebus* ($\times 1$), showing the similarity between these two types.

The comparative anatomy of the lemuroid skull is also of particular importance because it so clearly reinforces the evidence from other anatomical systems that the modern lemurs have become segregated in two rather well-defined and contrasted groups, the Lemuriformes and Lorisiformes. It is a matter of some theoretical interest to determine when, in the course of their evolutionary history, this segregation took place. If the common assumption is correct that the lemuriform type of auditory bulla with an enclosed tympanic ring is an aberrant and specialized modification, it can only be inferred that the two groups had already separated by the Lower Eocene. On the other hand, if the lemuriform type of bulla represents the condition to be expected in primitive Primates generally (for which, as we have indicated, there is certainly some highly relevant evidence), then it may well be that the lemuriform and lorisiform stocks became differentiated some time between the Eocene and the Lower Miocene.

TARSIOIDS

The skull of *Tarsius*, the only living representative of the infraorder Tarsiiformes, is characterized by its enormous orbital cavities the openings of which are directed forwards and outwards (figs. 61 and 62c). It has been suggested that certain of the cranial characters of *Tarsius* which resemble those found in higher Primates may really be secondary to this aberrant feature, and

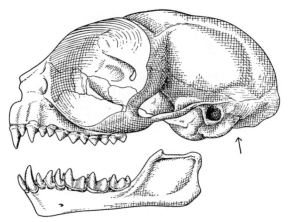

FIG. 61. Skull of *Tarsius spectrum* (×2).

this makes it difficult to determine whether some of the pithecoid resemblances of the tarsioid skull are, as it were, merely fortuitous, or whether they do in fact indicate a real structural affinity with the monkeys [153].

The reduction of the snout is at least partly illusory, for its posterior region is overlapped by the orbits. The latter are not only surrounded by a bony ring, but are also partly shut off from the temporal region of the skull by an incomplete bony wall formed from coalesced expansions of the frontal, zygomatic, and alisphenoid bones, approaching in this respect the normal condition in the Anthropoidea. As in the Lorisiformes, the ethmoid enters into the formation of the inner wall of the orbit, and the lacrimal is separated from the zygomatic by the maxilla. In the temporal region of the skull the alisphenoid articulates with the parietal bone. The brain-case is evenly rounded and is not marked by strong bony crests for muscular attachments. The fossa at the back of the skull which lodges the cerebellum (cerebellar fossa) is almost completely overlapped from above by the cerebral fossa. The lacrimal bone has a moderate facial extension, and the lacrimal foramen is placed outside the orbit. The foramen magnum is displaced on to the basal surface of the skull and looks almost directly downwards as it commonly does in the Anthropoidea (fig. 55B). The occiput is also prominent and well rounded, and the flexion of the basicranial axis is more marked than it is in the lemurs. The bony palate is triangular in shape, narrowing sharply in front towards the incisor teeth.

The tympanic bullae are relatively enormous, and extend in an antero-medial direction so that they approximate to the palatal region and almost meet in the mid-line (fig. 55B). The bulla wall is formed mainly from the petrous bone, the ectotympanic being applied to its outer surface superficially. Moreover, the latter element is produced laterally to form a very definite tubular auditory meatus, almost precisely as in the Old World monkeys. The entocarotid artery enters the tympanic cavity by piercing the under surface of the bulla near its centre, and its course through the cavity corresponds to that found in the Lemuriformes with the exception that the continuation of the artery (the *arteria promontorii*) is relatively large and enters the cranial cavity more directly.

The mandible in the tarsioids is lightly built, with a relatively low and broad ramus and a small coronoid process. The bodies diverge conspicuously from the region of the symphysis; such a divergence is correlated with (and thus an indication of) the relatively wide cranial base, for this determines the wide separation of the glenoid fossae which articulate with the mandibular condyles. In the modern *Tarsius* the symphysis menti remains un-ossified, but in the mandible of the Middle Eocene *Caenopithecus* (which, as we have noted, may be a tarsioid) it is synostosed, thus resembling the normal condition in the Anthropoidea.

A few of the extinct tarsioids are represented by skulls or frag-ments of skulls, and these remains are not only of importance for establishing the tarsioid nature of the fossils (which from the dentition alone might in some instances be considered doubtful), they also serve to emphasize the great antiquity of some of the specializations peculiar to the group. For example, the crushed skull of a specimen of *Nannopithex* and the front portion of the skull of *Pseudoloris* (both from the Upper Eocene) demonstrate that at this early period the enlarged orbits, the abbreviated muzzle and the triangular palate characteristic of the modern *Tarsius*, had already been acquired. The maxillary fragment of *Microchoerus* also provides an indication of spacious orbits. In the American genus *Tetonius* (fig. 62B), of Lower Eocene date, the orbits are not so large relatively as those of *Tarsius*, and they open widely into the temporal area of the skull behind. The facial part of the skull is also less reduced (though approximating to typical tarsioid dimensions), the brain-case is smaller and more constricted an-teriorly, the cranial roof flatter, and the occipital region less promi-nent and rounded. The tympanic bullae are large and elongated antero-medially. The skull of the European genus *Necrolemur*, of the Middle Eocene, is known in some detail from several speci-mens, and it offers a number of points of peculiar interest (fig. 62A). As compared with *Tetonius*, the orbits are yet further reduced and the snout is longer (though relatively narrow from side to side). The brain-case appears more elongated when viewed from the lateral aspect, while its breadth in dorsal view is partly exaggerated by the inflation of the mastoid air cells (as in the modern *Galago*). The cerebellar fossa of the skull is not so completely overlapped by the cerebral fossa, the foramen magnum looks downwards and backwards instead of almost directly downwards as it does in *Tarsius*, and the basicranial axis is more straightened out. On the other hand, the tympanic bulla is voluminous and extends antero-medially almost to meet its fellow of the opposite side as in *Tarsius*, and the auditory meatus is also extended in tubular form.

It has been generally accepted that the auditory meatus of *Necro-lemur* is constructed from the ectotympanic element as in *Tarsius*

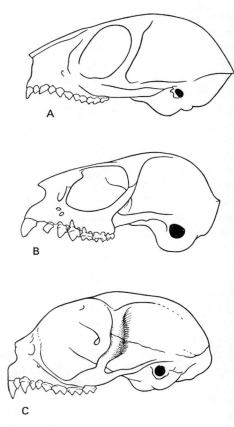

FIG. 62. Lateral view of the skull of (A) *Necrolemur* (×$\frac{5}{4}$), (B) *Tetonius* (×$\frac{3}{2}$), and (C) Tarsius (×$\frac{3}{2}$). Note the more primitive features of the fossil tarsioids (*a* and *b*), as indicated by the lesser reduction of the snout region, the smaller orbits, and the less voluminous brain-case.

and the higher Primates. However, a closer study has demonstrated that the ectotympanic is partly enclosed within the bulla [142]. But it is not a simple bony annulus as in the Lemuriformes; it is produced into a bony tube which extends outwards to form the meatal wall. The demonstration of an intrabullar part of the ectotympanic in *Necrolemur* led some authorities to dissociate this genus altogether from the Tarsiiformes and classify it either with the lemurs, or in an entirely separate group of its own. On the other hand, we have already seen that there is some reason to regard an intrabullar disposition of the ectotympanic as a primitive and characteristic feature common to the basal Primates, and, if this is so, it would not be surprising to find it still partially present in some of the primitive tarsioids of Eocene antiquity. It is true that the position of the entocarotid foramen in *Necrolemur* (close to the posterior margin of the bulla), and the tortuous course of the entocarotid artery within the bulla, have been adduced as further evidence of lemuriform affinities. But much the same position of the foramen is found in the marmoset, and the tortuous course of the artery is to be regarded as a primitive mammalian condition rather than a peculiarity which can be taken to be diagnostic of the Lemuriformes. The fact that the stapedial artery is a relatively small branch of the entocarotid, and that the continuation of the latter through the tympanic region is a relatively large vessel presumably supplying a considerable part of the vascularization of the brain (both non-lemuriform characters), suggests an approximation to the higher Primates. It is difficult, also, to overlook the implication of those features of the skull already mentioned which, taken in combination, indicate tarsioid (rather than adapid) affinities. Such a conclusion not only conforms with the evidence of the dentition, it is also consistent with highly significant features of the brain and the limb skeleton (see pp. 199 and 253).

We have noted the suggestion that some of the 'advanced' characters of the skull of the modern *Tarsius* may have been secondarily conditioned by the unusual development of the orbital cavities. No doubt this specialization has led to a certain degree of distortion of adjacent parts of the skull, but, if the fossil tarsioids are also taken into account, there is equally no doubt that the numerous similar tendencies in the evolutionary development of the skull which the tarsioids have shown in common with the Anthropoidea (even though superimposed upon a foundation of cranial specializations peculiar to the tarsioids) indicate a derivation from an ancestral stock with similar potentialities. In other words, there is reason to suppose that many of the tarsioids, including the modern *Tarsius*, have ultimately been derived from generalized precursors which also gave rise to the higher Primates. So far as the skull is concerned, such a precursor may have been not unlike *Necrolemur* (but presumably with a much more generalized type of dentition). From a skull structure of this kind, a progressive expansion of the brain-case, a forward rotation of the orbits without their specialized enlargement, a more complete bony partition separating the orbital cavities from the temporal region, a reduction of the bulla leading to the exposure on the surface of the whole extent of the ectotympanic element, a continued and marked retrocession of the facial skeleton, and the further displacement downwards to the cranial base of the foramen magnum (as in *Tarsius*) —comparatively moderate changes of this sort would have been adequate to transform the early tarsioid type of skull into a type such as must be presumed to have characterized the earliest evolutionary stages of the Anthropoidea.

MONKEYS

The simplest expression of the type of skull characteristic of the Anthropoidea is to be found in some of the New World monkeys (e.g. *Callitrix* and *Cebus*). Compared with the lower Primates, the

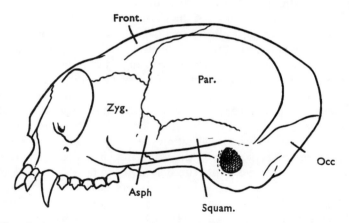

FIG. 63. Skull of a marmoset, *Callitrix jacchus* (× 2). Note the extension backwards of the zygomatic bone to articulate with the parietal and alisphenoid bones.

brain-case of monkeys is proportionately much larger and dominates the relatively small facial component of the skull (figs. 64 and 65). This enlargement of the brain-case is partly related to the pronounced expansion of the frontal and occipital areas of the cerebral cortex; in association with these factors, the frontal bone in the

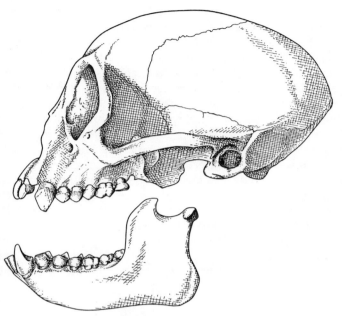

FIG. 64. Skull of *Cebus fatuellus* (×1).

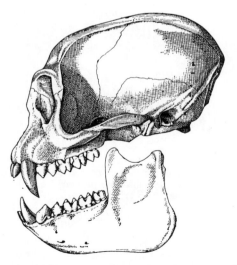

FIG. 65. Skull of a langur monkey *Presbytis* (×$\frac{2}{3}$).

153

Anthropoidea forms a proportionately large part of the cranial roof, and the occipital region of the skull is developed to the degree that the cerebral fossa of the cranial cavity entirely overlaps the cerebellar fossa. The relative extent to which different bony elements participate in the construction of the lateral wall of the brain-case varies in the different groups of monkeys. The parietal bone forms the major part and commonly articulates below and in front with an upward extension of the alisphenoid (as in the skulls of primitive mammals generally). This is the arrangement found in the New World monkeys and is also usual in the Colobinae of the Old World monkeys. In the Cercopithecinae, on the other hand, the squamosal commonly extends forwards to make sutural contact with the frontal, thus separating the parietal from the alisphenoid, but even in this subfamily there is considerable variation in the sutural pattern. In the New World monkeys the zygomatic bone is characteristically extensive and reaches back on the side wall of the skull to articulate with the parietal (fig. 63); this appears to be a specialization which has been avoided in the other groups of the

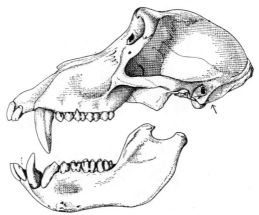

FIG. 66. Skull of a male baboon, *Papio*, showing the great development of the facial skeleton and the powerful canine teeth in these large Catarrhine monkeys (×⅓).

Anthropoidea. The lacrimal bone is confined within the orbital cavity (except for a very slight facial extension sometimes to be seen in a few genera such as *Cebus*, *Macaca* and *Papio*), and the lacrimal foramen is inside the orbital margin.

The recession of the snout which is foreshadowed to varying degrees in the Prosimii is carried to a further stage in monkeys generally, and there is a corresponding dwindling of the olfactory chambers. In fact, however, even among the Anthropoidea there is considerable variation in these proportional relationships (figs. 65 and 66). The baboons and mandrills, for example, appear to form conspicuous exceptions to the general rule, and in the larger individuals the massive, projecting muzzle is a very marked feature. The explanation is partly to be found in the fact that, in mammals at a corresponding level of cerebral development, the greater the body size the smaller proportionately is the brain and brain-case; on the other hand, the jaws and teeth, together with the masticatory muscles, increase their size in more direct proportion to the body weight as a whole. Thus, in the skulls of the larger monkeys the facial skeleton is usually more bulky as compared with the size of the brain-case, and in the skulls of small monkeys in which the jaws are markedly reduced in proportionate size the brain-case appears *relatively* large. These varying proportions in the size of the jaws and the brain-case are made very evident by the study of growth changes in the skull of individual animals; for example, in the baby baboon the jaws are relatively small, but with the growth of the animal as a whole they increase their proportionate size. This is simply an example of the well-known phenomenon of allometric growth, as an expression of which an increase in the general size of an animal at a certain over-all rate may be accompanied by a different rate of increase in a part of the same animal. But although such a differential rate of growth is not a *direct* relationship, it may be based on a more or less constant ratio which is expressed by the formula $y = bx^k$ (where y is the size of the part, x the size of the body, and b and k are constants). It is of the utmost importance to understand the implications of allometry in evolutionary development as well as in the growth of the individual, for it has not infrequently happened that statistical comparisons, by ignoring allometric phenomena, have grossly over-emphasized the taxonomic significance of differences in shape. As the result of allometric growth, the general shape and proportions of the skull (and also of other parts of the skeleton) may be very different in two quite closely related types. Such differences clearly have no great taxonomic value, since they may be related only to one major factor, i.e. body size. For example, the skull of a small macaque monkey differs greatly in general shape and proportions from that of a large baboon, even though they are members of the same subfamily; on the other hand, the general shape and proportions of the skull of distantly related monkeys of equivalent size, such as *Cebus* and *Macaca*, may show a much closer resemblance. It follows from these observations that, in order to effect valid comparisons between one type and another for the assessment of taxonomic affinities, the factor of absolute size must always be taken into account. Incidentally, since in the larger monkeys the jaws, teeth and masticatory muscles are as a rule more powerfully developed as compared with the size of the brain-case, the skull in these animals tends to develop more conspicuous bony crests for muscular attachments. These, likewise, may have little taxonomic significance. As examples of the traps into which the unwary statistician may fall by overlooking the factor of allometry, we may recall that a direct comparison of the massive skull of the extinct hominid *H. erectus* with that of the small gibbon led some authorities to the conclusion that the former was really a giant gibbon. And the comparison of *Australopithecus* with an immature ape's skull, or with the infantile skull of the pygmy chimpanzee, might also be used to support the argument that this primitive hominid was really an ape! It is worth noting, further, that even in modern human skulls there is a significant correlation between shape and absolute size.

The premaxilla of the upper jaw is of moderate development in the monkeys and apes and is never reduced to the extent seen in the modern lemurs. It usually extends up along the lateral border of the nasal bone, and in some of the Old World monkeys it may even reach up as far as the frontal bone.

The basicranial axis is flexed to the extent that the facial part of the skull comes to be more below the front of the brain-case than it is in the lower Primates. Similarly, the foramen magnum is displaced more completely on to the basal aspect of the skull. A curious exception is seen in the Howler monkey (*Alouatta*), in which the foramen magnum looks almost directly backwards; but it seems that this displacement is secondary to a change in the poise of the head which may be associated with the remarkable size of the hyoid bone and the lower jaw in this creature.

The orbits are relatively large and directed straight forwards and they are almost completely shut off from the temporal region of the skull by a bony wall composed predominantly of the alisphenoid and zygomatic bones (an advanced character which, as we have seen, is adumbrated in *Tarsius*). The orbital plate of the ethmoid forms a considerable part of the medial wall of the orbit, while the orbital process of the palatine is restricted in size.

In the New World monkeys the tympanic cavity is distended into a swollen osseous bulla, derived from an expansion of the petrous bone, and the ectotympanic element forms a simple ring which is adherent to its outer surface (fig. 55c). The entocarotid artery pierces the bulla posteriorly on its ventral surface, somewhat as it does in *Tarsius*; however, there is no persistent stapedial artery in any of the Anthropoidea (though this vessel is always present in foetal stages of development—even in the human embryo). In the Old World monkeys the tympanic cavity is encapsuled by the petrous bone, but there is no inflated bulla; the ectotympanic, also, is extended into a tubular meatus.

The mandible is stoutly built in most monkeys, and with the recession of the snout its body tends to be shorter in relation to the vertical ramus. The two halves of the mandible are firmly synostosed at the symphysis menti.

As with the dentition, palaeontological evidence bearing on the earliest evolutionary stages of the skull of the Anthropoidea is exceedingly sparse. One of the most ancient relics of a skull which definitely belongs to a catarrhine Primate is the frontal fragment of Oligocene age found in Egypt. This specimen gives evidence of a post-orbital closure characteristic of the Anthropoidea. It approximates in size to the skull of a marmoset and the orbit is relatively large. But, unlike lemuroids and tarsioids, the shape of the orbital aperture is rectangular rather than rounded and it is directed more forwards. The curvature of the frontal bone also indicates that the posterior part of the brain-case was somewhat expanded [134]. It is not known to which of the Oligocene Anthropoidea this frontal bone belongs. The skull of the Miocene ceboids from Colombia shows no fundamental differences from modern ceboids, and the skull of extinct cercopithecoids so far discovered (of Pliocene or Pleistocene antiquity) are closely similar to those of modern cercopithecoids. The skulls of *Mesopithecus* (fig. 67) and *Dolichopithecus* in their general proportion and structure suggest affinities with the subfamily Colobinae. The skull of the African *Libypithecus* is characterized by an unusually strong sagittal crest

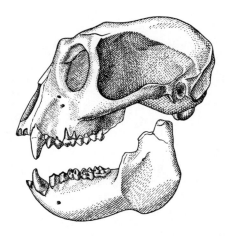

Fig. 67. Skull of a Pliocene monkey, *Mesopithecus pentelicus* ($\times \frac{4}{9}$).

and by a strong development of the snout region. The Pleistocene baboons of South Africa, so far as can be judged from their cranial remains, were highly variable, and included a giant type *Dinopithecus* which must have been a most formidable creature.

HOMINOIDEA

If the skull of one of the modern large anthropoid apes is compared with the skull of a modern man, the differences are very striking. The distinguishing characters in *H. sapiens* are in general related to (1) the much greater size of the brain-case, with which is associated a vertical forehead and a rounded and backwardly-projecting occipital region, (2) the greatly reduced size of the jaws and teeth, and (3) the poise of the head which, in correlation with the erect posture, comes to be balanced more evenly on the top of a vertical vertebral column. But it would be wrong to say (as is sometimes stated in careless writing) that these are the diagnostic characters whereby *man* differs from *apes*, for there are fossil representatives of the Hominidae in which the skull lacks such strongly contrasted features, or only displays them to an incipient degree (figs. 71 and 73). It is also the case that the skull characters of the Early Miocene apes (so far as they are known) differed in some species considerably from those of Recent apes and approximated to the cercopithecoid monkeys. In fact taking all this evidence into account, it is not easy to draw a firm morphological boundary sharply differentiating the Pongidae either from the Hominidae or from the Cercopithecidae on the basis of skull structure alone. We have already noted (in the first chapter) that major taxonomic categories such as these families are not to be defined by reference simply to the end-products of their evolutionary sequences, but by the main trends which have characterized them in the course of phylogenetic development. This, of course, applies to the skull as well as it does to other anatomical systems.

Of the skull of the modern anthropoid apes, that of the small gibbons (Hylobatinae), because of the relative proportion of the brain-case and facial skeleton, appears superficially to show the closest resemblance in general shape to the more primitive hominids (fig. 68). The relative abbreviation of the jaws may even be

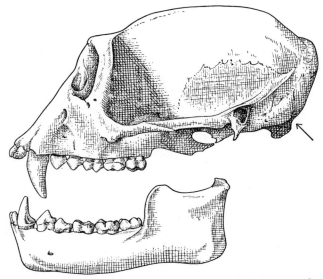

FIG. 68. Skull of a gibbon, *Symphalangus syndactylus* ($\times \frac{2}{3}$).

associated, in some individuals, with what appears to be an incipient mental eminence, or chin. The frontal bone characteristically extends far back on the top of the skull, intervening as a triangular wedge between the parietal bones anteriorly. On the side of the cranium the alisphenoid reaches up to articulate with the parietal. The occipital condyles are placed well behind the level of the auditory meatus, and the latter is elongated in tubular fashion. The brain-case is rounded and relatively smooth, but in occasional specimens the temporal ridges may extend up to the mid-line of the cranial vertex and even form a low sagittal crest.

Of the large anthropoid apes (Ponginae) the skulls of the African genera, the chimpanzee and gorilla (figs. 69 and 71A), show a general resemblance, their common features tending to be more exaggeratedly developed in the latter. The orbital apertures are

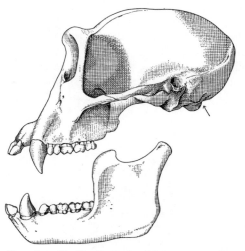

FIG. 69. Skull of a male chimpanzee ($\times\frac{1}{3}$).

relatively large and somewhat rectangular, and are surmounted by a continuous transverse ridge of powerful construction, the supra-orbital torus. Associated with the relatively larger jaws, the skull of adult male gorillas is usually surmounted by a conspicuous sagittal crest which extends back into direct continuity with the elevated nuchal crest. This coalescence of the sagittal and nuchal crests produces a beak-like, backward projection at the top of the occipital region of the skull, which makes the brain-case appear longer than it really is. The sagittal crest is formed during the period of growth to accommodate the growing temporal muscles which, as their fibres extend in length, migrate up the sides of the cranial wall to reach the mid-line of the roof of the skull. A sagittal crest is only rarely found in chimpanzee skulls, but the nuchal crest extends far up on the occiput as it does in all the large anthropoid apes. This high position of the nuchal crest results from the expansion upwards of the nuchal area of the occiput to provide for the attachment of the powerful neck muscles. On the side wall of the skull of the African apes, the temporal bone normally reaches forward to articulate with the frontal, so that the alisphenoid is separated from the parietal. In fully adult gorillas the mastoid region of the skull is sometimes produced to form a definite mastoid process, somewhat similar to that of the hominid skull; but this is not a consistent feature and is not found in immature skulls. The zygomatic arch is powerfully built, and a considerable proportion of it is formed by a backward elongation of the temporal process of the zygomatic bone. The occipital condyles are consistently placed well behind the transverse level of the auditory aperture and (in the adult) face backwards as well as downwards. In the infantile skull of the large apes, the occipital condyles and foramen magnum look more directly downwards (thus approximating to the adult hominid condition). But during growth, they progressively shift backwards to varying degrees. The premaxillary element of the upper jaw is relatively large (associated with the hypertrophy of the incisor teeth), but the suture between it and the maxilla becomes obliterated very early in post-natal life.* The palate is long and narrow, and of a general rectangular form. In the lower jaw of the large apes the symphysial region slopes downwards and markedly backwards, and the ascending ramus presents a broad surface for the attachment of a powerful masseter muscle.

The skull of the large Asiatic ape, the orang-utan, shows certain obvious differences from that of the African apes (fig. 70). There is no continuous supraorbital torus, and partly as a result of this the forehead region appears more fully rounded and thus more human in aspect. On the other hand, the occiput forms a flattened, vertical surface, and (viewed from the lateral aspect) the nuchal crest ascends steeply to a high level. In the adult male the jaws are quite massive and the cranial roof not uncommonly shows a high sagittal crest similar to that of the male gorilla.

* It has been claimed that the skull of *H. sapiens* is sharply distinguished from that of apes by the disappearance early in embryonic life of the facial component of the premaxilla. However, its occasional persistence has been recorded in the full-term human foetus. We have no evidence of its condition in early post-natal life in primitive hominids such as *H. erectus*.

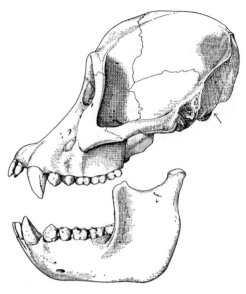

FIG. 70. Skull of a male orang-utan ($\times\frac{1}{3}$).

From this brief description it might appear that there is no great difference in general construction of the skull between the modern anthropoid apes and the Catarrhine monkeys. In fact, there is a considerable difference in the volume of the brain-case proportionately to the absolute size of the skull as a whole. This contrast is perhaps most obvious in the gibbons which, on the basis of body size, are more directly comparable with some of the monkeys. But it is also obvious enough in the large apes if due account is taken of the factor of allometry. There are also certain other differences, for example those which manifest themselves in the intracranial cavity in association with the relatively better regional development of certain parts of the brain, and there are differences in the shape of the nasal aperture and the conformation of the front part of the jaws. As we have seen, it is a characteristic of the modern large apes that the incisor teeth have undergone a relative enlargement; this has led to a general broadening of the muzzle (including the lower margin of the nasal aperture). The widening of the incisor region of the lower jaw has also resulted in a splaying apart of the two sides of the front part of the mandible, and the increased gap between them has become partly filled in below by a shelf of bone, the simian shelf, which extends back from the lower border of the symphysial region (fig. 42). As a secondary result of the widening of the incisor region, also, the post-canine teeth of the two sides are disposed in more or less parallel rows instead of diverging posteriorly, or they may even converge slightly towards the back.

Two almost complete skulls of early fossil apes are known, *Aegyptopithecus* and *Dryopithecus* (*Proconsul*). They are of unusual importance, for they indicate (as indeed was to be expected) that in Oligocene and Miocene times the pongid skull was far more primitive than that of the Recent anthropoid apes. In a number of features it resembles the skull of a catarrhine monkey more closely than does that of modern apes, for example in the relatively small size of the brain-case, the narrow nasal apertures, the less complete closure of the bony wall separating the orbital cavity from the temporal fossa, and the moderate dimensions of the front part of the jaws. These ceropithecoid traits make a strange contrast with the typically pongid pattern of the dentition. The frontal part of the cranial roof is smooth and the mandible lacks a simian shelf. It may be that the larger species of *Dryopithecus* when found (e.g. the African species *Dryopithecus major*) will show more strongly developed pongid features, but even in these the simian shelf of the mandible is not developed. In the later Miocene and the Early Pliocene pongids such as *Dryopithecus*, the scanty remains of jaws indicate that in these types the simian shelf, if present at all, was but slightly indicated. It seems likely, therefore, that the strong shelf which is so characteristic of the Recent large apes did not become fully established till the latter part of the Pliocene period. The skull of the Early Pliocene *Oreopithecus* is known from an almost complete, but severely crushed, specimen. In its general proportions it appears to have been very similar to the skull of a large gibbon, with an abbreviated muzzle, a rounded chin, and a relatively voluminous brain-case [60].

We have remarked that, while the modern ape skull is very strongly contrasted with the skull of modern *H. sapiens*, it is not possible to make such sharp distinctions when the much more primitive skulls of some of the extinct hominids are taken into consideration. In order to render comparisons of primitive types more certain, a great variety of measurements and indices have been contrived in attempts to give quantitative expression to distinctions which otherwise may not be very obvious. For example, various methods have been suggested for estimating the facial angle to allow comparisons of the degree to which the jaws project in front of the brain-case (prognathism); different angles for comparing the flexion of the several components of the cranial base have been devised; quantitative comparisons may be made of the cranial capacity as an absolute measurement, or in relation to some other measurement of the skull (e.g. the total area of the palate, or the gross dimensions of the teeth); indices may be applied for expressing the position of the condyles relative to the total skull length, the cranial height relative to the cranial length, the width of the postorbital constriction of the brain-case relative to its maximum width, and so forth. Not all indices of this sort are entirely satisfactory, if only because the points used for each series of measurements often vary independently of each other, so that the same indices employed on skulls of different configuration do not necessarily record identically equivalent morphological characters. However, it is not our intention here to recount the various methods which have from time to time been proposed for the quantitative comparison of cranial dimensions, for their details are readily available in text-books which more particularly deal with the technique and aims of craniometry. A brief descriptive comparison of the skull of a modern ape with that of the most primitive type of hominid so far known (*Australopithecus*) will serve to indicate some of the more general cranial features which, in the course of their divergent evolution, have come more and more to distinguish the hominid line of development from the pongid line. In this connection, it is important to note that (as the palaeontological evidence has now abundantly demonstrated) the expansion of the brain in the course of hominid evolution was a relatively late phenomenon; consequently, the differentiating and diagnostic characters of the skull of the *earlier* representatives of the Hominidae depend not so much on the relative enlargement of the brain-case as on modifications related to other factors such as the adoption of an erect posture and the acquisition of the hominid type of dentition.

In fig. 71 is shown the skull of *Australopithecus* compared with that of a female gorilla of comparable size. The distinguishing features of the australopithecine skull, wherein it reveals its

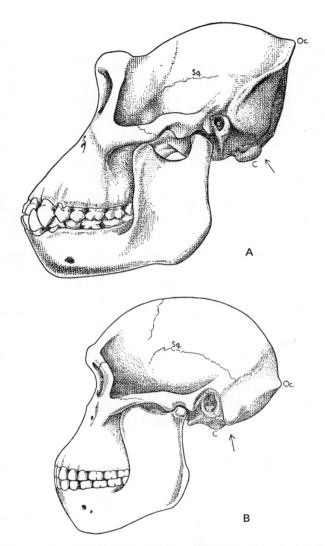

Fig. 71. Skull of an adult female gorilla (A) compared with that of
one specimen of the primitive hominid *Australopithecus* (B) (× ⅓).
Note the profound contrasts in the position of the occipital protuberance, *Oc.*, the inclination of the foramen magnum, the height of
the brain-case above the level of the supraorbital margin, the degree
of development of the supraorbital ridges, the relative position and
inclination of the occipital condyle, *C*, and the upward curvature of
the squamous suture, *Sq.* (By courtesy of the Chicago University
Press.)

essentially hominid character, are as follows. The cranial roof is elevated to a high level above the supraorbital margin; this is partly consequent on an increased upward bending of the basicranial axis which raises the brain-case to a higher level relative to the facial skeleton. Correlated with the increased flexure of the basicranial axis, and also with its abbreviation, the occipital condyles are placed further forwards in relation to the total skull length and also in relation to neighbouring structures of the cranial base. The nuchal crest (and its median eminence, the occipital protuberance) is situated low down on the back of the skull; this is related to the small size of the nuchal area of the occiput which, again, is correlated with a modest development of the nuchal musculature. The restricted extent of the nuchal area, it may be emphasized, is a particularly distinctive character of the hominid skull, and is related to the fact that, since in the erect posture the skull comes to be more nicely balanced on the top of the vertebral column than it is in the apes, the functional importance of the nuchal musculature for counterbalancing the weight of the front part of the skull is very much reduced. The forehead region lacks the huge supraorbital torus characteristic of the modern African apes. The mastoid process is consistently well developed in all the known australopithecine skulls, whether of young or fully adult individuals. The palate (like the dental arcade) is rounded and evenly curved. The jaws relatively to the size of the skull as a whole, though massive, do not project so far in advance of the frontal margin of the brain-case. The brain-case is not larger than that of the gorilla, but almost certainly it was larger relatively to the body-size so far as the latter can be inferred from remains of the trunk and limb skeleton. In the larger, or robust, variety of *Australopithecus* (ascribed by some authorities to a different genus of the Australopithecinae, *Paranthropus*), the small size of the brain-case relatively to the large jaws has commonly led to the development of a low sagittal crest. But it differs from the crest seen in anthropoid apes in the fact that it does not extend back into continuity with a high nuchal crest, and by itself it has no great taxonomic significance. It is a secondary development related to the extension upwards on the side of the skull of the relatively large temporal muscle. Finally, the australopithecine skull differs from the pongid type of skull, and conforms to the hominid type, in a number of apparently minor features which, taken in combination, constitute

a morphological pattern of considerable taxonomic significance—
for example, the conformation of the glenoid fossa and its immediate relationships, the disposition of the auditory meatus and the
petrous bone, the contour and extent of the squamosal, the relative
shortness of the temporal process of the zygomatic bone, the position of the mental foramen in the mandible, the shape of the
symphysis menti, the depth of the pituitary fossa, and so forth.
Not all these characters are readily amenable to statistical analysis
(except, perhaps, by prolonged studies based on the complicated
technique of multivariate analysis). But some of those dimensional
differences which are evidently related to erect bipedalism* have
been checked biometrically and are shown in fig. 72. Thus, the
nuchal area height index AG/AB brings into sharp contrast the

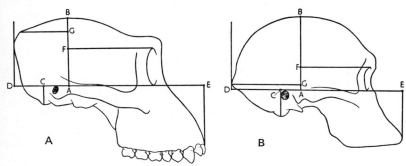

FIG. 72. Outlines of the skulls of (A) a female gorilla, and (B)
Australopithecus. The indices illustrated here are: nuchal area height
index, AG/AB; supraorbital height index, FB/AB; condylar
position index CD/CE. (By courtesy of Chicago University Press.)

relative height of the nuchal crest in apes and hominids, the supra-
orbital height index FB/AB records the elevation of the roof of the
brain-case, and the condylar position index CD/CE expresses the
relatively forward position of the occipital condyles in *Australopithecus*.

It will be understood that the typical hominid characters of the
australopithecine skull, some of which at this evolutionary level
are in little more than an incipient stage of development, become

* It is not to be assumed that these dimensional characters, taken *by themselves*, prove that *Australopithecus* was an erect, bipedal creature. But, when the
first skulls were discovered, they provided a strong presumption that this was
the case, and the inference was fully confirmed later by the discovery of remains
of the pelvis, lumbar vertebrae, and limb bones.

greatly accentuated in more advanced types of the Hominidae. In the skull of *H. erectus*, for example, the brain-case had already undergone considerable expansion by the beginning of the Middle Pleistocene, though the cranial capacity was still small by modern standards. This expansion particularly affects the occipital region of the skull, and it is partly due to this that the occipital condyles

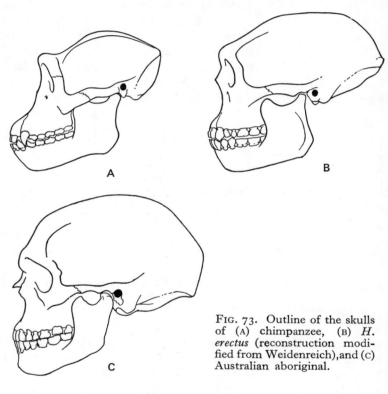

A

B

C

FIG. 73. Outline of the skulls of (A) chimpanzee, (B) *H. erectus* (reconstruction modified from Weidenreich),and (C) Australian aboriginal.

come to occupy a position still further forwards in relation to the total skull length. The jaws and teeth, though large, are of more modest proportions than in *Australopithecus*. The forehead region is still retreating and poorly developed, the nuchal ridge is often quite strongly marked, and the mandible lacks a prominent chin. Although the skull of *H. erectus* represents a considerable advance on the australopithecine skull, in its general appearance it does show a number of features in which it occupies a position more or less intermediate between the skulls of anthropoid apes and *H. sapiens* (fig. 73). By the end of the Middle Pleistocene, the hominid skull had attained a degree of development very similar to modern man; indeed, except for the rather strongly developed supraorbital ridges, some of the cranial remains of this date are hardly to be distinguished from modern man (see p. 358). So far as the skull is concerned, then, there is a close gradation—both as a morphological series and a time sequence—linking *Australopithecus* through *H. erectus* with *H. sapiens*. Detailed accounts of these later stages in the evolution of the hominid skull (including the aberrant skull of the more extreme type of Neanderthal man) are to be found in publications which deal specifically with palaeolithic prehistory.

5

The Evidence of
the Limbs

APART from the obvious modifications which occur in certain
groups of Primates in relation to functional specializations, the
general structure of the limbs of the order as a whole is remarkably
primitive. From a consideration of all the evidence based on a
comparative study of the more generalized types of mammal and
of their reptilian forerunners, there is good reason to assume that
in the early land vertebrates the limbs were initially capable of
some degree of grasping or clinging power, in addition to their
main function as components of a locomotor apparatus for propel-
ling the body forwards. These prehensile capabilities were associated anatomically with relatively free movements in all directions
at the shoulder joint, with a well-developed clavicle, with completely separate bones of the forearm (radius and ulna) between
which some degree of rotation was possible, with a separate tibia
and fibula, and with pentadactyl extremities in which the individual digits could be spread well apart or flexed in a converging
movement. Even in the most primitive vertebrates the clambering
activities of the forelimbs and hindlimbs are associated with
different functional requirements, for while the forelimbs may be
used from time to time for seeking and securing a grasp by which
the body weight is suspended and hauled up from above, the
hindlimbs serve more as struts which intermittently support the
body weight from below and push it upwards. These functional
differences are reflected in anatomical differences, particularly in
the constructional details of the limb girdles. Thus the shoulder
girdle is designed to permit a high degree of mobility at the
shoulder joint, and pelvic girdle (through which the whole weight
of the body may on occasion be transmitted to the hindlimbs) is
designed to permit a greater degree of mechanical stability at the hip
joint. For the same reason, the mobility between the tibia and fibula
is very much more limited than that between the radius and ulna.

In mammals which during the course of evolution have adopted a wholly terrestrial habitat with a quadrupedal gait the limbs come to serve more exclusively as props or supports to sustain the weight of the body in an approximately horizontal, or pronograde, position, and as instruments of speed for rapid propulsion in running and galloping. The adoption of a terrestrial mode of life, therefore, leads to modifications the effect of which is to increase the stability of the limbs at the expense of much of their primitive pliancy. In other words, limbs which are required mainly for support and progression on the ground lose to a greater or lesser degree their prehensile functions. Structurally, such terrestrial adaptations are manifested in a limitation of movement at the scapulo-humeral or shoulder joint, a tendency to the atrophy or complete disappearance of the clavicle, a fusion of the radius and ulna with consequent loss of any rotatory movement between them, a fusion of the tibia with the fibula, and an atrophy or loss of one or more digits in both the anterior and posterior extremities.

Structural adaptations to a fully terrestrial type of quadrupedalism have been almost entirely avoided in the Primates, and this is evidently due to the fact that, from the time when they first began to undergo evolutionary differentiation as a separate order from the basal stock which gave rise to placental mammals generally, they have in almost all cases maintained an arboreal mode of life. Thus, in the Primates as a whole the forelimbs, and to a considerable extent the hindlimbs also, have preserved an ancient simplicity of structure and function. Indeed, the capacity for grasping and clinging, which characterized the limbs of the earliest representatives of the basal mammalian stock, has not only been maintained, it has been further enhanced by a wider range of movement at the shoulder joint, a greater freedom of the rotatory movements between the radius and ulna, a more flexile ankle joint, an increased mobility of the digits, and the development on the latter of highly efficient friction pads.

It is one of the outstanding characters of almost all Recent Primates that one or more of the digits are covered on the terminal phalanx by a flattened nail (ungula) instead of a sharp and curved claw (falcula). Falculae are commonly found in all primitive and generalized mammals. They are also a characteristic reptilian feature, not only in modern but in extinct forms (as evidenced by the shape of the terminal phalanx), including the Palaeozoic group

of theromorph reptiles from which it is certain that mammals
were originally derived. It may be assumed, therefore, that in the
basal mammalian stock from which, in Palaeocene times, the
Primates took their origin, the digits were armed with claws.

A claw is a specialized epidermal appendage employed for
attack and defence, scratching, scraping, digging or climbing. In
its typical form it is strongly compressed from side to side, acutely
curved, and closely moulded on the terminal phalanx (which itself
assumes a claw shape). A longitudinal section of a claw shows that
its horny substance is disposed in two layers (fig. 74). The deep
stratum is the mechanically important part, being composed of

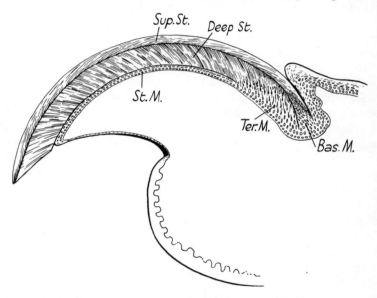

FIG. 74. Diagram of a longitudinal section of a claw. The horny
substance consists of a deep stratum which gives the claw its mechanical strength, and a superficial stratum which functions as a
protective layer. The germinal matrix of epidermis from which the
claw grows forwards is correspondingly divided into the terminal
matrix and the basal matrix. The greater part of the epidermal bed
of the claw is formed of a passive or sterile matrix. In the flattened
nails of most Primates, the deep stratum disappears, leaving only
the superficial stratum.

horny lamellae which are directed upwards and forwards towards
the tip; on it depends the maintenance of a sharp and strong point.

The superficial stratum is much thinner and, being composed of lamellae which lie flat in relation to the surface, does little more than provide a protective cover for the main underlying deep stratum. A typical flat nail must be regarded in itself as a degenerative formation, a retrogression from the more elaborate structure of the claw. It is curved laterally to a variable degree in different Primates and, while it may be used for scratching or picking, it is otherwise little more than a mechanical support for the digital pad of the terminal phalanx. In some Primates the nail becomes so small and insignificant that it is hardly more than a vestigial structure—and it may disappear altogether [76].

The structure of a flat nail represents only the superficial stratum of a claw—the deep stratum has disappeared entirely. But in the living Primates all stages of transformation from claw to nail occur, with an increasing retrogression of the deep stratum. In the tree-shrews all the digits are furnished with typical claws, as were also the plesiadapids of Palaeocene times. In *Daubentonia* and the marmosets sharp, curved claws are also characteristic of all the digits except the hallux. In these creatures they are not so sharply compressed from side to side as in the tree-shrews, and the groove on the under surface is relatively wider. But the deep stratum characteristic of true claws is quite strongly developed in *Daubentonia*, and is still present as a thin layer in the marmosets.

The evolutionary development of a flattened nail in the Primates is associated with an increasing functional importance of the terminal digital pads. Compared with claws these provide a much more efficient grasping mechanism for animals which find it necessary to indulge in arboreal acrobatics, for by their greater pliability they can be adapted with much more precision to surfaces of varying shape, size, and texture. They also come to be richly supplied by sensory nerves and thus to form tactile organs with a high degree of sensitivity. But, although this elaboration of the digital pads is clearly a progressive trend, the associated transformation of the specialized claw into a flattened nail is no less a degenerative change. It is of some importance to recognize this fact for it has sometimes been argued that the sharp-pointed claws in some of the Primates are retrogressions from a stage of evolution in which only flattened nails were present, a complete reversal of what was evidently the real direction of evolutionary development. This point will be referred to again in connection with the different

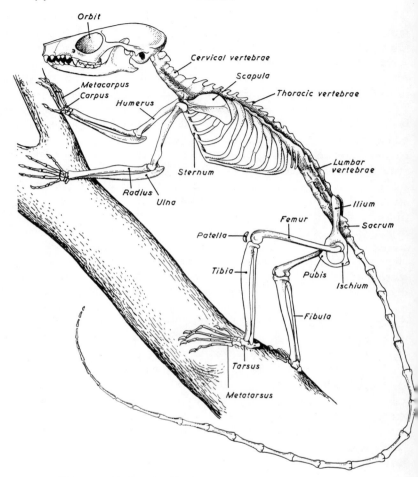

FIG. 75. Skeleton of the pen-tailed tree-shrew, *Ptilocercus*.
(Natural size.)

subdivisions of the Primates, but it may be remarked that it is
precisely in those Primates in which the digital pads have attained
a most conspicuous development that the nails show the most
advanced degree of degeneration; for example, in *Tarsius* on most
of the digits they are reduced to minute horny plaques to which
it would be difficult to assign any function. It is interesting to note,
also, that in some of the arboreal marsupials similar changes have

also occurred. Thus in the little *Tarsipes* the digits of the manus are provided with flat, scale-like nails, and in several marsupial groups the hallux has lost the relics of its claw altogether.

TREE-SHREWS

The limb structure of the living tree-shrews is in many respects of a very generalized character. The hindlimbs exceed the length of the forelimbs, as in primitive mammals generally, and all the digits of the manus and pes are equipped with typical claws. The joints connecting the different segments of each limb are capable of ranges of movement which permit a remarkable degree of agility in climbing, running and leaping.

In order to facilitate and make intelligible subsequent references to modifications of the limb structure in different groups of Primates, it is convenient here to recount briefly the main features of the generalized limb skeleton of the pen-tailed tree-shrew, *Ptilocercus* (fig. 75). In the shoulder girdle there is a relatively stout clavicle with a slight forward convexity. This bone, articulating with the sternum medially and the scapula laterally, serves as a strut which, by providing for appropriate leverage and also for muscular attachments, allows free movements of the upper limb away from and towards the body, as well as movements fore and aft. The clavicle is therefore an essential requisite for a full range of prehensile activities, and in terrestrial mammals in which the shoulder joint is used for little more than antero-posterior move-ments (for example, artiodactyls and perissodactyls), this bony ele-ment is absent.* The scapula of *Ptilocercus* has a characteristic triangular shape, the vertebral border (which forms the base of the triangle) being relatively short (fig. 76). Laterally, at the apex of the triangle, is the glenoid fossa for articulation with the head of the humerus. The shallowness of the socket formed by the fossa allows a very wide range of movement at the joint. Overhanging the joint is the acromion process of the scapula which gives attach-ment to the deltoid muscle, and from this process the spine of the scapula extends inwards to the vertebral border.

The humerus (fig. 77) has a rounded articular head facing directly backwards, adjacent to which are two small eminences,

---

* That the absence of the clavicle is the result of a secondary retrogression is made clear by the fact that in some species it develops, and actually undergoes ossification, in the embryo but is later resorbed.

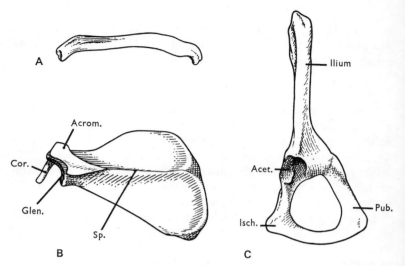

FIG. 76. Left clavicle (A), scapula (B), and right os innominatum (C) of *Ptilocercus* (×3).

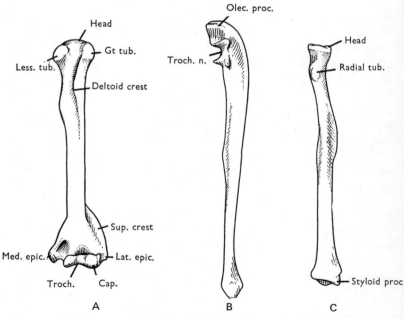

FIG. 77. Left humerus (A), ulna (B) and radius (C) of *Ptilocercus* (×3).

176

the greater and lesser tuberosities, which receive the insertion of the short scapulo-humeral muscles. The deltoid muscle is attached to a crest on the anterior aspect of the upper third of the shaft; this crest gives a slight angulation to the shaft as viewed laterally (fig.

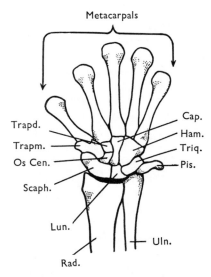

FIG. 78. Skeleton of left forefoot of *Ptilocercus*, seen from in front. The phalanges are not included (× 4). (*Proc. Zool. Soc.* 1934). The carpal bones are labelled: capitate, hamate, triquetral, pisiform, lunate, scaphoid, os centrale, trapezium, and trapezoid.

82). The lower extremity of the humerus has a rounded articular surface, the capitulum, for the head of the radius, and a pulley-like surface, the trochlea, which articulates with the ulna. The capitulum and trochlea are not sharply demarcated from each other. A bluntly-pointed medial eminence, the medial epicondyle, provides for the attachment of the flexor muscles of the forearm; immediately above it is the medial supracondylar ridge under cover of which the entepicondylar foramen transmits the brachial artery and the median nerve. At the outer margin of the lower extremity of the bone is a small lateral epicondyle above which is a conspicuous flange, the lateral supracondylar ridge or supinator crest. This crest attaches some extensor muscles of the wrist and (in some tree-shrews of the genus *Tupaia*) a powerful flexor of the elbow joint, the brachioradialis muscle.

The ulna (fig. 77) has a strongly concave articular surface at its upper end, the trochlear notch, and above the latter projects the olecranon process to which is attached the tendon of the triceps muscle. The radius has a disc-shaped articular head which allows the bone as a whole to pivot about the longitudinal axis of its shaft. It is this rotatory movement which permits pronation and supination of the forearm, a movement which is lost in those terrestrial mammals in which the radius and ulna become fused to provide for a greater stability. Near the upper end of the shaft is the radial tuberosity which receives the insertion of the biceps muscle. The lower extremity of the radius is slightly expanded to provide an articular surface for the wrist joint, and laterally projects to form the styloid process. In *Ptilocercus* the radius is but slightly longer than the humerus, the percentage ratio (called the brachial index) being about 105. The wrist joint is formed by the radius and ulna above and the carpal bones below. These small bones are illustrated and labelled in fig. 78. Those forming the distal row of the carpus articulate with the metacarpal bones of the five digits.

The manus (or hand) of the tree-shrews is furnished with friction pads on its palmar surface, corresponding in number, arrangement, and distinctness to the generalized mammalian condition (fig. 79). Thus, proximally there are two pads (thenar and hypothenar), and distally four interdigital pads. All these pads are covered by a fine pattern of papillary ridges, and the skin is richly supplied with sweat glands. On the ulnar side of the forearm, immediately above the wrist, is a small skin papilla in which are inserted the carpal vibrissae. Such vibrissae are found in primitive arboreal mammals of many types (see p. 284). The five digits (of which the middle is the longest) can be widely spread apart, and observations on the living animal have shown that by a converging flexion movement they can secure a strong grip of small objects. So far as the pollex, or thumb, is concerned, this type of movement no doubt provides the elementary basis of opposability, a function which becomes well developed in many of the higher Primates. In *Tupaia* the disposition of some of the short muscles of the hand does suggest a minor degree of functional differentiation of the thumb [101], but this is not to say that the tupaioid pollex is opposable in the usual meaning of this term, for the phrase 'opposition of the thumb' is commonly limited to the perfected movement whereby the thumb can be rotated and circum-

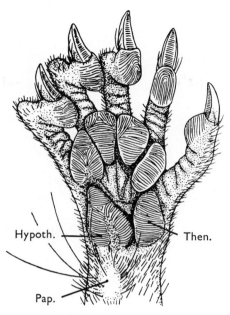

FIG. 79. Right forefoot of *Ptilocercus*, showing the palmar pads, and the carpal papilla to which the carpal vibrissae are attached (×3). (After R. W. Haines, *Proc. Zool. Soc.* 125, 1955.)

ducted at the carpo-metacarpal joint so that its palmar surface is brought into direct and firm contact with the palmar surfaces of the other digits. As already noted, all the digits of the tree-shrews are equipped with sharp, curved claws, and it is interesting to note that, in feeding, they convey their food in small pieces to the mouth by the use of the hand.

Because the hindlimb is required to resist the sudden stresses engendered by the forward propulsion of the body in running and leaping, and also, in arboreal creatures, to sustain its whole weight intermittently in the action of climbing, the mammalian pelvic girdle is more rigid and more firmly hafted to the trunk skeleton than the shoulder girdle. Thus, the three bony elements of the pelvis—ilium, ischium and pubis—are welded together to form the os innominatum, the ilium articulates firmly with the sacrum, and the pubic bones of the two sides are united ventrally along a strong symphysial joint. In the tree-shrews, the os innominatum retains the form characteristic of primitive arboreal mammals generally (figs. 75 and 76). The ilium is elongated and rod-shaped, but it is also slightly spatulate with a hollowed lateral surface which provides an area for the attachment of the gluteal muscles of the buttock region; it articulates with the first two sacral vertebrae. The pubic symphysis is short. The ischium is slightly expanded where it forms the ischial tuberosity; the latter underlies the region exposed to pressure in a 'sitting' position. The socket (acetabulum) for the reception of the head of the femur is deep and cup-shaped.

The femur (fig. 80) has a straight cylindrical shaft. At its upper end are the spherical articular head (which is connected to the

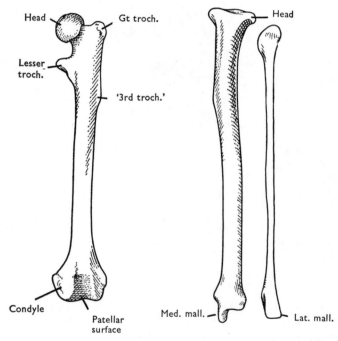

FIG. 80.   Left femur, tibia and fibula of *Ptilocercus* (×3).

shaft by a short neck), the greater trochanter to which is attached some of the gluteal musculature, and the lesser trochanter which receives the insertion of a powerful flexor muscle of the hip joint, the ilio-psoas. Towards the lateral aspect of the back of the upper end of the shaft is the gluteal ridge for the attachment of the

gluteus maximus; this ridge is rather pronounced and is sometimes termed the third trochanter. The lower extremity of the femur
has a complicated joint surface composed of the lateral and medial
condyles which articulate with the tibia, and a grooved patellar
surface. The condyles are separated by the intercondylar notch,
which accommodates the strong cruciate ligaments binding the
lower extremity of the femur to the upper surface of the tibial
head. The patella or knee-cap is a small sesamoid bone embedded
in the tendon of the quadriceps extensor muscle.

The shaft of the tibia has a sharp anterior border. It broadens
above into the head of the bone, and below it has an articular
surface for the talus. This surface is prolonged medially on to the
medial malleolar process. The fibula is a slender bone articulating
above and below, by synovial joints, with the tibia. Its lower extremity is extended to form the lateral malleolus. A slight degree
of play is permitted between the tibia and fibula at the ankle-joint,
and so accentuates the plasticity of movement at this joint.

The tarsal bones of the foot of *Ptilocercus* are illustrated in
figure 81. The talus (=astragalus) forms the tarsal component of
the ankle-joint, and is gripped by the malleolar articular surfaces.
The calcaneus has a prominent heel process and articulates in front
with the cuboid bone. The talus transmits stresses from the ankle
joint to the fore part of the foot through the navicular bone, and
the latter in turn transmits them to the three inner toes through
the three cuneiform bones. Of the metatarsals, the third is the
longest. The first metatarsal (of the hallux), in contrast with the
first metacarpal in the manus, is more obviously differentiated
from the other elements of the series. Thus, not only is it conspicuously shorter, but in *Ptilocercus* (not, however, in *Tupaia*) it
has a stouter shaft. Its base is also marked by a rounded peroneal
tubercle for the attachment of the tendon of the peroneus longus
muscle, and the facet by which it articulates with the entocuneiform is oblique and somewhat saddle-shaped. These structural
features of the first metatarsal are reflected in a certain degree of
functional specialization, for the hallux in the tree-shrew can be
widely abducted (more so than the pollex), and to a slight extent
can be brought nearly into opposition with the other digits and thus
used for grasping purposes (see fig. 16). This mobility of the hallux
is associated with a conspicuous differentiation of the intrinsic
muscles of the foot which are attached to this digit. Like those of

the manus, all the pedal digits are equipped with sharp, curved claws. Finally, it may be noted that on the sole of the foot the plantar pressure pads in *Ptilocercus* are disposed in the same pattern as the palmar pads and are all separate; in the genus *Tupaia*, on the other hand, the thenar and first interdigital pads are fused together as is usually the case in the lemurs.

The limb structure of the tree-shrews conforms closely with that of primitive arboreal mammals in general, and at the same

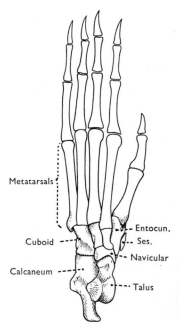

FIG. 81. Skeleton of left hind-foot of *Ptilocercus* (×3). (*Proc. Zool. Soc.* 1939.) The tarsal bones are labelled, and also a small sesamoid bone embedded in ligaments.

time permits a degree of mobility which, so to say, adumbrates the still greater pliancy of limb movement common to higher Primates. Its archaic character is emphasized by the fact that the limb skeleton of the plesiadapid Primates of Palaeocene antiquity shows a number of resemblances to that of the modern tree-shrews (fig. 82). Taken by themselves, these resemblances do not necessarily betoken a close phylogenetic relationship between the Plesiadapidae and the Tupaioidea, but they have been taken to indicate a somewhat similar degree of limb mobility. The distinctively Primate characters of the tree-shrew limbs are not, at first

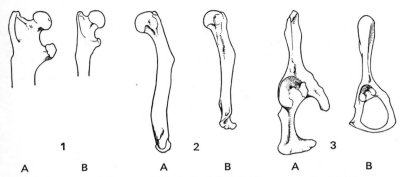

1 A B   2 A B   3 A B

FIG. 82. Limb bones of *Plesiadapis* (A) (×1) compared with those of *Ptilocercus* (B) (×$\frac{3}{2}$). 1. Posterior aspect of the upper end of the left femur. 2. Left humerus seen from the medial aspect. 3. Right os innominatum seen from the lateral aspect.

sight, very obvious, and undoubtedly the fact that the digits are armed with sharp claws (and not with flattened nails) has been partly responsible for the reluctance of some authorities to recognize the Tupaioidea taxonomically as Primates, even though the clawed plesiadapids are included in this order.

The muscular system has not been considered systematically in this book because it only occasionally offers evidence applicable to the study of the interrelationships of the various Primate groups. But it is worth while noting that there are a number of significant features in which the limb myology of the Tupaioidea shows resemblances to that of the lemurs and at the same time departs from the Insectivora with which they were originally classified. Thus, the following conditions are normally found in *Tupaia* and the lemurs, but are usually absent in the Insectivora:

1. M. piriformis present.

2. M. caudo-femoralis present.

3. M. brachioradialis usually present (but absent in *Ptilocercus*).

4. The elements of the M. quadriceps extensor are relatively well differentiated at their insertion.

5. The insertion of M. popliteus does not extend down beyond the upper quarter of the tibial shaft.

6. M. extensor digitorum brevis provides a tendon for the hallux, and in *Ptilocercus* this part of the muscle is even partly differentiated from the main mass to form an independent M. extensor hallucis brevis.

7. A well developed M. abductor hallucis is present.
8. Differentiation of M. peroneus digiti quarti.

The following conditions are also frequently lacking among the Insectivora, though normally found in tupaioids and lemurs:

1. M. teres minor present.
2. M. sartorius present.
3. M. coraco-brachialis fully developed.
4. Coracoid head of M. biceps present.
5. Differentiation of M. flexor pollicis brevis and M. flexor digiti minimi brevis.

LEMURS

The modern lemurs are all arboreal creatures, either moving rapidly by running and leaping from branch to branch as in the genera *Lemur*, *Propithecus* and *Galago*, or moving slowly with great deliberation by means of a crawling gait as in *Nycticebus*. In either case both anterior and posterior extremities are used for grasping the branches firmly, and the manus and pes are modified accordingly. The forelimbs are always shorter than the hindlimbs (a proportion which, as we have noted, is characteristic of primitive types of mammal), but in the Lorisiformes they approximate quite closely in size (see fig. 83 for the skeleton of a modern lemur, *Propithecus*).

The lemuroid clavicle is relatively stout and has a slight curvature, and the scapula in its proportions is similar to that of most lower mammals, that is, the vertebral border is relatively short. The humerus in Recent lemurs has a straight and slender shaft with a well-marked (but not very obtrusive) deltoid ridge; with the exception of *Arctocebus* the medial supracondylar ridge is pierced by an entepicondylar foramen transmitting the brachial artery and median nerve. The trochlear surface and capitulum at the lower extremity of the humerus are rather sharply differentiated. The radius and ulna are always separate, relatively slender, and usually fairly straight bones, and their articulations allow comparatively free movements of pronation and supination.

The slenderness and gracility of the limb bones in modern lemurs are related to the lightness of their body-build and, in general, to their agility of movement. It is interesting to compare the long bones of the upper extremity of the Eocene lemur, *Notharctus*. In

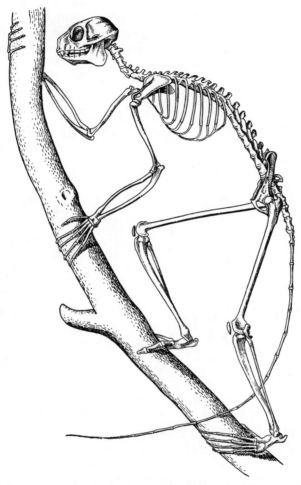

FIG. 83. Skeleton of a lemur (*Propithecus*) ($\times\frac{1}{4}$). (Adapted from Milne Edwards' Atlas publ. in 1875.)

this extinct genus the humerus is of more robust and coarse construction, and it approaches more closely in its shape to the generalized mammalian condition in its relatively short and curved shaft, its small articular head, and its prominent supinator ridge (fig. 84). As in primitive mammals, the supinator ridge extends further up

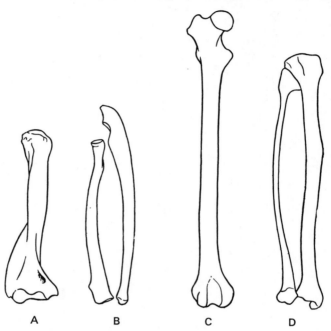

Fig. 84. Limb bones of *Notharctus osborni* (×⅔); (A) humerus, (B) radius and ulna, (C) femur, and (D) tibia and fibula.

the humeral shaft than in Primates generally, and was evidently associated with a powerful brachio-radialis muscle. The less extensive origin of this flexor muscle of the forearm in modern Primates is probably related to the capacity for a more complete extension of the elbow joint, such as is required for a greater mobility of the arm in arboreal activities. The humerus of *Notharctus* shows definitely Primate features in that the medial epicondyle is rather small (associated with a less strong development of the flexors of the fingers than is usually the case with clawed mammals), the trochlear surface is separated from the capitulum by a low lip (whereas in most primitive mammals the capitulum and trochlea merge into a more continuous articular surface), and the capitulum is globular (allowing a free rotatory movement of the head of the radius in pronation and supination). The radius and ulna of *Notharctus* are stoutly constructed, with flattened shafts which show a marked curvature.

In so far as *Notharctus* can be taken to be representative of the Eocene lemurs, it appears from the limb bones that these early Primates were more heavily built in relation to body size than Recent lemurs. Presumably, also, they had not acquired the agility of movement found in their successors. This conclusion is still more obviously applicable to the earlier genus, *Plesiadapis*, whose limb bones indicate that if these creatures were indeed arboreal, they must have been rather clumsy climbers.

In all modern lemurs the manus shows modifications from the generalized mammalian condition, some of which are of a distinctly aberrant nature. In certain species, e.g. some of the Galaginae, the palmar pressure pads retain the primitive pattern—that is to say. there are two proximal (thenar and hypothenar) pads and four interdigital, all of which preserve their individuality. In other forms, such as *Lemur*, there is some degree of fusion between adjacent pads, for example between the first and second interdigital, the first interdigital and thenar, or the fourth interdigital and hypothenar pads, and the hypothenar pad may show a tendency to split into two parts. Finally, in some lemurs, e.g. *Nycticebus*, the individual pads are merged into each other to such an extent that their outlines are but feebly indicated; in this feature an approximation is shown to the higher Primates.

As regards the digits, a characteristic of all Recent lemurs, and even of the Eocene *Notharctus*, is the relative elongation of the fourth digit both in the manus and the pes (fig. 85). Such a distortion of the primitive digital formula (in which the third is the longest) may be related to the fact that when the hand grasps a branch of relatively large diameter, a more secure and encompassing hold is obtained if the palm is placed obliquely across it with the thumb opposed to and meeting the outer digits round the branch, in this way increasing the span of the grasping hand. Thus (it has been argued) the early lemurs—which were relatively small animals in relation to the size of the branches among which they moved—tended to develop a functional specialization of the first (thumb) and fourth digits which together act somewhat as the two blades of a pair of forceps. It is of some importance, as emphasizing its far-reaching character, to note that the specialization of the fourth digit in lemurs includes more than a mere proportional difference in length, for it also involves a rearrangement of the small interossei muscles of the manus and pes which have become disposed about the fourth digit as their functional axis, whereas in other Primates generally (as in primitive mammals) the third digit characteristically forms this axis. In some lorisiform lemurs the specialization of the first and fourth digits is still further accentuated by retrogressive changes in the intervening digits (fig. 85). Thus, in *Perodicticus* and *Arctocebus* the index finger is reduced to

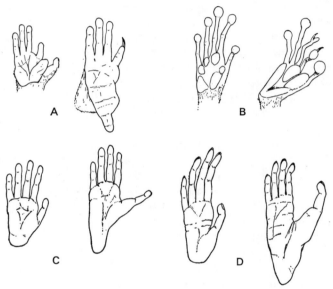

FIG. 85. Manus and pes of (A) *Nycticebus*; (B) *Tarsius*; (C) *Macaca* and (D) *Hylobates*.

a small nail-less stump containing only two vestigial phalanges, while the third digit is considerably shortened. Obviously, the typical lemuroid digital formula is of very ancient origin, for, as we have seen, it was already established during Eocene times.

In all modern lemurs except the aye-aye (*Daubentonia*) the digits of the manus are provided with flattened nails. In the aye-aye on the contrary, all the fingers are sharply clawed. It has been suggested that the claws are really curved and laterally compressed nails of secondary origin, but it has already been noted (p. 173) that they show the intrinsic structure characteristic of true claws, even though they are marked on their palmar aspect by a wider groove than is commonly found in the more typical claws of lower mammals. The retention of claws on all the fingers of *Daubentonia* is quite in harmony with the fact that this aberrant lemur retains in a number of other anatomical features persistent evidence of its derivation from a mammalian group of primitive placental type.* It is interesting to note that in *Notharctus* the shape of the only available ungual phalanx of the manus indicates that the nails in this Eocene genus were longer and somewhat more laterally compressed than in modern lemurs (except *Daubentonia*). This provides some confirmatory palaeontological evidence of the derivation of the flattened nail of Primates from a narrow claw.

In the carpus of Recent lemurs several modifications from the primitive carpal pattern are to be noted, which are seemingly related to the degree of divergence of which the thumb is capable, as well as to its functional differentiation (figs. 86 and 87). In the

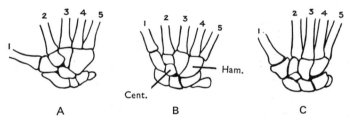

FIG. 86. Carpal skeleton of (A) *Lemur*; (B) *Tarsius*; and (C) *Macaca*. The numbers refer to the digits. Note in *Lemur* the large size of the hamate bone, the reduced capitate, and the displaced os centrale.

proximal row of carpal bones the scaphoid is wide, but the lunate is small and narrow. In the distal row, the capitate is constricted by a lateral compression while the hamate is typically large. The shrinkage of the capitate may prevent it from reaching proximally to gain contact with the lunate bone, thus losing an articulation which is characteristic of the generalized mammalian carpus (compare *Ptilocercus*, fig. 78). There is an os centrale, but it has been displaced medially from its primitive position to such an extent that it establishes contact with the hamate instead of being

---

* As emphasizing the aberrant nature of the aye-aye, it may be noted that the third finger is peculiarly attenuated, apparently to allow the creature to probe fine crevices in search of grubs.

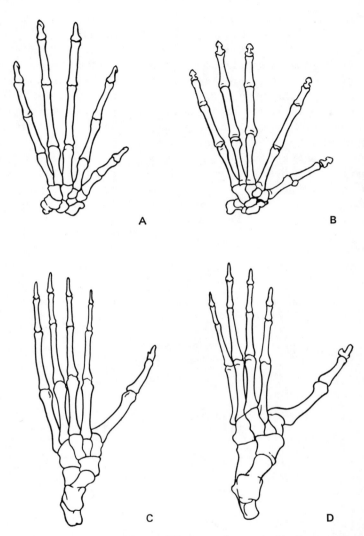

FIG. 87. Skeleton of the manus and pes of *Notharctus* (A **and** C) and of *Lemur* (B and D) ($\times\frac{3}{4}$ approx.).

separated from that bone by the capitate (as in the generalized condition which has been retained in higher Primates). It has been suggested that these specializations of the carpal pattern may all be secondarily related to the counterpressure exerted by a widely divergent pollex, which has resulted in a crowding of the carpal elements towards the ulnar side.

Notharctus preserves a more primitive carpal pattern than modern lemurs (fig. 87), for in this fossil type the capitate is relatively less compressed, the lunate is not so reduced in size, and the os centrale probably had little, if any, contact with the hamate.

The pelvic girdle in the lemurs differs but little from that of primitive mammals generally (fig. 83). The ilium is elongated (especially in *Daubentonia*) and only slightly flattened to form a 'blade'; its outer surface usually presents a shallow concavity over the area for the attachment of the gluteal musculature. The iliac surface (which gives origin to the iliacus component of the iliopsoas muscle) is relatively narrow and faces ventrally or, in some cases, even slightly laterally. The ilium articulates only with the first sacral vertebra in some genera, e.g. *Lemur* and *Notharctus*, and in others with two vertebrae, e.g. *Nycticebus* and *Propithecus*. The pelvis as a whole is narrow and, associated with this, the iliac crests diverge rather conspicuously where they serve to give attachment to the abdominal muscles (fig. 88). The ischial tuberosities are not markedly divergent as they commonly are in monkeys, correlated with the fact that lemurs are not so much accustomed to sitting upright on their haunches. The symphysis pubis is rather short.

The femur in modern lemurs is slender and has a straight, cylindrical shaft. The greater and lesser trochanters are well developed, and a third trochanter of moderate size is commonly present. The tibia and fibula are always separate, and are only slightly bowed. In *Notharctus* the femur is relatively of more robust construction and the shaft thus appears relatively shorter; the tibia and fibula are stoutly built and show a marked degree of bowing (fig. 84).

An outstanding characteristic of the lemuroid foot is the enlarged and widely abducting hallux (fig. 87). The fourth digit is always the longest (as in the manus), and the second and third toes may undergo degenerative changes (e.g., in the pottos). All the toes of the modern lemurs have flattened nails on their terminal phalanges with the exception of the second, which (being less

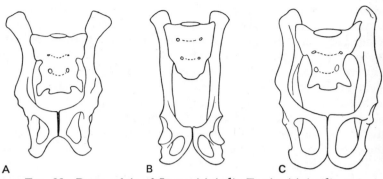

A   B   C

FIG. 88. Bony pelvis of *Lemur* (A) (×½), *Tarsius* (B) (×1½), and *Callitrix* (C) (×¾).

specialized and important functionally for grasping) retains a modified type of claw. The latter is used for scratching purposes, and
the second pedal digit is therefore sometimes termed the 'toilet
digit'. In the aye-aye, all the pedal digits are clawed except the
hallux, which has a flattened nail. As in the manus of this creature,
the retention of claws is probably a primitive feature.

The marked hypertrophy and structural differentiation of the
hallux in the lemuroid foot contrasts with the less modified form
of the pollex in the hand. This is simply another expression of the
fact, already remarked upon, that in climbing activities the hindlimb is required to serve rather different purposes, and to meet
more severe stresses, in comparison with the forelimb. Not only
must the prehensile function of the pes be adequate for sustaining
the weight of the body intermittently in climbing as the forelimbs
reach upwards in search of hand-holds, it must also provide a
stable *point d'appui* for the forward propulsion of the body in
running or leaping among the branches. Thus, the biramous or
'forceps' action of the hallux in opposition to the other digits needs
to be sufficiently powerful to secure a very firm grip of supporting
branches. Since in this action the hallux alone serves as a counter-
blade to all the other toes, its opposability is highly developed and
the digit itself acquires considerable strength. In association with
these requirements, the metatarsal bone and phalanges of the hallux
are much more robust than those of the other digits as compared
with the pollex, the joint surface whereby the metatarsal articulates
with the entocuneiform bone of the tarsus is modified to permit

a high degree of opposability at this joint, and the friction pad on the terminal or ungual phalanx is broadened for securing a firm foot-hold. It seems probable, from the eivdence of comparative anatomy, that in the early evolutionary differentiation of the Primates from a generalized placental mammalian stock, the primary modification of the digits was manifested in the specialization of the hallux for such arboreal activities. The relative stoutness of the hallucial metatarsal in the pen-tailed tree-shrew suggests the very earliest expression of such a developmental tendency. As we have already noted, also, in *Daubentonia* the hallux alone bears a flattened nail, and this is also the case in the marmosets (p. 206). The modification of the other digits of the Primate manus and pes, with the development of the terminal tactile pads and the replacement of claws by nails, presumably followed subsequently. Incidentally, the importance of the early hypertrophy of the hallux in Primate phylogeny has a direct reference to the evolution of the peculiar features of the human foot (see p. 219).

The basic mechanisms of the Primate foot have been the subject of many detailed investigations, some of which are noted in the references in the bibliography. Of outstanding importance, both for the factual evidence presented and also the logical interpretation of this evidence, is the series of papers by D. J. Morton which appeared a number of years ago [94-97]. According to this author, the Primates may broadly be divided into two main groups in regard to the mechanism of the foot. In the more thoroughly arboreal types the foot is adapted to maintain a clinging grasp to the branches, and it is this requirement which leads to a specialization of the hallux with its capacity for wide abduction from the other digits and its free and strong opposability. An arboreal animal which employs its hindlimb in this way will, in climbing and leaping movements, use the anterior tarsal segment (i.e. that part of the tarsus in front of the tibial articular surface of the talus) as a fulcrum for the leverage of the foot. In the other group— which comprises most of the higher Primates—the gait is more cursorial, the hallux is not ordinarily used as an independent digit for purposes of locomotion, and the fulcrum of the foot in forward progression is supplied by the heads of the metatarsal bones. To these contrasting mechanisms the somewhat cumbrous (but conveniently expressive) terms 'tarsi-fulcrumation' and 'metatarsi-fulcrumation' have been applied. The structural modifications which are associated with these different uses of the foot are important.

All the lemurs employ the foot to a greater or lesser degree in maintaining a clinging or perching grasp and thus belong to the 'tarsi-fulcrumating' group. This specialization reaches its most advanced expression in the galagos, in which the hallux can be very widely abducted; the movement is partly effected by the peroneus longus muscle and the first metatarsal is thus provided with a prominent peroneal tubercle for its insertion. The galagos adopt a springing gait, and in correlation with this saltatory activity they have developed a great lengthening of the anterior tarsal segment by a forward extension of the front part of the calcaneus and elongation of the navicular. Thus the tarsal pattern has been considerably modified by a pushing forwards of the calcaneo-cuboid joint well beyond the level of the talo-navicular joint. In this structural specialization of the tarsus, the galagos resemble the condition in *Tarsius* (compare fig. 89B), though not to such an extreme degree.

In the slow-moving Lorisinae the most primitive type of tarsus and metatarsus among the lemurs is preserved; the calcaneus in these creatures is quite short, for movements are made by a slow crawl and do not require much leverage at the heel. In the lemuriform lemurs (fig. 89A) there is a partial transition to the 'metatarsi-fulcrumating' mechanism, the metatarsals being somewhat lengthened, but the hallux is specialized as in the Lorisiformes, and the lengthening of the front part of the calcaneus with the forward displacement of the calcaneo-cuboid joint is well marked (though by no means to such a degree as in the Galaginae). The functional specialization of the hallux in all the lemurs is reflected in the tarsus, as is shown by a lateral compression of the mesocuneiform (which is somewhat analogous to the constriction of the capitate bone in the carpus, to which reference has already been made).

In *Notharctus* the hypertrophy and functional specialization of the hallux are already well established (fig. 87). On the other hand its pedal skeleton is in general rather more primitive than in Recent lemurs; in this Eocene genus, for example, the metatarsals are relatively shorter and stouter, and the mesocuneiform has not yet become distorted by lateral compression. Only a few, rather uninformative, fragments of limb bones of the European fossil lemur *Adapis* are known, and there is even some doubt whether they have in all cases been correctly attributed to the genus. But as far as this dubious evidence goes, the calcaneus by the abbreviated

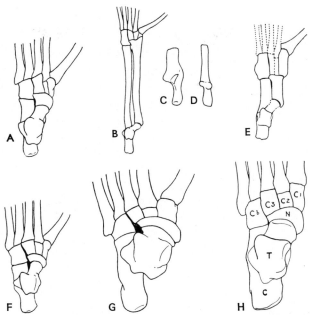

FIG. 89. Tarsal skeleton of (A) *Lemur*, (B) *Tarsius*, (E) *Hemiacodon*, (F) *Macaca*, (G) *Gorilla*, and (H) *Homo sapiens*. (Partly adapted from D. J. Morton, *Amer. J. Phys. Anthr.*, 1924.) Also illustrated are the calcaneus (C) of *Teilhardina*, and (D) of *Necrolemur*. Individual bones are indicated in (H)—Talus (T), Calcaneus (C), Navicular (N), Entocuneiform (C1), Mesocuneiform (C2), Ectocuneiform (C3), and Cuboid (cb).

development of its anterior part suggests a slow mode of progression similar to that of the modern lorisiforms. No fossil material is available to provide information on the hindlimb skeleton in the other Eocene lemurs, or in the Miocene genus *Progalago*.

There is a structural peculiarity in the mode of vascularization of the limbs of the slow-moving lorises whose significance is still obscure. The main arteries of the limbs, instead of continuing down to the wrist and ankle as individual trunks, break up into leashes of small vessels, forming the so-called 'rete mirabile'. Such a vascular arrangement is found in the limbs of sloths, and it has been conjectured that it is a device for providing a larger reservoir of oxygenated blood for muscles which are held in sustained contraction in the slow movements and the clinging postures of these animals.

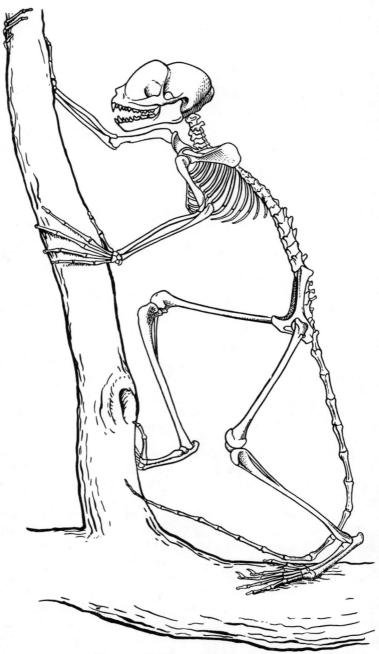

FIG. 90. Skeleton of *Tarsius* (×¾).

The living tarsier, *Tarsius*, is highly specialized in its mode of progression. It is entirely arboreal, maintains its hold of the branches by grasping them with both hands and feet, and moves about by leaping from one position to another with great rapidity and accuracy. The hindlimbs have become extensively modified to serve these saltatory functions, showing some resemblance in general proportions to those of other jumping mammals such as the jerboa and kangaroo, but with considerable differences in the detailed pattern of their skeletal construction. The resemblances to *Galago* are particularly obtrusive, but they are only to be explained on the basis of convergent evolution. The few fragmentary remains of limb bones that have been ascribed to fossil tarsioids suggest that the peculiar specializations of the hindlimb characteristic of the modern tarsier may be of very ancient origin, and had perhaps already become established to a moderate degree even as far back as the Eocene.

The forelimb of *Tarsius* is much shorter than the hindlimb (fig. 90). The clavicle has a slight S-shaped curve. The scapula is elongated transversely, with a short vertebral border. The humerus is unusually abbreviated in proportion to all the other components of the limb skeleton, particularly in relation to the length of the forearm bones; it has a well-rounded articular head, a strong supinator crest, a globular capitulum, and a trochlear surface which on its anterior aspect is not grooved and only faintly demarcated from the capitulum. The medial supracondylar ridge is perforated by an entepicondylar foramen. The radius and ulna are separate bones, both elongated, slender, and fairly straight.

In the manus the proportions of the digits characteristic of generalized mammals are maintained; that is to say, the third digit is the longest of the series (fig. 85B). The proportionate length of the pollex is also less reduced than it is in many types of lemur. The terminal digital pads are curiously elaborated to form disc-like expansions covered with fine papillary ridges, by whi h the animal is enabled to maintain a firm grip of smooth surfaces of stems and branches. All the digits are provided with tiny flattened nails of triangular shape, scale-like and rudimentary in appearance. The palmar friction pads are of a generalized pattern, except that the thenar pad is partially fused with the first interdigital pad.

In the carpus (fig. 86B) there is an os centrale placed in its typical position between the proximal and distal row of carpal bones, and

separated from the hamate by the capitate. The pisiform articu
lates with the ulna, the hamate is the largest carpal bone, and the
capitate is not laterally compressed as it is in the Recent lemurs.
The skeletal elements of the thumb show no very evident structural differentiation, and in the living animal this digit appears to
possess very little functional individuality in grasping movements
of the manus. But while the thumb is strictly non-opposable, it does
possess an unusual degree of mobility at the metacarpo-phalangeal
joint.

In the pelvic girdle (fig. 88B) the ilium is elongated in the form
of a prismatic rod and articulates with two sacral vertebrae. The
gluteal surface shows a slight concavity, and the iliac surface faces
directly ventrally. The pubic symphysis is quite short and the
ischio-pubic arch very attenuated. The femur is unusual for its
relatively great length, and its shaft is straight and slender. It has
a small but prominent third trochanter for the attachment of the
gluteus maximus muscle.

The tibia is of approximately the same length as the femur, and
the fibula (which is extremely slender) is completely fused with it
in the distal half of its extent. The head of the fibula is also commonly synostosed with the tibia by ossification of their connecting
ligaments. This tibio-fibular fusion is a highly specialized trait in
which *Tarsius* stands out among all the other living Primates;
clearly it provides for a very high degree of stability and rigidity
such as is required in the sudden propulsive movements by which
the animal leaps from branch to branch. But it is in the foot skeleton that the tarsier exhibits the most remarkable specializations
(fig. 89B), for the tarsus is extended to a very unusual degree by the
elongation of the navicular and the front part of the calcaneus (resembling but surpassing the tarsal modifications in the Galaginae).
The extreme lengthening of these tarsal bones is correlated with
the fact that in its foot leverage *Tarsius* has carried 'tarsi-fulcrumation' to an advanced stage in association with its saltatory habits.
Of the cuneiform bones, the medial is noteworthy for its relative
shortness. The metatarsal of the hallux is stouter than those of the
other digits in relation to its strong and powerful grasp in opposition. The fourth digit is the longest, resembling in this feature the
lemurs. The second and third toes bear long, pointed claws, while
the other digits are provided with minute, flattened nails. The two
claws are weak in construction, and are employed only for the

toilet of the fur. As in the fingers, all the terminal digital pads are expanded into discoid formations.

It appears that the saltatory specializations of *Tarsius* may have been characteristic (though not to the same degree) of the tarsioid group as a whole. At any rate, in those fossil tarsioids whose limb skeleton is partly known, the latter has provided some evidence of similar modifications. For example, the femur that has been ascribed to the Middle and Upper Eocene genus *Necrolemur* is like that of *Tarsius* in its general conformation (though the shaft is somewhat more robust), the tibia and fibula are synostosed below, and the calcaneus is markedly elongated anteriorly (figs. 89D and 91). The calcaneus of the Lower Eocene *Teilhardina*, on the other hand, shows only a comparatively slight lengthening of the calca-

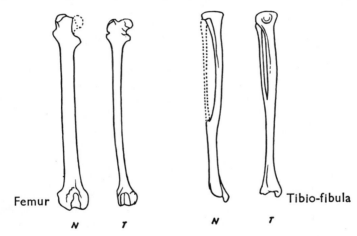

FIG. 91. Femur and tibio-fibula of *Necrolemur* (*N*) and *Tarsius* (*T*). Approximately natural size.

neus, which is not very much more marked than in many of the Recent Lemuriformes and in the primitive Notharctinae (fig. 89C). Nevertheless, it appears to indicate the preliminary stages which led to the ultimate production of the elongated calcaneus found in later tarsioids. In *Nannopithex* the resemblances to the modern tarsier which have been described (by Weigelt) appear to be quite striking, particularly in the morphological details of the pelvis and femur, the elongation of the calcaneus, and in the fusion of the tibia and fibula at their lower end. It is to be noted, however, that the evidence for the tibio-fibular synostosis has been questioned [137]. In the American fossil tarsioid, *Hemiacodon*, the elongation of the tarsus affects not only the calcaneus and navicular, but also the cuboid and cuneiform bones; thus, while it shows a broad resemblance to the condition in *Tarsius*, the relative proportions of the tarsal elements are not identical (fig. 89e).

It has to be recognized, of course, that this fossil evidence of the limb structure of extinct tarsioids is very meagre indeed, but, so far as it goes, it leads to the inference that the characteristic tarsioid specialization of the hindlimb was probably initiated near the beginning of the Eocene period, that it became definitely established by the Middle Eocene, and by the Upper Eocene had reached a stage comparable with, though still less advanced than, the final condition represented in *Tarsius*.

MONKEYS

All the Platyrrhine monkeys are entirely arboreal in their mode of life, and so are most of the Catarrhine monkeys. Many of the latter, however, frequently descend from the trees and move quite freely on all fours on the ground, as they also do along the larger branches; they are thus typically quadrupedal in their gait. The baboons have adopted a fully terrestrial habitat, living in open country and being capable of scampering over the ground with great rapidity.

In arboreal progression all types of monkey except the marmosets make use of the forelimbs for securing handholds above the head and swinging the body forwards from one branch to another; this mode of progression is termed brachiation. But in most types, it is a modified brachiation that is only practised to a minor degree —running along the branches on all fours and leaping from branch to branch are the usual activities. Some of the Platyrrhine monkeys (in particular *Ateles*) have become arboreal specialists to the extent that they normally rely always on brachiation to pass from one branch to another, but they can also walk along branches quadrupedally to a limited extent. A characteristic feature of the arboreal specializations related to brachiating habits is a tendency towards lengthening of the forelimbs (which may be very considerably longer than the hindlimbs, as in the spider monkeys and the anthropoid apes), and atrophy of the thumb. The latter is associated with the functional specialization of the hand as a suspending 'hook' rather than a grasping mechanism, allowing a very rapid release in swinging from one bough to another. In monkeys which are not such advanced arboreal acrobats, and which run and leap among the larger boughs, more primitive proportions of the limbs are retained—that is to say, the forelimbs and hindlimbs approximate closely to each other in total length (see fig. 92 for the skeleton of a Catarrhine monkey, *Cercopithecus*).

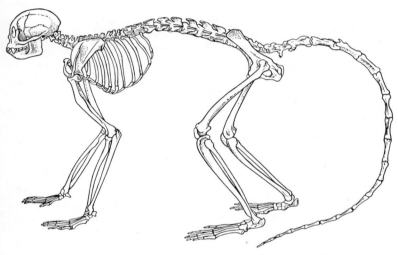

FIG. 92. Skeleton of *Cercopithecus* ($\times\frac{1}{6}$ approx.).

The clavicle in monkeys is always a well-developed and relatively stout bone, usually bent into an S-shaped curve. The scapula has a relatively short vertebral border, especially in the marmosets (Callitrichidae) in which it resembles closely that of *Tarsius*. In the more arboreal genera, however, the vertebral border is relatively longer in proportion to the other borders; in these types, therefore, the scapula as a whole becomes more extended in a cranio-caudal direction and makes some approach to the general shape of the bone characteristic of the Hominoidea. The humerus characteristically has a slender straight shaft and a well-rounded articular head which faces mainly in a backward direction. The supinator crest is strongly developed in some genera, particularly in *Papio* and some of the Platyrrhine monkeys (e.g. *Cebus*). The capitulum is globular and the trochlear surface sharply defined. These features of the humerus are associated with great freedom of movement at the shoulder joint, and with a considerable range (at least 90°) of pronation and supination of the forearm, which can be effected independently of flexion at the elbow joint. The primitive entepicondylar foramen is absent in all Catarrhines, but is commonly preserved in the Platyrrhines except for the marmosets and the woolly spider monkey (*Brachyteles*).

The radius and ulna are always separate and have slender shafts which are bowed to a slight degree. The articular surface of the head of the radius is approximately circular and thus well adapted for free rotatory movements. The olecranon process of the ulna is relatively long in most genera, particularly so in *Papio* and *Alouatta*. The lower end of this bone articulates directly with the carpus, commonly with both the triquetral and the pisiform bones.

The manus as a whole is typically rather elongated and slender (fig. 85c), but becomes shorter and broader in the more terrestrial types. The palmar tactile pads do not show the definition or distinctness found in many lemuroids and in most lower mammals. In association with the development of a closer intimacy between the palms and the surface of the branches which they grasp, and also correlated with the enhancement of the use of the hands as tactile organs, the primitively separate pads tend to lose their individuality and to merge with each other. The whole of the palm, also, becomes covered with a continuous system of fine papillary ridges. In the marmosets, and to a lesser extent in *Aotus*, palmar pads of a typical mammalian pattern are more evident than in other monkeys (fig. 93). Thus, there are to be recognized thenar and hypothenar pads separated by a wide furrow, and four interdigital pads of which the first and fourth blend proximally with the thenar and hypothenar pads respectively. In monkeys generally, the thenar pad tends to undergo atrophy and disappear altogether.

In all the Old World monkeys the terminal phalanges are provided with flattened nails which project but little beyond the finger tips. On the other hand, in many New World monkeys (e.g. Cebinae) the nails are elongated and laterally compressed, thus making some approximation to a fully developed claw. In the marmosets all the digits of the manus are armed with sharp, projecting and recurved claws (fig. 93). They are very similar in general appearance to typical claws, such as those of the modern tree-shrews, but in the presence of a more conspicuous groove on their under surface they show what might be taken for an incipient tendency to open out and thus to mark the earliest morphological stage in the transformation of the specialized claw into the less

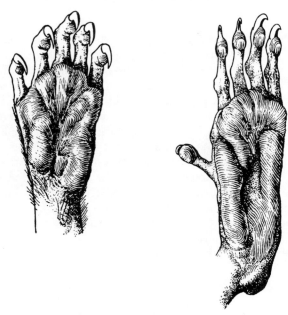

FIG. 93. Manus and pes of a marmoset (*Callitrix*) (× $\frac{4}{3}$ approx.). (After J. Beattie, *Proc. Zool. Soc.* 1927.)

elaborate nail which is characteristic of most modern Primates. If it is proper to speak of the so-called retention of claws in the marmosets, this is of considerable interest in the problem of pithecoid ancestry. It remains possible, of course, that the marmoset claw is a secondary modification of a true nail such as is found in the Cebidae, and that it merely simulates a true claw in its generat shape and funct ons. But microscopic examination shows that there still persists a thin stratum of the deep layer of horny substance which is the mechanically important element of a true claw (see p. 172), a stratum which seems to have disappeared entirely in the nails of other groups of monkeys. Whether the marmosl-claw is a primary or secondary morphological element in the phyleo genetic sense will only be determined as further palaeontological

evidence accrues. From the functional point of view the development of claws in the marmosets is associated with their special
arboreal habits, for they can run spirally up and down large tree
trunks in the manner of squirrels, which would not be possible if
the digits were provided with naked terminal pads as in other
monkeys.

In the carpus an os centrale is consistently present in all genera
of Old and New World monkeys (fig. 86c). The carpal pattern
contrasts with that of the lemurs, and resembles that of *Tarsius*,
in the relatively large size of the lunate and capitate bones. The
latter intervenes between the os centrale and the hamate, and thus
prevents the contact between these bones which is characteristic
of the modern lemurs. The thumb is variably developed. In the
Platyrrhines it is commonly rather short and only slightly opposable or even, in the marmosets, non-opposable. In the spider
monkeys it is absent altogether, except for a residual stump of the
metacarpal bone. In the Catarrhines the thumb can be incompletely opposed to various degrees. Opposability is best developed
in the more generalized arboreal and predominantly terrestrial
genera (e.g. *Papio* and *Macaca*). The whole question of the definition
of opposability has been discussed by Napier [100, 101] who distinguishes between the true opposability that occurs at a saddle-
shaped carpo-metacarpal joint at the base of the thumb, and the
pseudo-opposability of a prehensile hand where the movement
occurs at the metacarpo-phalangeal joint. While the thumb of the
Lemurs, *Tarsius* and the New World monkeys belongs to the latter
category, true opposability is limited to the Old World monkeys,
apes and man, reaching its ultimate expression in the human hand.
In the more thoroughly arboreal types of Catarrhine monkey the
thumb is much reduced in size. Indeed, in *Colobus* it has shrunk
into little more than a small nodule on which the nail, if present at
all, is quite rudimentary. In the higher Primates generally, the
movement of opposition of the thumb is enhanced by a modification
of the carpo-metacarpal joint, the articular surface of which, besides being obliquely set in relation to the surface of the palm, has
become markedly concavo-convex in shape.

In the smaller Platyrrhine monkeys the pelvis preserves a primitive type of construction (fig. 88c), that is to say, it is narrow transversely, the pubic symphysis is rather elongated and sometimes
involves the ischial as well as the pubic element, the ilium is long

and slender with a very short iliac crest, its gluteal area is restricted, and there is but little flattening to form an iliac surface. Articulation is practically limited to the first segment of the sacrum. In the larger Platyrrhines (e.g. *Cebus* and *Alouatta*), and also in the Catarrhines generally, the blade of the ilium is slightly expanded, the iliac surface faces more medially, and the bone articulates with two sacral vertebrae. In the more fully arboreal monkeys, the iliac surface is relatively more extensive, and in *Ateles* the ilium articulates with three vertebrae. The expansion of the iliac blade in the larger monkeys is no doubt partly related to its supporting functions in the erect or orthograde position of the trunk in climbing activities, but it also provides larger areas for the attachment of more powerful iliac and gluteal musculature. As might be expected, these areas increase in extent in correlation with increasing body weight.* In all monkeys the ischial tuberosities tend to be splayed out so as to provide a broad surface for maintaining an upright posture in a sitting position. In the Catarrhines they show a conspicuously broad, roughened surface with an everted lateral margin, which forms a base to support the overlying ischial callosities which characterize these animals. In this particular feature the Old World monkeys contrast strongly with New World monkeys, for in the latter callosities are consistently absent. It has been suggested that the development of ischial callosities is related to the habit of sleeping in a sitting position on the branches of trees.

The shaft of the femur is typically long, straight and slender, and the greater and lesser trochanters are prominent processes. In the Callitrichidae there is also a distinct third trochanter as in lower Primates, but this is rudimentary or absent in other Platyrrhines, and normally absent in the Catarrhines. The tibia and fibula are always separate, and have slender shafts. They articulate with each other below by means of a synovial joint which permits considerable freedom of movement at the ankle for eversion and inversion of the foot.

The digital formula of the foot is of a primitive type, the third digit being the longest, with the exception of *Callitrix* in which

---

* It should be remarked that, contrary to a generally held impression, the blade of the ilium, even in modern man, does not, to more than a minor degree, provide a *direct* support from below for the abdominal viscera—at any rate, not in the manner of a basin (which is what the word pelvis means). In the quadrupedal monkeys, and also the anthropoid apes, the flattened iliac blade is mainly dorsal or posterior to the lower abdomen.

the fourth digit is slightly longer than the third (thus paralleling the normal lemuroid condition). The plantar pads—like those on the palm—merge with each other to a considerable extent. In the marmosets the two proximal pads are distinctly separated from each other, but the first interdigital pad runs into continuity with the thenar pad and the other interdigital pads are not sharply individualized.

The distinctive characters of the pedal skeleton of monkeys are associated with the fact that most of these animals adopt to some extent a quadrupedal, cursorial gait on the ground or on the larger boughs of the trees which they inhabit. In so far as they use the heads of the metatarsal bones as the fulcrum in the leverage of the foot required for forward propulsion, they belong to the 'metatarsi-fulcrumating' group of Primates, a group which (in contrast to the 'tarsi-fulcrumating' group) is structurally characterized by a lengthening of the metatarsals. On the other hand, the bones of the mid-tarsal region not only preserve their primitive proportions but may even undergo a degree of relative shortening. Except for some of the arboreal specialists among the New World monkeys, the functional axis of the foot leverage corresponds to the line of the third digit; this is reflected in the greater length of this digit and particularly of its metatarsal element (fig. 94). In the larger and more exclusively arboreal Platyrrhines, and (as we shall see) in the anthropoid apes, this axis is shifted medially to a position between the first and second digits, i.e. the axis of the grasp of the foot as a whole; this change is accompanied anatomically by an increase in the length of the second metatarsal which now becomes the longest of the series. The toes are all provided with nails (sometimes rather strongly compressed from side to side); in the case of the Callitrichidae, however, only the hallux bears a flat nail, the other digits having sharp, curved claws of the same type as those of the manus. The grasping power of the hallux is well developed in the monkeys (but somewhat limited in the marmosets); structurally, this functional differentiation is associated with a distinct torsion of the shaft of the first metatarsal bone, and also with a characteristic curvature of the articular surfaces between this bone and the entocuneiform. In the predominantly terrestrial baboons and the plains-living genus *Erythrocebus*, the phalangeal components of the digits are short relatively to the metatarsals, and the hallux tends to be reduced in size though retaining its power of grasping. In the

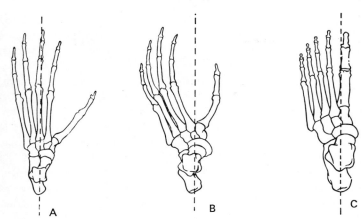

FIG. 94. Foot skeleton of (A) *Macaca*, (B) Chimpanzee, and (C) *Homo sapiens*, showing in each case the functional axis. (Adapted from D. J. Morton, *Amer. J. Phys. Anthr.* 1927.)

more specialized arboreal forms, the digits are longer and the hallux can be more strongly opposed to the other digits.

In general, the tarsal pattern in monkeys is more primitive than that of the modern lemurs, for they have avoided the high degree of functional specialization of the hallux with its wide abduction, the lateral compression of the middle cuneiform, and the tendency (except in the slow-moving Lorisinae) for an elongation of the mid-tarsal region. The marmosets preserve the primitive proportions of the tarsal, metatarsal and phalangeal elements to quite a remarkable degree. The hallux in these small creatures has evidently undergone a moderate reduction, but it is capable of grasping movements and, in conformity with this, the first metatarsal bone shows a characteristic torsion of the shaft and on the terminal phalanx the claw has been replaced by a flattened nail. The generalized type of tarsus is but little modified in the less exclusively arboreal Catarrhines.

HOMINOIDEA

On casual inspection, the contrasts between the limbs of the anthropoid apes and those of *Homo* are quite obtrusive, and it may thus seem inappropriate to consider them together in one section.

In fact, however, the structural contrasts are by no means so fundamental as some comparative anatomists in the past have supposed,
and a proper understanding of the evolutionary relationships of the
Pongidae and Hominidae is more readily obtained if the divergent
functional adaptations of the limbs in the two groups are discussed
as two facets of the same problem. In the anthropoid apes the
limbs have been modified for brachiation; in the Hominidae they
have been modified for erect, bipedal progression. But the evidence
of comparative anatomy and palaeontology is strongly in favour
of the assumption that the hominid pattern of limb structure is
secondary to the pongid pattern, with the necessary reservation
that the latter, as seen in the *modern* apes, has been accentuated to
an advanced stage of specialization in association with an extreme
development of brachiation. It is not to be assumed, therefore,
that the hominid pattern of limb structure would have been derived
from the exaggerated type of brachiating pattern seen in modern
apes. But there is no theoretical reason why it should not have
been derived from the more generalized brachiating pattern found
in the extinct apes of Miocene times.

The brachiating adaptations of the Pongidae, as compared with
the quadrupedalism of most Catarrhine monkeys, are reflected in
certain modifications of the upper limb related to the freedom of
movement required at the shoulder joint, and the use of the hands
as suspending hooks for swinging from branch to branch. Associated with the more habitual orthograde position of the trunk, also,
the thorax has become relatively broad and flattened from front
to back, instead of being (as in the monkeys and in pronograde
mammals generally) relatively narrow and flattened from side to
side. This modification of the thorax, again, is reflected in a
widely expanded breast-bone and in certain orientations of the
pectoral girdle. For example, the scapula is placed against the
back, rather than the side, of the chest.

In all the anthropoid apes the clavicle is strongly developed.
The scapula has become extended in a cranio-caudal direction, so
that the vertebral border is relatively long. Such a change of shape
(already adumbrated to a moderate degree in the more arboreal
monkeys) provides for a more effective leverage (by the action of the
rhomboid and serratus anterior muscles) in rotating the scapula
and allows elevation of the arm to its fullest extent. The articular
head of the humerus is orientated more medially than posteriorly,

to make contact with the scapula in its new position. The deltoid tuberosity is situated relatively lower on the humeral shaft than in the cercopithecoid monkeys (in relation to the amplification of the movement of abduction of the arm), but it is less prominent and this makes the shaft appear straighter. The radius and ulna are proportionately long, particularly in the gibbon and orang in which the brachial index (that is, the percentage ratio of the forearm to the upper arm) is about 100 or more. This relative lengthening of the forearm is one of the structural manifestations of brachiating habits (fig. 95). The somewhat lower indices in the chimpanzee 91, and gorilla 80, are probably the result of a secondary shortening following the partial abandonment of active brachiation. The olecranon process does not project appreciably beyond the level of the articular surface, thus permitting more complete extension at the elbow-joint and, by bringing the attachment of the triceps muscle close to the fulcrum of the joint, increasing the velocity with which the movement can be effected. The carpus normally includes a separate os centrale in the gibbon and orang, but not in the chimpanzee, gorilla or *H. sapiens*. However, this is in no way a fundamental distinction. The os centrale can be recognized as a separate element in the human embryo, but it normally fuses with the scaphoid bone during the third month of foetal life, and sometimes (but very rarely) it may persist as an independent bone into adult life. In the chimpanzee and gorilla it merges with the scaphoid only late in foetal life, and in the orang and gibbon, though normally persistent, fusion does sometimes occur in advanced age. It thus appears that there is a common tendency for fusion of the os centrale in the Hominoidea, the various genera differing only in the age at which this tendency finds expression.

A striking feature of the manus of the modern anthropoid apes is the relative shortness of the thumb, and this is particularly obtrusive in the gibbon and orang in which brachiation as a mode of progression has been highly developed (fig. 96). In the gorilla and chimpanzee it is also abbreviated to the extent that its tip does not reach the level of the proximal crease of the index finger. As compared with the Catarrhine monkeys, this abbreviation of the thumb is to some extent relative rather than absolute, for it is due to the elongation of the hand and finger as a whole (a brachiating specialization) that the thumb appears to be so much shorter than in many of the monkeys. But there can be little doubt that its relative

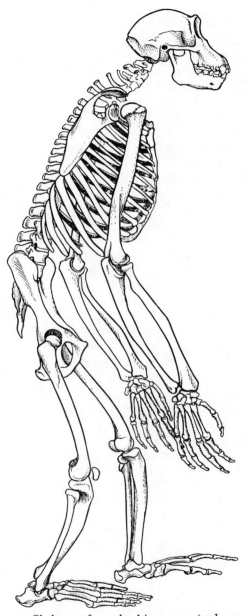

FIG. 95. Skeleton of a male chimpanzee ($\times \frac{1}{6}$ approx.).

shortness also involves a degenerative factor. It may be noted, for example, that the tendon of the flexor pollicis longus muscle is reduced or absent in many gorillas and orangs, and also in a large percentage of chimpanzees.

In *H. sapiens* the upper limb is constructed on the same general pattern as that of the large anthropoid apes, approximating most closely in the general proportions of its skeletal elements to the

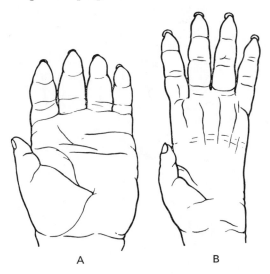

FIG. 96. Hand of a forest gorilla (A) and an orang-utan (B). (Adapted from A. H. Schultz, *Amer. J. Phys. Anthr.* 1942.)

chimpanzee. The orientation of the scapula and the articular head of the humerus, the free rotatory movements at the radio-ulnar joints, and the mobility of the hand, are all consonant with the thesis that the Hominidae were originally derived from an arboreal Primate in which brachiation had been developed to a moderate degree, that is to say, before brachiating specializations comparable to those of the Recent Pongidae had become manifested. Indeed, the construction of the human upper limb by itself provides a sound argument for the derivation of the Hominidae from an anthropoid ape ancestry, for there is good reason to suppose, from the evidence of comparative anatomy and palaeontology, that its essential features were primarily developed in relation to brachiating habits. But the upper limb in *H. sapiens* preserves more primitive proportions than the Recent pongids; the mean inter-membral index (i.e., the percentage ratio of the upper to the lower limb) is only 88, whereas in the modern apes it ranges from 136 in the chimpanzee to 178 in the gibbon. The mean brachial index is only 76. The human thumb, also, has undergone no apparent reduction, and in its mobility and degree of opposability it is a far more efficient and delicately adjusted grasping mechanism than in any of the modern apes. The contrasting linear proportions of the limbs in modern

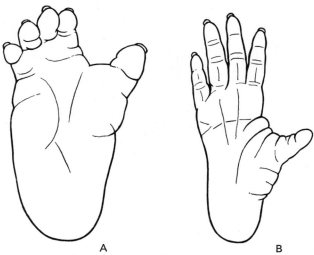

A   B

FIG. 97. Foot of a mountain gorilla (A) and an orang-utan (B). (Adapted from A. H. Schultz, *Anthr. Anz.* 1933 and *Quart. Rev. Biol.* 1926.)

man and the modern apes are less marked in early post-natal life, and still less so in the foetus,* and they are evidently the result of differential rates of growth; in other words, it need not be assumed that they represent fundamental differences of a genetic kind such as could only indicate a distant relationship.

Of critical importance for the problem of the phylogenetic relationship of the Hominidae and the Pongidae are the features of the limbs of the generalized Miocene apes in which structural specialization associated with extreme degrees of brachiation had not proceeded so far as they have in the *terminal* products of pongid

---

* In the full-term human foetus the mean intermembral index is 104, and the brachial index 80.

evolution, that is to say in the modern apes. In addition to an incomplete humerus of (probably) *Dryopithecus fontani* discovered many years ago in Miocene deposits in France and two forelimb bones of an allied, but perhaps identical, genus found in Austria, our knowledge of the upper limb bones of extinct apes relates mainly to those of the extinct Miocene apes of East Africa, *Dryopithecus* (*Proconsul*) and *Pliopithecus* (*Limnopithecus*), and of the Lower Pliocene *Oreopithecus*. An almost complete upper-limb skeleton of a young specimen of *Dryopithecus africanus* (except for the clavicle and scapula) has provided evidence of exceptional interest, for it combines features which are typical of the arboreal quadrupedal monkeys with characters no less distinctive of brachiating anthropoid apes (fig. 98). For example, the humeral articular head is directed more or less backwards as in the monkeys, the radius is relatively short and the brachial index low, the thumb is less reduced than in the modern apes, and the phalanges of the fingers lack the marked dorsal curvature characteristic of advanced brachiators. On the other hand the shaft of the humerus is long, straight and slender, and the lower articular extremity shows evidence of the free movements associated with some degree of brachiation. As a whole, the upper limb skeleton of *Dryopithecus* resembles most closely that of the quadrupedal monkey *Presbytis*, but it also adumbrates the sort of modifications to be expected in a group ancestral to the modern genera of large apes. In other words, the genus *Dryopithecus* forms a most important connecting link between the quadrupedal Cercopithecoidea and the Recent Pongidae. Of still greater significance is the fact that the generalized type of upper limb in *Dryopithecus* could theoretically have provided a basis for the evolutionary development of the hominid type. In the Miocene genus *Pliopithecus*, the bones of the upper limb show a similar combination of cercopithecoid and pongid features, the latter approaching those of the modern gibbon but by no means so advanced in the degree of their specialization [82]. They show a rather remarkable resemblance to those of *Ateles* —an interesting comparison because it has long been recognized that the spider monkey has developed, by a process of convergent evolution, a number of anatomical characters which resemble those of the gibbon but in less extreme form. In both *Dryopithecus africanus* and *Pliopithecus* the slender, straight shafts of their limb bones indicate that they must have been agile and lightly built creatures.

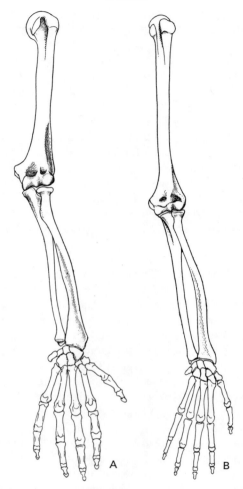

Fig. 98. Forelimb skeleton of a chimpanzee (A) and *Dryopithecus africanus* (B). Both have been reduced to the same total length. That of *Dryopithecus* is based on a reconstruction by J. R. Napier and P. R. Davis.

No doubt they were arboreal in habitat, but it seems equally certain that, like many of the Catarrhine monkeys but unlike the modern anthropoid apes, they were also capable of scampering, quadrupedal activities on the ground as well as along the larger boughs of the trees.

In *Oreopithecus* the limb proportions indicate an advanced stage in the development of brachiating habits which is well emphasized by the estimated index of 122 relating the humerus length to the femur length [132]. In cercopithecoid monkeys this index is below 100, and above 100 in most of the modern anthropoid apes. Thus *Oreopithecus* seems to have been as specialized for brachiation as the modern apes—a strong argument against the suggestion that the genus had any place in the ancestry of the Hominidae (as has been suggested).

The characteristic features of the lower limb in the Pongidae are partly related to the fact that their mobility for prehensile functions has been enhanced at the expense of their supporting functions, and partly to the habitual orthograde position of the trunk in climbing activities. Although anthropoid apes can rear up and balance themselves on their hind feet, except in the gibbon this is only an occasional attitude under natural conditions. The large apes on the ground normally walk clumsily in an oblique quadrupedal position, taking the weight on the soles of the feet and on the bent knuckles of the fingers (fig. 7). This has been termed semi-erect quadrupedalism, to emphasize that the posture of the large apes actually contrasts rather strongly with the erect bipedalism which is so distinctive of *Homo*. The oblique inclination of the trunk in moving on all fours results from the preponderating length of the forelimbs. The small and lightly built gibbon employs a method of terrestrial locomotion which is more closely comparable with that of man; on the ground, and also along horizontal boughs, it can walk upright, holding out its long arms to act as balancers. On the ground the gibbon walks on the flat of the foot, that is to say in a plantigrade manner like the human gait. In the trees, however, it does not rest on the heel but on the fore part of the foot which grasps the bough. It is to be noted that walking in an upright position on the ground is really an artificial mode of progression, for gibbons normally do not descend from the trees. As Carpenter [14] notes, 'Gibbon locomotion should be visualized as occurring in a network of branches rather than on the ground', and even in the trees these creatures only rarely 'walk' on their hindlimbs along the larger boughs.

The pelvis of the large anthropoid apes contrasts with that of the monkeys in the greater expansion of the ilium (fig. 99). As already noted, there is some correlation between the degree of this expansion and body size, and it reaches its maximum in large male gorillas. From the functional point of view, the increased area of the iliac blade provides for a more powerful development of the gluteal musculature required for hip joint movements, and the

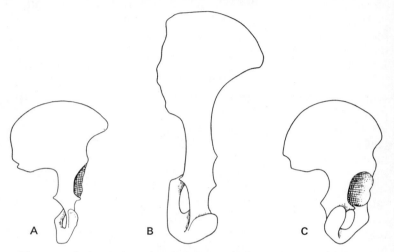

FIG. 99. Os innominatum of *Australopithecus* (A), chimpanzee (B), and *Homo sapiens* (C).

lengthening of the iliac crest increases the linear attachment of the abdominal muscles which, in the orthograde position of the trunk, are increasingly important for supporting the abdominal viscera by maintaining intra-abdominal pressure. In the general morphology of the pelvis the Hylobatinae form a connecting link between the large anthropoid apes and the quadrupedal monkeys, for in this subfamily the ilium is relatively narrower.

In spite of the expansion of the ilium, the os innominatum of the large anthropoid apes is to be regarded as a modification of the quadrupedal type rather than an approximation to the hominid type. Thus the ilium is elongated in a cranio-caudal direction, the sacral articular surface is relatively small in extent and distant from the acetabulum, the body of the ischium is long so that the ischial tuberosity is well removed from the acetabulum, the anterior inferior iliac spine can hardly be said to be present at all, and the axis of the pelvic canal is more or less parallel to the axis of the vertebral column. In the pelvis of *Homo* profound morphological changes have occurred in direct relation to the assumption of an erect bipedal gait. These are evidently adaptations to the mechanical requirements of bipedalism, though (in the female) the general broadening of the pelvic cavity is also related to the necessity for accommodating the relatively large foetal head during parturition. The distinctive features whereby the hominid contrasts with the pongid type of pelvis include (1) the backward extension of the ilium, which brings the gluteus maximus and gluteus medius muscles into a different alignment with the hip joint—an alignment which adapts their actions to the requirements of bipedal walking; (2) the great backward extension of the iliac crest, which provides a more extensive attachment for the muscles used in supporting the trunk in the erect posture, and in particular for the powerful back muscle called the sacrospinalis; (3) a rotation of the sacral articulation associated with a reorientation of the sacrum, so that the axis of the pelvic canal becomes disposed almost at a right angle to that of the vertebral column; (4) a great reduction in the relative height of the ilium; (5) the formation of an angulated and relatively deep sciatic notch associated with an accentuation of the ischial spine; (6) the development of a conspicuous and stoutly built anterior inferior iliac spine for the attachment of a strong rectus femoris muscle, and also of a powerful ligament (iliofemoral) which stabilizes the hip-joint when fully extended in the erect posture; (7) an abbreviation of the body of the ischium with a corresponding approximation of the ischial tuberosity to the acetabulum, which enhances the extensor action of the hamstring muscles in maintaining the position of full extension of the hip by bringing their upper attachments to a position behind, rather than below, the hip joint in the erect position; and as a corollary of the previous items (8) a relative approximation of the sacral articular surface to the acetabulum, which makes for greater stability in the transmission of the weight of the trunk to the hip joint (see figs. 99 and 104).

The contrasts in the morphology of the femur between the Pongidae and Hominidae, although not so immediately obtrusive as those of the pelvis, are quite as definite. In the lightly built Hylobatinae the femoral shaft is relatively straight and slender; in the large apes it is more robust and usually shows a pronounced antero-posterior curvature (fig. 100). At the lower extremity, the condyles diverge widely and are separated by a relatively broad

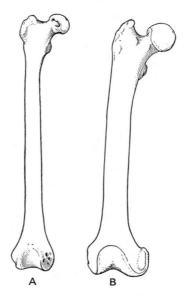

FIG. 100. Femur of *Dryopithecus* (A) and a chimpanzee (B).

and shallow intercondylar notch. These features are related to a considerable degree of mobility at the knee joint in a lateral, as well as in an antero-posterior, direction. In association with the fact that in the erect standing position the load line of the body passes through or medial to the medial condyle, the latter presents a much larger articular surface than the lateral condyle. In the East African Miocene ape, *Dryopithecus (Proconsul)*, the femur has a straight, slender shaft, and is altogether more delicately built than in the modern large apes. In its general proportions, therefore, it shows more resemblance to the femur of the modern gibbon. A femur of similar type, found at Eppelsheim in 1892, was at one time assumed to be that of a giant gibbon to which the name *Paidopithex* was given, but there can be little doubt that it is more properly ascribed to *Dryopithecus*.

In the hominid femur, the articular head is relatively larger, and the neck relatively longer, than in the anthropoid apes. At the lower extremity the condyles are more closely parallel and, since the load line in the standing position passes more nearly through the lateral condyle, the disparity in width between the two condyles is not so marked as it is in the apes. The parallel alignment of the condyles is related functionally to the limitation of free rotatory movement at the knee joint. But a small degree of rotation occurs as part of what is termed the 'locking' movement of the joint in full extension, and this is reflected in a slight angulation of the axis of the medial condyle which is very characteristic of the hominid femur. Finally, it may be noted that the intercondylar notch is narrower and, in order to accommodate the cruciate ligaments of the knee joint in full extension, extends further forwards.

While the functional axis of the foot leverage in the more generalized types of monkey corresponds to the line of the third digit, in the anthropoid apes (as also in some of the larger Platyrrhine monkeys) this axis is shifted medially to a position between the first and second digits, that is, to the axis of the grasp (fig. 94). This change is reflected anatomically in an increase in the length of the second metatarsal bone, which now becomes the longest of the series. Further, in the gorilla the interosseous muscles of the foot become disposed about the axis of the second digit as in man. The fact that the medial position of the functional axis is characteristic of the human foot, together with the relative size and stoutness of the hallux, is very difficult of explanation unless it be supposed that these features have been determined by their evolutionary derivation from an arboreal type of foot approximating to that of the large apes. While in terrestrial mammals generally there is a common tendency for the hypertrophy of one or two of the toes and a reduction (or even disappearance) of others, the human foot is quite unique in the fact that it is the *first* toe which has undergone hypertrophy. But such a unique character would naturally follow in the evolutionary specialization of the human foot if the first digit had already become enlarged in direct relation to prehensile functions. That prehensility of the hallux was a feature of human ancestry seems also to be indicated by the fact that the human foot still retains all those intrinsic muscles which in the apes are used for grasping movements of the big toe, and also by the torsional characters of the metatarsal bones. The latter point may be made clear by reference to fig. 101, which shows outlines of the bases of the metatarsal bones superimposed on those of the heads. It will be noted that in the anthropoid apes the right angle formed by the plane of the hallucial plantar surface with that of the other digits is not conditioned by the orientation of the *bases*

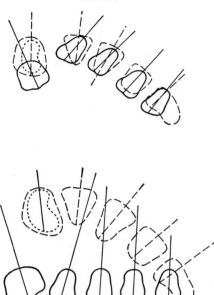

FIG. 101. Transverse sections through the bases (interrupted lines) and the heads (continuous lines) of the metatarsal bones of a young gorilla (above) and *Homo sapiens* (below). The longitudinal axes of the articular surfaces are indicated. (Adapted from D. J. Morton, *Amer. J. Phys. Anthr.* 1922.)

of the metatarsals, but by the orientation of the *heads*. This results from a torsion of the shafts such that the head of the first metatarsal is rotated in one direction and those of the outer four bones in the other direction, the flexor-extensor axes of the four outer metatarsals being twisted so that their plantar aspects present towards the hallux, and that of the hallux being rotated in the opposite direction. In other words, it is this opposing torsion which orientates the toes to the extent that the hallux can be brought into full opposition with the rest of the foot. It is an interesting fact that an appreciable amount of just the same kind of torsion still exists in the first and second metatarsals of the human foot, even though movements of opposition are not possible.* The evolutionary im-

* It has been claimed by some anatomists that the configuration of the joint between the first metatarsal and the medial cuneiform is fundamentally very different in man and apes. But this argument against the derivation of the Hominidae from a pongid ancestry has been exploded by A. H. Schultz [131] who has demonstrated that 'the exact formation of the joint . . . differs between man and ape only in degree but not in principle.'

plications of this comparative anatomical evidence are considerably reinforced by a comparison of the feet of the chimpanzee and gorilla with the human foot, shown in fig. 102. This illustration demonstrates a rather remarkable gradation in which the progressive diminution of the outer toes appears to be associated with a relative hypertrophy of the hallux. In the mountain gorilla the latter is actually more closely approximated to the other toes and is reported to be not quite so freely opposable as it is in the forest gorilla. But this series makes it clear that, so far as the relative size of the toes is concerned, the human foot is distinguished not so much by a larger big toe as by smaller lateral toes.

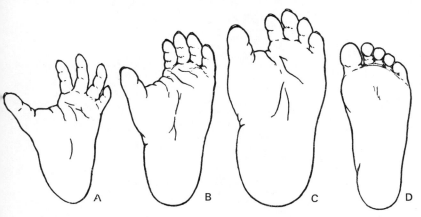

FIG. 102. Foot of (A) chimpanzee, (B) forest gorilla, (C) mountain gorilla, and (D) *Homo sapiens*. (After D. J. Morton and D. D. Fuller, *Human Locomotion and Body Form* 1952.)

It may perhaps be said that the characteristically large size of the human hallux was implicit even in the earliest phases of Primate evolution, for a differentiation and hypertrophy of the hallux was certainly one of the more consistent features among the earlier Primates so far as this part of their limb skeleton is known (fig. 103).

In a number of anatomical features, such as the details of the intrinsic muscles and the relative proportions of the digits, the gorilla foot shows the closest resemblance to the human foot. This does not necessarily imply that man was derived from a pongid ancestor of gorilloid dimensions, for there are certain arguments against such a proposition. For example, an increase in body-size

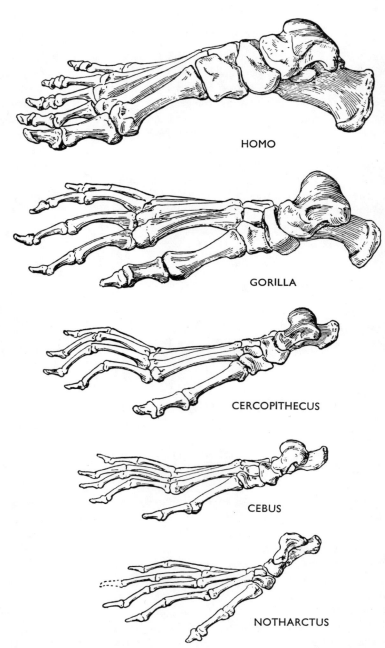

HOMO

GORILLA

CERCOPITHECUS

CEBUS

NOTHARCTUS

FIG. 103. Foot skeleton of a series of Primates. (After W. K. Gregory *Mem. Amer. Mus. Nat. Hist.* 1920.)

in brachiating apes is accompanied by the development of progressively more powerful arms and more massive shoulders, and this has the effect of raising the centre of gravity of the trunk upwards and forwards, and further away from the fulcrum of the hip joint. As a result (and also because of the relatively weaker hind limbs), movements for elevating and maintaining the trunk in a vertical position are performed at a greater mechanical disadvantage. So it has happened in the case of the modern large apes that, while in the trees they maintain an upright (or orthograde) position of the trunk in their climbing movements, as soon as they reach the ground they revert to a quadrupedal type of gait, whereas the small hylobatine apes, with their lighter build, normally retain the erect posture on the ground by assuming a bipedal gait. The effect of a forwardly displaced centre of gravity is further demonstrated in the tarsal pattern, for in the large anthropoid apes (orang, chimpanzee and gorilla) the anterior tarsal elements, i.e. cuneiforms, cuboid and navicular, have undergone a conspicuous shortening (fig. 89G). This appears to be related to the fact that, with an arboreal type of foot, the weight of the body in large animals with a forwardly displaced centre of gravity is transmitted to the anterior tarsal segment and the latter becomes better adapted by the flattening of its components to withstand the stress imposed upon them. In the human foot the anterior tarsal elements have retained a more primitive pattern, which is also found in the gibbon's foot. On the basis of evidence of this sort it has been argued that the early ancestors of the Hominidae must have forsaken the trees and adopted a terrestrial mode of life at a stage when they were not much larger or heavier than the modern gibbon, and that they increased in body size *after* they had assumed a truly erect posture—using the heel more for the support of the body weight and so avoiding the concomitant distortion of the anterior tarsal elements which has evidently occurred secondarily in the modern large apes. Two tarsal bones of the foot of *Dryopithecus* (*Proconsul*) are known (talus and calcaneus), from which it is evident that the tarsal skeleton of this Early Miocene ape still retained some characters similar to those of the more generalized Catarrhine monkeys.

The evidence which we have been discussing for the derivation of the erect bipedalism of the Hominidae from a brachiating ancestry is of course only indirect evidence. The more direct evidence which might be provided by the palaeontological record is unfortunately very meagre indeed, but, even so, it provides information of considerable interest. The extinct anthropoid apes of Miocene times certainly had a more generalized limb structure than Recent apes, and we have noted that some of them were evidently lightly built and agile creatures. In the absence of extreme degrees of specialization for brachiating habits, and also of structural changes related to increasing body weight which would more and more limit the capacity to assume an erect posture on the ground, there is no theoretical difficulty in supposing that the evolution of the hominid type of locomotor apparatus may have proceeded from the basis of the generalized limbs which were evidently characteristic of some of the Miocene apes. But there remains a long hiatus in the fossil record between these extinct apes and the earliest known hominid, *Australopithecus*, for in this Early Pleistocene genus an erect bipedalism had already developed to an advanced stage. This is demonstrated most clearly by the anatomical features of the australopithecine pelvis in which all the fundamental and distinctive characters of the hominid pelvis are present (figs. 99 and 104). Thus, the ilium is short and broad, and is extended backwards over a deep sciatic notch, the anterior inferior iliac spine is as prominent and robust as it is in modern man, the ischial tuberosity is approximated to the acetabulum, the sacral articular area has become rotated as in man and so forth. All these characters are to be regarded as essential elements of a total morphological pattern which is evidently related to the mechanical requirements of an erect posture, even though in certain features the australopithecine pelvis still retained some archaic characters which suggest that the erect posture had not been developed to the degree of perfection seen in *H. sapiens*. Significant features of the upper end of the femur are known from three fragmentary specimens. The relative length of the neck exceeds both that of the modern apes and that of modern man though closer to the extremes of the latter, and the size of the articular head is relatively small. A groove on the back of the neck made by the tendon of the obturator externus muscle (a groove not present in modern apes) provides convincing evidence for the hyperextension of the hip joint that occurs in relation to upright bipedalism. On the other hand, the absence of the rough ridge called the trochanteric line, which in *Homo* serves for the attachment of a powerful ligament in front

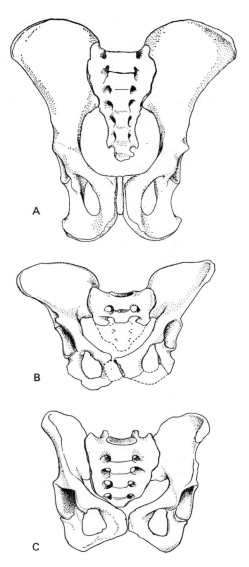

Fig. 104. The pelvis of a chimpanzee (A), *Australopithecus* (B), and a Bushman (C). All drawn to the same scale from a photograph kindly supplied by Dr J. T. Robinson. It is to be noted that, except for the lower half of the sacrum, the australopithecine pelvis (found at Sterkfontein in South Africa) is virtually complete, for those parts which were missing on one side of the fossil specimen have fortunately been preserved on the other side.

of the hip joint, suggests that the maintenance of the erect standing posture in the australopithecines did not depend as much on the stabilizing tension of this ligament as it does in modern man [20]. The lower end of the femur in *Australopithecus* also carries convincing evidence that it was adapted for an upright gait, for the alignment of the condyles is similar to that of modern man (and very different from the anthropoid apes), and the intercondylar notch extends far forwards. It is to be noted that there are certain differences in their morphological details that distinguish the pelvis of the more lightly built, or gracile, type of *Australopithecus* from that of the heavier or robust type, suggesting that in the latter the mechanism for a fully erect posture and bipedal gait was less well developed than in the former. Napier [103] has advanced a plausible explanation for these differences by supposing that, subsequent to their derivation from a common ancestral australopithecine, the robust type remained in a forest environment where the stimulus for a more perfected development of bipedalism was less intense, while the gracile type adventured on a savannah-living habitat where the competition for evolutionary improvement was more severe. To some extent this hypothesis finds support in differences of dental morphology, for it has been inferred that these betoken dietary contrasts, i.e. that the gracile type tended to be more carnivorous and that the robust type, with its larger molars and relatively smaller canines, was predominantly vegetarian [121].

From the palaeontological record, then, it seems clear that the anatomical correlates of an erect bipedal posture had reached an advanced stage of development by the beginning of the Pleistocene, while the pongid precursors of the Hominidae which existed in the Miocene possessed limbs which (it seems) could readily have been adapted to the functional changes demanded by a transition from generalized brachiating habits to bipedal progression on the ground. It is to be anticipated that the connecting links objectively demonstrating the postulated sequence will eventually come to light when the Pliocene phase of hominid evolution is revealed in the fossil record.

6

The Evidence of
the Brain

UNDOUBTEDLY the most distinctive trait of the Primates, wherein
this order contrasts with all other mammalian orders in its evolutionary history, is the tendency towards the development of a
brain which is large in proportion to the total body weight, and
which is particularly characterized by a relatively extensive and
often richly convoluted cerebral cortex. It is true that during the
first half of the Tertiary epoch of geological time the brain also
underwent a progressive expansion in many other evolving groups
of mammals, but in the Primates this expansion began earlier,
proceeded more rapidly, and ultimately advanced much further.

Apart from its relative size and general elaboration, the Primate
brain shows certain features of its intrinsic organization which also
distinguish it from that of other mammals. We shall refer to these
in more detail later, but it is convenient at this juncture to make a
brief reference to some of them. For example, the neural apparatus
of vision becomes highly developed in relation to the need for a
high degree of visual acuity in arboreal life, and with this is associated a corresponding enhancement of tactile sensibility. Coincidentally, the olfactory regions of the brain undergo a retrocession,
because the sense of smell is of far less immediate importance for
the tracking of prey or the avoidance of enemies in the trees than
it is in terrestrial life. The olfactory cortex in the Primates is thus
relatively reduced in extent, while the non-olfactory areas, and
particularly the visual cortex, become progressively expanded. As
a whole, the cerebral cortex is basically concerned with the reception, analysis and synthesis of sensory impressions of diverse kinds,
and with transforming the resultants of these activities into behavioural reactions appropriate to the immediate environmental
situation. Thus the progressive elaboration and differentiation of
the cortex in the evolving Primates have led to increasing powers
of apprehending the nature of external stimuli, a greater capacity

for a wider range of adjustments to any environmental change, and an enhancement of the neural mechanisms for effecting more delicately co-ordinated reactions.

The basic functions of the cerebral cortex in mammals generally are reflected anatomically in the differentiation of a number of sensory areas which receive incoming impulses related to different sensory modalities, a motor area which is predominantly concerned with the emission of impulses whereby voluntary movements are initiated and controlled, and intervening 'association areas' whose functions are largely concerned with the interrelation of the activities of the sensory and motor projection areas. Stated thus, this is of course a gross over-simplification of the subject, for there are, in fact, many other functions which can be ascribed to the cerebral cortex; the functions of even a single area are very highly complex and closely integrated with those of every other area. However, it is important to recognize the general concept that the cortex is by no means uniform in function and structure and is made up of a mosaic of different areas, for the reason that the differentiation, disposition, and relative extent of these areas show certain features which may be regarded as typical of the Primate brain.

The main sensory and other cortical areas which can be demarcated in the Primate brain may also be identified in the brains of lower mammals, but in the latter they are not so sharply differentiated in their microscopical structure, and the boundaries between adjacent areas are thus less sharply defined. With the progressive refinement of the analytical functions of the cortex in the Primates (more so, of course, in the higher than the lower members of this order), and with the increasing powers of sensory discrimination served by the several cortical areas, the latter undergo a progressive structural differentiation so that their extent and boundaries become more clearly definable by histological examination. Thus, for example, in the Primates the visual cortex (which receives impulses relayed from the retina) has an intrinsic structure which is so distinctive that the extent of the area can be mapped on a chart of the cerebral cortex with the greatest precision simply by the microscopic examination of suitably stained sections. In a similar manner, though not in all cases with the same degree of precision, the other areas can be defined and charted. By constructing such maps, or cyto-architectonic charts as they are commonly termed, it is possible to compare the relative extent and degree of differentiation of the various areas in the brains of different species. Examples of cyto-architectonic charts are shown in figs. 109, 120, 123, and 125.

The progressive differentiation and structural contrast displayed by the cyto-architectonic areas in the higher Primates are closely related to the type of convolutional pattern which characterizes their cerebral cortex. Broadly speaking, it may be said that there are two main factors by which the pattern of convolutions (gyri) and intervening sulci is determined—one is a general mechanical factor associated with the stresses to which the brain as a whole is exposed during its development;* the other depends on the growth and expansion of individual cortical areas in relation to each other. These two factors may interact in determining the precise location of an individual sulcus within the general pattern. In the mechanical category are many of the gyri and sulci which contribute to the convolutional pattern characteristic of lower mammals such as carnivores. The cerebral hemisphere during foetal development becomes buckled ventrally to form a notch called the pseudo-sylvian sulcus, and this is followed by the appearance of a series of

FIG. 105. Diagram of the cerebral hemisphere of a cat from the lateral and medial aspects illustrating the basic pattern of sulci and convolutions commonly found in lower mammals. *Ps.* Pseudosylvian sulcus. *C.C.* Corpus callosum.

arcuate gyri which are disposed concentrically about the ventral notch. Such a convolutional pattern is seen in the cat's brain (fig. 105). On the medial surface of the brain, similar series of sulci are formed in concentric relation to the development of the massive commissure, the corpus callosum, which bridges across from one cerebral hemisphere to the other and coordinates their functional

* It would perhaps be more correct to say that these sulci are developed in conformity with the stresses imposed on the developing and expanding cerebral hemisphere, than to suppose that these stresses *directly* produce them. The relationship of the sulci to the mechanical stresses may only be indirect, but the relationship is obviously there.

activities. Sulci having the same mechanical basis also contribute in part to the convolutional pattern of the brain in the higher Primates, including man; for example, the Sylvian fissure (lateral sulcus) and the concentric pattern of sulci disposed about the corpus callosum.* But in the higher Primates generally, more of the sulci are developed as foldings of the cortex along the boundary lines between adjacent cortical areas (limiting sulci) or as foldings along the middle of a specific area (axial sulci). The convolutional pattern thus comes to have a much closer topographical relationship to the various cortical areas, a relationship which, as already noted, apparently results from the sharper contrasts in their intrinsic structure already noted. It is this factor which is responsible for the characteristic convolutional pattern of the Anthropoidea (see fig. 116).

The expansion in various directions of the cerebral cortex in the Primate brain leads to a conspicuous development of certain of its lobes (see fig. 106). Of the latter, the occipital lobe becomes enlarged to accommodate the expanding visual cortex; it is one of the earliest features to become established in the evolution of the Primates, and even in the tree-shrews it is already obtrusive (fig. 109). With its progressive increase in size and its backward extension, the occipital lobe comes to overlap almost entirely the rest of the brain, a feature in which the higher Primates contrast strongly with lower mammals. The rapid development of the temporal lobe is another distinctive character of the Primate brain, and its precocious evolutionary enlargement is one of the most striking characters of the Eocene lemur, *Adapis* (fig. 111). The significance of this expansion of the temporal lobe is not clear, but there is some evidence of a clinical nature that (apart from the auditory cortex which forms only a small part of it) it constitutes a mechanism for the "storage" of visual as well as other memories. If this is so, its progressive development *pari passu* with the increasing elaboration of the visual powers in Primates becomes readily intelligible, for it enhances the ability to profit from experience on the basis of visual cues—an attribute which is obviously of the greatest importance for active arboreal creatures. Anatomically, there is also evidence that the temporal lobe receives impulses relayed, directly or

* In this connection, it is interesting to note that in those anomalous human brains in which the corpus callosum fails to develop, the sulci and gyri on the medial surface of the cerebral hemisphere are arranged quite differently—in a radial and not in a concentric pattern.

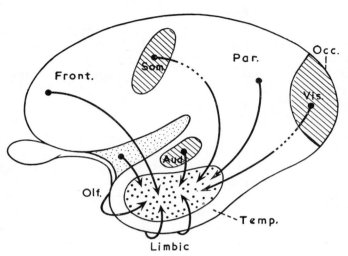

FIG. 106. Schema of the mammalian cerebral hemisphere indicating the main lobes (frontal, parietal, occipital and temporal) and some of the main sensory projection areas—general sensory (Som.), visual, and auditory. The cortex of the temporal lobe receives impulses (directly or indirectly) from almost all the other cortical areas, including the olfactory cortex and the limbic cortex (which borders the corpus callosum on the medial surface of the hemisphere).

indirectly, from almost all other areas of the cortex (fig. 106); in other words, it appears to provide the means for a final integration of the resultants of activity of the cortex as a whole.

The frontal lobes of the brain are commonly assumed to represent the highest functional level of cortical activity. Such an assumption is not altogether correct, but it is a fact that in the ascending scale of the Anthropoidea, from the monkeys to the apes and finally to man, these lobes show a progressive expansion. In man their large size is responsible for the development of a vertical forehead. There is every reason to suppose that the functions of the cortex of the frontal lobes are highly complex, but, in a general way, it can be said on the basis of experimental and clinical observations to form an essential part of the neural mechanism by which attention can be focused and concentrated on a particular objective, and in some manner it is instrumental in supplying the initiative to embark on a task as well as the urge to pursue it to completion without distractions.

Coincidentally with the expansion and differentiation of the cerebral cortex in Primate evolution, the cerebellum also undergoes an elaboration. The latter involves particularly the middle lobe of the cerebellum, which is intricately linked by fibre connections with most areas of the cerebral cortex and is to a great extent concerned with the co-ordination of the activity of the voluntary musculature. On either side the middle lobe expands to form the lateral lobes (or hemispheres) of the cerebellum. The anterior and posterior lobes of the mammalian cerebellum are more directly concerned with the automatic regulation of the equilibrium of the body as a whole, on the basis of information supplied to them partly from the semicircular canal system and partly from nervous pathways (spino-cerebellar tracts) ascending in the spinal cord.

There is good reason to suppose that the comparative anatomy of the central nervous system is of particular value in the elucidation of phylogenetic relationships. By the very nature of cortical activity, it must be quite exceptional for the higher functional levels of the nervous system to undergo retrogressive changes during the course of evolution; in other words, the general concept of the irreversibility of evolution might be expected here to have considerable validity. For if it be accepted that a progressive elaboration of the higher functional levels of the brain permits a greater facility in adapting behaviour with more precision to environmental needs, it is difficult to suppose that when such an advantage has been gained during evolutionary development it can be dispensed with, unless this is accompanied by very extreme forms of specialization in other systems of the body.

THE BRAIN OF A PRIMITIVE MAMMAL

As a preliminary to a survey of the comparative anatomy of the brain in Primates, it is convenient first to give a brief description of its main features in a primitive and generalized eutherian mammal, partly to indicate the arrangement and organization of the main subdivisions of the brain, and partly in order to emphasize the contrast which it shows with the more highly developed brain of the modern Primates. For this purpose we may refer to the archaic type of mammalian brain which is still preserved in the Madagascan insectivore *Centetes* (the Tenrec). In general proportions the brain of *Centetes* (fig. 107) resembles quite remarkably those of certain primitive Eocene mammals. The olfactory bulb (which receives the terminals of the olfactory nerves) is very large, projecting well in advance of the cerebral hemisphere and attached to the latter by a broad pedicle. A lateral view of the brain shows that the greater proportion of the cerebral cortex is made up of the piriform lobe—the cortical layer of which in its front part receives impulses from the olfactory bulb conveyed to it by the olfactory tract. This region of the piriform lobe thus constitutes the olfactory cortex. The cortex of the piriform lobe as a whole is a phylogenetically ancient part of the forebrain, and it is sometimes referred to as the archipallium. The dorsal surface of the cerebral hemisphere, which is separated from the piriform lobe by a horizontal groove, the rhinal sulcus, is composed of a layer of cortex concerned with the reception and interpretation of sensory impulses other than those of smell. This area is not present in submammalian vertebrates and, because of its relatively late appearance in vertebrate phylogeny, it is termed the neopallium. In general, it may be said that the progressive elaboration of the neopallium in mammals provides an index of the increasing influence of tactile, visual and auditory stimuli in controlling behavioural reactions. In an animal like *Centetes* the neopallial cortex is of very limited extent and is quite devoid of sulci or gyri, and it is evident, therefore, that smell is the dominant sense. As already noted, however, an arboreal habitat tends inevitably to lead to a reduction of olfactory mechanisms, while it also demands a greater acuity in other senses; it is because of this that in the evolution of the higher Primates the neopallium becomes increasingly elaborate and convoluted and the olfactory area of the piriform lobe more reduced, so that ultimately the latter is almost completely overshadowed by the former and becomes pushed entirely on to the medial aspect of the cerebral hemisphere.

On the base of the brain in *Centetes* is a large olfactory tubercle, in which are incorporated other olfactory receiving centres. The medial view of the brain (fig. 107) shows a very small corpus callosum and a big anterior commissure. Both these commissures serve to link across the midline the cortical areas of the two hemispheres, but the anterior commissure also contains fibres which connect the two olfactory bulbs as well as certain other primitive elements of the forebrain. The corpus callosum is only found in eutherian mammals, being absent in submammalian vertebrates

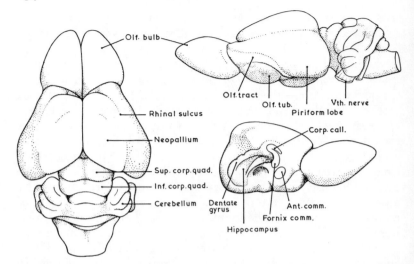

FIG. 107.   Brain of the insectivore *Centetes* from the dorsal, lateral
and medial aspects (×2).

and also in monotremes and marsupials. In the latter the anterior
commissure is thus the only neopallial commissure, and it is
correspondingly large. The proportionate size of the two commissures in *Centetes* is a measure of the lowly status of this primitive insectivore; in higher mammals, and particularly in most of
the Primates, the corpus callosum increases in size and the anterior
commissure dwindles. Below the corpus callosum is the fornix
commissure which connects across the midline the two curved
strips of cortex comprising the hippocampal formation (hippocampus and dentate gyrus). The hippocampal formation was for
many years erroneously assumed to be in its entirety a component
of the olfactory apparatus of the brain; its function still remains
obscure, but since it is consistently present in all vertebrates from
the highest to the lowest, and is linked by fibre connections with
a diversity of important centres of the forebrain, it presumably
plays an essential part in the more fundamental activities of the
cerebrum as a whole.

While olfactory impulses are conveyed directly to the archi-
pallial cortex by fibres of the olfactory tract, all other types of
sensory impulse have to be relayed to the neopallial cortex through
a central mass of grey matter called the thalamus. This large sen-

sory mechanism is concerned with sorting the incoming impulses as well as relaying them to the cortex, and on its sorting activities largely depend the analytical functions of the sensory areas of the cortex. Consequently, as the neopallium increases in extent and differentiation, the thalamus itself becomes larger and more complex. Ventral to the thalamus, and exposed on the base of the brain in close relationship to the pituitary body, is the hypothalamus, a zone of grey matter through which many of the visceral functions of the body are subjected to nervous control (see fig. 110).

The olfactory bulbs, cerebral cortex and thalamus, together with certain other important structures embedded in the hemispheres, comprise the forebrain. Behind it is the midbrain, in the roof of which are two pairs of rounded eminences, the anterior and posterior corpora quadrigemina. The anterior pair receive some of the terminal fibres of the optic tract and are concerned (*inter alia*) with mediating ocular reflexes. The posterior pair perform a similar function as auditory reflex centres.

The cerebellum, which is developed as an excrescence in the roof of the hindbrain, is of a very simple type in *Centetes*. It has a large median part (the vermis) which is subdivided by two deep fissures into anterior, middle and posterior lobes. The last named is extended laterally on each side to form a conspicuous lobule, the flocculus. The middle lobe bulges out on each side to form the main mass of the cerebellar hemispheres (lateral lobes); in the primitive brain of insectivores these are relatively small, but in higher mammals, and particularly in the Primates, they expand to a considerable size.

Besides the cerebellum the hindbrain is made up of the pons and medulla, both of which contain vital centres of autonomic function, the motor and sensory nuclei of most of the cranial nerves, and numerous ascending and descending tracts of nerve fibres through which the forebrain is linked with the hindbrain, midbrain, and spinal cord.

TREE-SHREWS

There is a marked difference between the brain of the Ptilocercinae (represented by *Ptilocercus*) and that of the other tree-shrews, the Tupaiinae. In most of its anatomical systems *Ptilocercus* is a much more primitive creature and, in contrast to the diurnal

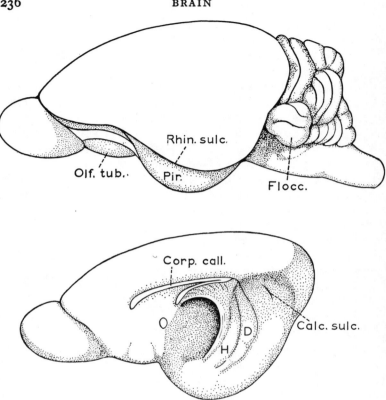

FIG. 108. Brain of *Tupaia minor* from the lateral and medial aspect. *H*, Hippocampus; *D*, Dentate gyrus (×4).

habits of tree-shrews in general, it is crepuscular. These differences are reflected in the general features of the brain. We shall here confine our attention to the brain of *Tupaia* (fig. 108), partly because its structure has been studied in some detail, and partly because it displays a number of features which clearly indicate affinities with the lemuroid brain. These affinities may most clearly be brought to view by comparing *Tupaia*, on the one side with the insectivores (with which the tree-shrews were at one time grouped in zoological classifications), and on the other side with the primitive Madagascar lemur *Microcebus* (see figs. 109, 110 and 111). Quantitative data are meagre, but so far as they are available they demonstrate that the brain of *Tupaia* is relatively large in com-

KEY TO AREAS:
1–3 general sensory,
4 motor, 5 and 7 parietal,
6 and 8 frontal, 13–16 insular,
17 and 18 visual, 19 occipital,
20-22 temporal, 28 piriform.

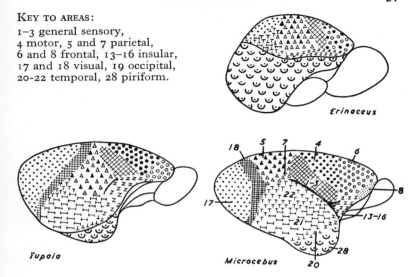

FIG. 109. Cerebral hemisphere of *Erinaceus* (hedgehog), *Tupaia* (tree-shrew), and *Microcebus* (mouse lemur), showing the cytoarchitectonic charts of the cortical areas. The numbering of the areas is that introduced originally by Brodmann in 1904.

parison with sub-Primate mammals of an equivalent size. For example, a specimen of *Tupaia javanica* with a body weight of 80 grams was found to have a brain weight of 2·3 grams. The corresponding figures for a mole (*Talpa*) have been reported as 77 grams and 1·5 grams and for *Microcebus* 62 grams and 1·9 grams. Compared with all insectivores, the olfactory bulbs of *Tupaia* are relatively much smaller (though larger than those of any other Recent Primates), the olfactory tubercle is reduced in extent and flattened, and the piriform cortex has been displaced downwards and medially by the expanding neopallium so that not more of it is to be seen on the lateral aspect of the hemisphere than is the case with *Microcebus*. The corpus callosum is elongated and straight as in small lemuroids, while the anterior commissure is quite small. The hippocampus and dentate gyrus form continuously exposed strips of cortex which extend from the corpus callosum down to the piriform lobe (fig. 108).

The neopallial cortex shows a relative expansion in the occipital

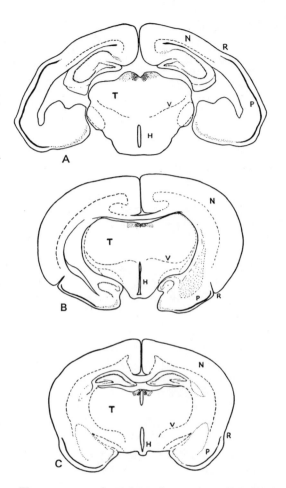

FIG. 110. Transverse sections through corresponding levels of the brain of (A) *Centetes*; (B) *Tupaia minor*; and (C) *Microcebus* (all ×4). *H*, hypothalamus; *N*, neopallium; *P*, piriform cortex; *R*, rhinal sulcus; *V*, ventral margin of the thalamus (T). Note the resemblance between *Tupaia* and *Microcebus*, and the corresponding contrast with the insectivore brain.

and temporal regions, but the frontal lobe remains small and attenuated. The temporal lobe projects downwards to form a broad and blunt temporal 'pole', and in front of this the hemisphere is excavated by the large orbital cavities of the skull to form a broad

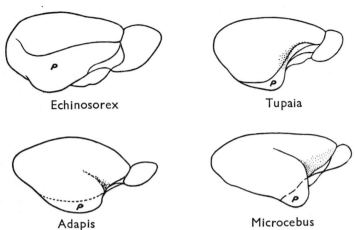

Echinosorex   Tupaia

Adapis   Microcebus

FIG. 111. Cerebral hemisphere of the insectivore *Echinosorex* compared with *Tupaia*, *Adapis* (based on an endocranial cast) and *Microcebus*. *P*, piriform lobe.

and shallow Sylvian fossa (as in *Tarsius*, but less extensive). With the expansion of the temporal lobe, the rhinal sulcus becomes strongly deflected downwards instead of pursuing a relatively straight, horizontal course as it does in the insectivore brain. A projecting occipital pole is present which overlaps the whole of the midbrain, as well as the anterior surface of the cerebellum. The only neopallial sulcus which is visible macroscopically is a short calcarine sulcus on the medial surface of the occipital lobe; it marks the anterior boundary of the visual cortex here.

The histological structure of the neopallial cortex is considerably more elaborate and differentiated than it is in insectivore brains; indeed, most of the cortical areas are as clearly circumscribed and appear to be as richly cellular as those of *Microcebus*. In comparison with the latter, however, the frontal and parietal association areas are less extensive and the frontal cortex is less differentiated. The visual cortex of *Tupaia* is remarkably well developed. In extent it forms a relatively large part of the neopallium, and on the medial aspect of the hemisphere it reaches downwards and forwards practically as far as the corpus callosum. In its microscopical structure it shows a degree of differentiation (particularly in the definition of its cell laminae) which, in mammals of equivalent size, is only paralleled by the other Primates. It may be noted that, while in the

Insectivora an increase in the size of the brain as a whole is accompanied by an expansion of the piriform lobe and the retention of a limited area of neopallial cortex (as, for instance, in *Echinosorex*, the largest of the living insectivores), in *Tupaia*—as in Primates generally—the increase in size of the brain is marked by a developmental trend in precisely the contrary direction, that is, by an expansion of the neopallium and a relative retrogression of the piriform lobe.

In conformity with the elaboration of the neopallial cortex in *Tupaia*, the thalamus is expanded to accommodate the various sensory nuclei of which it is composed, and the latter show a considerable degree of intrinsic differentiation. The expansion of the thalamus as a whole is made evident in transverse sections of the brain by comparing its size with that of the hypothalamus which lies ventral to it (fig. 110). The relay nucleus of the thalamus through which visual impulses are transmitted to the cortex, the lateral geniculate nucleus, merits particular attention, for not only is it relatively large, it also shows a laminar arrangement of its constituent cells which in its more sharply defined lamination very closely approaches that characteristic of the higher Primates (fig. 113A). The elaboration of the lateral geniculate nucleus in *Tupaia* is correlated with the relatively large size of the optic nerve, as well as with the high grade of development of the visual cortex.

The midbrain of *Tupaia* appears to be remarkably primitive, for the anterior corpora quadrigemina form great rounded prominences which are relatively larger than in any other mammal (with the possible exception of the so-called 'flying lemur'—*Cynocephalus* = *Galeopithecus*). Their unusual size even gives them a superficial resemblance to the optic lobes of the reptilian brain. Undoubtedly this feature is also (like the visual cortex) related to a high degree of visual acuity, and it is interesting to note that the latter is served by an elaborate midbrain mechanism as well as by a highly differentiated cortical mechanism. The posterior corpora quadrigemina, by contrast, are rather small.

The cerebellum of *Tupaia* is of a simple and generalized mammalian type, with a well developed floccular lobe. In the degree of its fissuration it is more complicated than that of insectivores of comparable size, and is actually rather more complex than the cerebellum of *Microcebus* and *Tarsius*.

Since the structure of the brain offers some of the most cogent evide nce for the affinities of the tree-shrews with the Primates, as well a s demonstrating their distinction from the Insectivora with which they were formerly grouped, the following summary is given of those characters in which the tupaioid brain contrasts with the latter.

1. The relative size of the brain as a whole.
2. The expansion of the neopallium, accompanied by a displacement downwards of the rhinal sulcus.
3. The formation of a distinct temporal pole of the neopallium.
4. The backward projection of the occipital pole.
5. The presence of a calcarine sulcus.
6. The well-marked lamination and cellular richness of the neopallial cortex in general, and the degree of differentiation of the several cortical areas.
7. The pronounced elaboration of the visual apparatus of the brain, particularly at the cortical level.
8. The advanced degree of differentiation of the nuclear elements of the thalamus.
9. The well-defined cellular lamination of the lateral geniculate nucleus.

On the other hand, the brain of *Tupaia* still remains more primitive than that of any of the other *living* Primates in the following features:

1. The relative size of the brain as a whole.
2. The relative size of the olfactory bulb.
3. The absence of a retrocalcarine sulcus (an axial folding within the area of the visual cortex).
4. The small extent of the parietal and frontal association areas of the cortex.
5. The large size of the anterior corpora quadrigemina.

The brain of the pen-tailed tree-shrew, in which the olfactory regions are relatively more extensive and the visual apparatus only poorly developed, provides a sort of transitional stage between the advanced type of brain found in *Tupaia* and the lowly type characteristic of the Insectivora. Unfortunately, there is no certainty about the size and proportions of the brain in the early fossil Primates of the family Plesiadapidae; the most we can say is that the skull of *Plesiadapis* suggests that the brain was even more primitive than that of *Ptilocercus*.

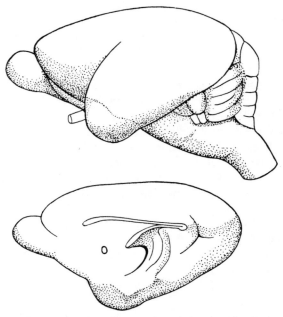

FIG. 112. Brain of *Microcebus murinus* from the lateral and medial
aspect (×3).

LEMURS

Among the living lemurs the most primitive type of brain is that
of *Microcebus*, the small mouse-lemur of Madagascar. In this genus
the olfactory bulbs, though reduced in comparison with *Tupaia*
and lower mammals, are relatively better developed than in other
lemuroid genera. A considerable part of the piriform cortex remains exposed on the lateral surface of the cerebral hemisphere to
form the projecting tip of the temporal lobe, and it is delimited
above by a shallow but quite distinct rhinal sulcus (fig. 112). The
neopallium is relatively simple, showing sulci in two regions only.
On the lateral surface is a deep lateral sulcus (Sylvian fissure), while
on the medial aspect of the occipital lobe is a triradiate sulcus, the
calcarine complex. The posterior limb of the latter, which extends
back horizontally to form an axial folding within the visual cortex,
is commonly termed the retrocalcarine sulcus. It is highly charac-

teristic of the Primate brain and, indeed, is found consistently in all Recent Primates with the exception of the Tupaioidea. Nor does it occur in any other mammalian group except sporadically in the Carnivora. The development of a retrocalcarine sulcus allows for the accommodation of a greater expanse of the visual cortex.

The formation of a deep Sylvian fissure is related to the great size of the temporal lobe which, as already noted (p. 230), is a very distinctive feature of the typical Primate brain. The occipital lobe of the cerebral hemisphere is also large, but the frontal lobe remains relatively inconspicuous. The corpus callosum is elongated and slender, showing slight thickenings at the anterior and posterior extremities (genu and splenium), while the fornix commissure is relatively reduced. As in *Tupaia*, the hippocampus and dentate gyrus are exposed as two separate strips of cortex from the corpus callosum downwards.

In fig. 109 is shown a cyto-architectonic chart of the cerebral cortex of *Microcebus*. The cortical areas demarcated are numbered according to the notation introduced by K. Brodmann who, over sixty years ago, completed a pioneering study of the comparative anatomy of cortical structure [10]. In comparison with the brain of *Tupaia* it will be seen that the cortex in *Microcebus* is further elaborated by an expansion of the parietal association areas and the temporal areas, and also (but to a lesser degree) of the frontal areas. In particular the appearance and incipient expansion of the area frontalis (area 8) may be noted; this area, which undergoes still further differentiation and assumes much greater importance in the higher Primates, is too poorly defined in *Tupaia* to be represented on its cyto-architectonic chart. All the cortical areas just mentioned come within the category of what are commonly termed 'association areas'—that is to say, they are not concerned so much with the direct reception and emission of impulses from or to lower levels of the central nervous system, but rather with correlating the functions of the sensory and motor projection areas and thus providing the anatomical substratum of mental processes such as the interactions of sensory impressions, the coordination of motor reactions in relation to the resultant of these interactions, the association of ideas, memory and so forth.

The thalamus in *Microcebus* shows distinctive Primate features in the enlargement and increased differentiation of most of the principal nuclear elements, and particularly in the elaboration of the lateral geniculate nucleus. The configuration and detailed cell-ular structure of this important visual centre present character-istic features in the main groups of Primates which are of some importance because of their possible taxonomic significance. In the illustration in fig. 113 is shown schematically a diagram of a trans-verse section through the lateral geniculate nucleus and adjacent part of the thalamus of a tupaioid, a lemur and a monkey. In such a section, the nucleus in the tupaioid brain has a vertically disposed

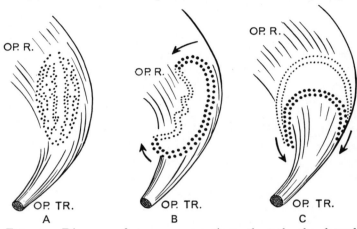

FIG. 113. Diagram of transverse sections through the lateral geniculate nucleus showing the different dispositions of the cell laminae in the Primates.

In (A) is shown a generalized condition such as is seen in the tree-shrew in which the laminae are not quite so sharply separated and are arranged more or less vertically; the optic tract fibres enter from below and laterally, and the optic radiation emerges from the medial aspect. In the lemurs and in *Tarsius* the laminae become inverted (B), while in all the Anthropoidea they become everted (C).

lentiform shape, and the lamination of its cells, though very distinct, may be somewhat irregular. The fibres of the optic tract enter the nucleus mainly through its lateral surface, while the fibres of the optic radiation, which project from the nucleus to the visual cortex, emerge from the medial aspect. In the higher Primates, the lami-nation becomes much more distinct and clear cut, and the cells which form the superficial laminae are conspicuous for their rela-tively large size (for further details see p. 278). Moreover, with the expansion of the rest of the thalamus the geniculate nucleus as a whole becomes rotated and displaced ventrally, and the laminae become folded. But the manner of this folding shows a rather marked contrast in the lemurs and the Anthropoidea. In the former the laminae become incurved or inverted, the fibres of the optic tract entering them from the lateral aspect, and the fibres of the optic radiation issuing from the medial concavity. In the Anthropoidea the laminae become everted, leading to the formation of a concavity which is directed downwards; the optic tract fibres in this case penetrate the geniculate nucleus mainly from below, and the fibres of the optic radiation emerge from the dorsal convexity. The inverted type of nucleus which, as just noted, is a lemuroid characteristic, is often accompanied by an irregular crinkling of the most medial laminae.

The midbrain of *Microcebus* is proportionately rather large as compared with most lemuroid brains, and both pairs of corpora quadrigemina form conspicuous elevations. The cerebellum is relatively simple; the lateral lobes are small and the median vermis and the floccular lobe quite well developed.

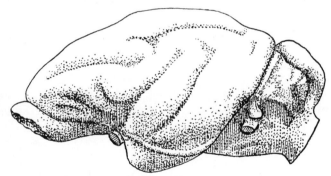

FIG. 114. Lateral aspect of an endocranial cast of *Lemur catta* ($\times 1\frac{1}{2}$).

In the larger lemurs the neopallial cortex is more richly convoluted and the olfactory bulb and olfactory area of the cortex relatively more reduced. The basic convolutional pattern characteristic of the lemuroid brain in general is shown in figs. 114 and 115. From these illustrations it will be noted that the sulci in the frontal and parietal regions run more or less longitudinally, and (as demonstrated in fig. 116) they tend to cut across the series of cortical areas which can be defined histologically. For example, the sulcus rectus intersects the middle part of the frontal area of cortex and also the motor cortex, while the intraparietal sulcus likewise cuts across the association areas in the parietal lobe and extends back into the front part of the visual cortex. None of the cortical areas outlined in fig. 116A is demarcated by limiting sulci. By contrast, in the Anthropoidea the sulci over the dorsal surface of the hemisphere

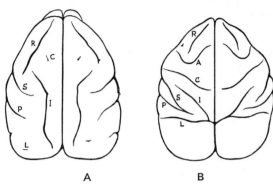

A   B

FIG. 115. Dorsal aspect of the cerebral hemispheres of (A) *Lemur nigrifons* (×1), and (B) *Cercopithecus mona* (×⅔), showing the contrasting dispositions of the sulci. *A*, arcuate sulcus; *C*, central sulcus; *I*, intra-parietal sulcus; *L*, lunate sulcus; *P*, parallel sulcus; *R*, sulcus rectus; *S*, Sylvian fissure (lateral sulcus).

are disposed more or less transversely, for they develop as limiting sulci in relation to the margins of the cortical areas here. Thus, for example, one of the most consistent sulcal elements in the brains

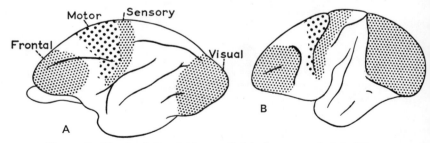

FIG. 116. Cerebral hemisphere of (A) *Lemur* and (B) *Macaca*, showing the contrasting relationships of the sulci to the motor, sensory, frontal and visual areas of the cortex.

of monkeys, apes and man is the sulcus centralis, which extends across the hemisphere to form a boundary line between the motor and general sensory areas of the cortex (fig. 116B). In the larger

lemurs the sulcus centralis is commonly represented by no more
than a very small and irregular dimple in the cortex near the upper
end of the boundary between these two areas. But there are occasional exceptions, for in the brain of *Perodicticus* there is a well-
developed sulcus centralis, and the latter may also be sporadically
found in other lemurs. Thus, while the absence of limiting sulci
over the frontal and parietal regions, and the presence here of longitudinally disposed sulci, are the prevailing characters which distinguish the convolutional pattern of lemurs from that of the
Anthropoidea, the potentiality for the development of a transverse
pattern of limiting sulci in occasional species or varieties does serve
to emphasize the affinities between these two groups of Primates.

The brain of *Daubentonia* (like many other details of the anatomy
of this curious creature) shows aberrant features which by no
means conform with those of the typical lemuroid brain. The
olfactory bulbs are relatively larger, the rhinal sulcus is well defined, and the configuration of the sulci shows a strong resemblance

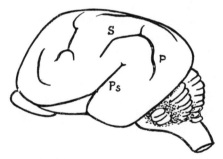

FIG. 117.   Brain of the aye-aye (*Daubentonia*) (×1). *P*, post-sylvian
sulcus; *Ps*, pseudo-sylvian sulcus; *S*, supra-sylvian sulcus.

to that of lower mammals such as the carnivores and certain small
ungulates (fig. 117). Thus, there is a relatively short 'pseudo-
sylvian' sulcus surmounted by an arcuate gyrus, and the latter is
bounded above by a continuity of the supra-sylvian and post-sylvian
sulci. The other sulci on the lateral aspect of the cerebral hemisphere show such a high degree of individual variation that it is
difficult to determine their homologies with certainty. In some
specimens, indeed, their disposition shows a curious resemblance
to the sulci of a cat's brain. On the medial surface there is a triradiate calcarine sulcus with a retrocalcarine element of the

AOM R

Primate type, and the corpus callosum is relatively massive. The temporal lobe is poorly developed—another obtrusively non-lemuroid feature. It is by no means easy to assess the significance of this unexpected combination of features in the brain of *Daubentonia* for, so far as can be ascertained, they seem to bear no relationship to habitus. Perhaps the most that can be said is that the aberrancy of the brain of this genus is in some manner a reflection of the aberrancy of much of the rest of its anatomy—for example, the dentition, certain elements of the skull, the clawed digits and so forth. But the primitive features of the cerebrum do lend some support to those who have contended, on other grounds, that *Daubentonia* represents the terminal product of a side-line of lemuroid evolution which must have become segregated from other groups of lemuroids at a very early geological date.

Endocranial casts of *Adapis* show that the brain of this Eocene genus was in several respects more primitive than any modern lemuroid brain (figs. 111 and 121). It was much smaller in proportion to the body size of the animal, the olfactory bulbs were relatively larger, and the cerebral hemispheres were abbreviated to the extent that the whole of the cerebellum seems to have been exposed above. On the other hand, the cortex shows some elaboration in the presence dorsally of a longitudinal (? intraparietal) sulcus, and the temporal lobes are relatively very large. As already noted, this precocious expansion of the temporal lobe is a characteristic feature in the evolution of the lemuroid brain, and it is also evident in the endocranial cast of the Early Miocene genus *Progalago*. The endocranial casts of the large Pleistocene lemurs of Madagascar show a profusion of convolutions which in most types are disposed in a generally longitudinal pattern. But an interesting exception is to be seen in *Archaeolemur*, a genus whose skull, as already noted, shows certain simian proportions. In this extinct lemur the brain is globular and richly convoluted, and the cerebral hemispheres overlap the cerebellum to a considerable extent; apart from these features, the sulci on the dorso-lateral aspect of the cerebral hemisphere are mostly transversely disposed (as they are in monkeys), and there is a well marked sulcus which corresponds in position and direction to the sulcus centralis. In general, then, the brain of *Archaeolemur* has a remarkably pithecoid appearance, though the resemblances to the brain of a monkey are offset by the poor development of the frontal lobes. The endocranial cast of *Megaladapis*

shows that the olfactory bulbs extended well in front of the cerebral hemispheres and were attached to the latter by stout, elongated peduncles. This feature, together with the longitudinal disposition of the convolutional pattern, gives to the brain a curiously
ungulate appearance.

TARSIOIDS

The brain of the modern *Tarsius* (figs. 118 and 119) comprises
a most remarkable mosaic of advanced and primitive characters,
which in their combination pose a difficult problem because of
their apparent incongruity from a functional point of view. In
general appearance the cerebral hemispheres show somewhat simian proportions. Thus they are broad and rounded when viewed
from above, while from the lateral aspect they are prolonged backwards to such an extent that the occipital lobes overlap most of the
cerebellum as in the Anthropoidea. The olfactory bulbs are much

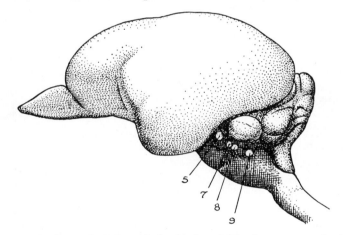

FIG. 118. Lateral aspect of the brain of *Tarsius spectrum* ($\times$3).
Some of the cranial nerves are indicated. (After H. H. Woollard,
Proc. Zool. Soc. 1925.)

reduced, and the olfactory tract and olfactory cortex are correspondingly small. It is to be noted, however, that the olfactory area
of the piriform lobe appears on the lateral surface of the hemisphere
forming almost as much of the tip of the temporal lobe as it does
in the lemuroid brain. A glance at the cyto-architectonic chart of
the cerebral hemisphere of *Tarsius* (fig. 120) makes it apparent

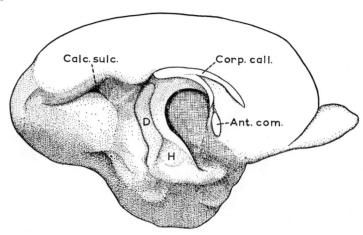

FIG. 119. Medial aspect of the brain of *Tarsius spectrum* (×4).
D, dentate gyrus; H, hippocampus.

that almost half the cortex is made up of the visuo-sensory cortex
(area 17) and the functionally associated area 18. It is the great
expansion of these cortical areas which has led to such a pronounced development of the occipital lobe and gives so pithecoid
an appearance to the brain as a whole. Moreover, the occipital lobe
is hollowed out by a posterior horn of the lateral ventricle, a diverticular extension of the ventricular cavity which is present in
the Anthropoidea but absent in the lemurs.

Not only is the visual cortex extensive in area, it has a complexity of structure which is only paralleled in certain of the Platyrrhine
monkeys (e.g. *Cebus*). Thus, the cell laminae are multiplied in
number and they are also very sharply differentiated. It is difficult
to offer an interpretation of this elaboration of the visual cortex for,
although the eyes are of enormous size and the retina is provided
with a macula (see p. 274), the retinal epithelium is constructed
for night vision and presumably, therefore, is not designed for a
high degree of visual discrimination. Apart from the visuo-sensory
and immediately associated areas, the cortex of *Tarsius* does not
show an advanced degree of differentiation. In the general development of the other areas it approximates to *Microcebus*, but the
temporal cortex is much more restricted and the frontal cortex is
also hardly more extensive.

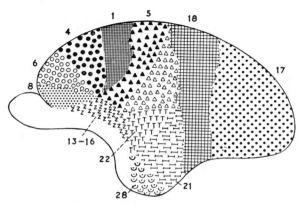

FIG. 120. Cyto-architectonic chart of the cerebral cortex of *Tarsius spectrum*. (Adapted from H. H. Woollard, *Jour. Anat.*, Vol. 60, 1925.) See Key to Fig. 109, p. 237.

The hemispheres show no fissuration on their exposed surface, except for a short oblique sulcus which may be present in the Sylvian fossa and which is of doubtful homology. The Sylvian fossa (which can be regarded as a very broad pseudo-sylvian sulcus) is excavated by the large orbits over the roof of which the cerebral hemispheres are moulded. Thus *Tarsius* lacks a fully formed Sylvian fissure (lateral sulcus) such as is found in the brains of all other living Primates except the tree-shrews. On the medial surface of the occipital lobe is a deep triradiate calcarine sulcus, similar to the corresponding sulcus in *Microcebus* and lying in a deep depression which fits over the large corpora quadrigemina of the midbrain. The appearance of the hippocampal formation (hippocampus and dentate gyrus) resembles that of the higher Primates in the fact that the hippocampus is only exposed at its lower extremity (fig. 119). On the other hand, the corpus callosum is of very primitive construction, rather like that of an insectivore such as *Erinaceus* and quite different from the elongated commissure which is so characteristic of Primates generally. The anterior commissure is correspondingly large as in most lower mammals.

The thalamus of *Tarsius* is similar in general proportions to that of *Microcebus*, except that the lateral geniculate nucleus is relatively much larger. This is also the case with the adjacent nuclear element (also concerned with visual functions) which projects to area 18 of the cerebral cortex; it is this thalamic element which in the

Anthropoidea becomes extended posteriorly, by the expansion of the thalamus as a whole, to form a projecting pole called the 'pulvinar'. The geniculate nucleus consists of well-defined layers of cells which are strongly inverted in lemuroid fashion. The midbrain shows a primitive feature in the strong development of the corpora quadrigemina. The cerebellum is astonishingly simple, its lateral lobes being even smaller and less complicated by fissuration than in *Microcebus*. Its elementary simplicity in *Tarsius* seems paradoxical in relation to the extent of the neopallial cortex; it is also surprising in view of the creature's arboreal agility, for it might be supposed that the accuracy with which it can leap considerable distances from branch to branch would demand a very precise control and co-ordination of muscular activity through the co-operative functions of the cerebellum and cerebrum.

We have remarked on the unexpected mixture of archaic and advanced features in the tarsioid brain. From a consideration of all the individual details, it now seems clear that it is really an extremely primitive brain on which has been superimposed a very specialized and 'one-sided' development of the visual centres accompanied by a reduction of the olfactory centres. But while the visual centres seem designed by the degree of their structural differentiation to deal with complex patterns of visual stimuli, it is puzzling to understand to what use the animal puts all this sensory information. For the non-visual areas of the neopallium are of limited extent and the effector mechanisms of the central nervous system—including the motor cortex, the cerebellum and the motor pathways of the brain-stem and spinal cord—are quite poorly developed. That the primitive features of the brain are truly primitive, and not the result of secondary retrogressive changes, seems assured because it is difficult to suppose that, once the advantages of a more precise control of behavioural reactions had been conferred by the acquisition of a more elaborate cortex, cerebellum and motor pathways, these advantages would be discarded by any process of natural selection. At any rate, this argument must surely apply to the cortex of the temporal lobe if, as we surmise, it plays so important a part in increasing the efficiency of cortical activity as a whole, for, as we have noted, the temporal cortex is but weakly developed in *Tarsius*.

The peculiar and distinctive tarsioid type of brain, in its combination of characters so unlike the brain of any other group of

Primates, is of high antiquity. The endocranial cast of the Eocene tarsioid *Necrolemur* shows a brain whose proportions are almost identical with that of *Tarsius* except that the olfactory bulbs are slightly less reduced, and the cerebral hemispheres do not so completely cover the cerebellum. The brain of the American fossil tarsioid, *Tetonius*, appears to be even more nearly identical with *Tarsius* in its size, shape, and the relative development of its several components. The contrast between the brain of *Necrolemur* and the Eocene lemuroid, *Adapis*, is made evident by reference to the illustrations in fig. 121, from which it will be seen that the adapid cerebral hemispheres were more elongated and were furrowed dorsally by a longitudinal sulcus, while the cerebellum was relatively

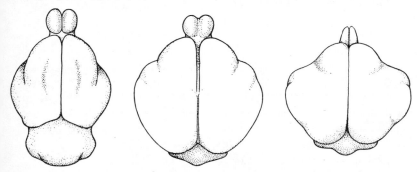

FIG. 121. Dorsal aspect of the endocranial cast of *Adapis* ($\times\frac{4}{3}$) and *Necrolemur* ($\times 2$) compared with the brain of the modern *Tarsius* ($\times\frac{5}{3}$). Note the close similarity in general proportions between *Necrolemur* and *Tarsius*.

much larger. We have already noted, also, that the temporal lobes of *Adapis* were precociously developed. This fossil evidence makes it clear that the tarsioid type of brain from its earliest development was very different from the lemuroid type of brain, and that it had almost reached its modern level of development during the Eocene. It also serves to emphasize one of the most significant trends of Primate evolution—that the brain began to expand in size at an earlier stage of evolution than in mammals generally.

The pithecoid appearance of the tarsioid brain has attracted the attention of many authorities, but the question needs to be considered whether this appearance betokens a phylogenetic affinity, or whether it may not be, in a sense, illusory. The large size of the occipital lobes and the structural differentiation of the visual cortex in *Tarsius* may be—at least in part—associated with a peculiar mode of nocturnal vision, and thus not entirely comparable with the development of the visual centres in the monkey's brain. Again, the reduction of the olfactory bulbs (and other olfactory centres) may be indirectly conditioned by the enormous expansion of the orbital cavities which has led to an extreme compression and limitation of the nasal cavities. In any case it should be noted that, in spite of the apparent pithecoid features such as the extensive occipital lobes and the relative breadth of the cerebral hemispheres, the brain of *Tarsius* as a whole is not appreciably larger in proportion to body weight than that of a lemur of equivalent size (e.g. *Galago*); in other words, the cerebral status (or cephalization coefficient) of tarsioids and lemurs is of an equivalent order.

ANTHROPOIDEA

The cerebral development of the modern monkeys shows a marked advance beyond the prosimian level. Thus in the marmosets, which in body size are comparable with *Tarsius* and *Galago*, the brain weight is about three times greater. These small Platyrrhine monkeys possess the most elementary and primitive type of brain seen in the Anthropoidea, and it therefore provides a convenient comparison with the prosimian brain.

In *Callitrix* (fig. 122) the cerebral hemispheres are voluminous with conspicuously large frontal lobes, and with a backward extension of the occipital lobes which almost completely overlap the cerebellum and medulla. Among the Platyrrhine monkeys this latter feature is particularly pronounced in the squirrel monkey (*Saimiri*). As in all the Anthropoidea, the occipital lobe contains a diverticular extension of the ventricular cavity—the posterior horn of the lateral ventricle. The olfactory bulbs are greatly reduced (more so than in *Tarsius*), and there is a corresponding diminution in the size of the olfactory tubercle and the olfactory area of the piriform lobe. Following on the expansion of the neopallial cortical areas, the piriform cortex is almost entirely displaced on to the basal and medial surface of the cerebral hemisphere, and is limited by a shallow rhinal sulcus which can be traced without difficulty throughout its usual course. The corpus callosum (in correlation with the neopallial development) is relatively large and elongated, with a well-marked splenium and genu, while the anterior com-

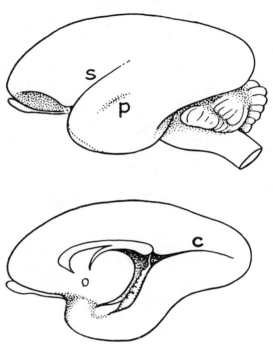

FIG. 122. Lateral and medial aspects of the brain of *Callitrix* (× 2).
S, Sylvian sulcus; *P*, parallel sulcus; *C*, calcarine sulcus.

missure and the fornix commissure are both very small. The dentate gyrus is a narrow, crinkled band of cortex reaching from the splenium of the corpus callosum down to the piriform lobe, and the hippocampus is only exposed at its lower extremity (as in *Tarsius*, but to a slightly less extent).

The neopallial cortex is smooth except for a deep Sylvian fissure which is related to a bulky temporal lobe, and the calcarine sulcus which extends back (as the retrocalcarine element) deeply into the middle of the visual cortex. There may also be a faintly indicated parallel sulcus on the temporal lobe; this sulcus is quite short and very shallow, so that it usually forms little more than a dimple on the surface of the cortex. The relative smoothness of the neopallium of the marmoset brain—as compared with the brains of most other Primates—can hardly be attributed to the small size of the animal, for in small lemuroids such as *Galago* the cortex may be quite well convoluted. It may thus be assumed to be essentially a primitive feature.

An analysis of the cortical structure of *Callitrix* (fig. 123) shows that, compared with lower Primates, there is a considerable expansion of the association areas, particularly in the frontal and

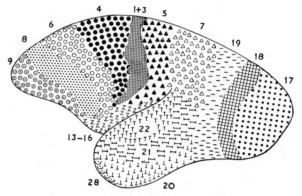

FIG. 123. Cyto-architectonic chart of the cortical areas of *Callitrix*. (Adapted from Brodmann.) See Key to Fig. 109, p. 237.

parietal regions, an expansion which is accompanied by an increased differentiation of their intrinsic structure. The expansion of the parietal cortex has thrust apart the visual and general sensory areas so that these are much more widely separated than they are in *Tarsius*. The visual cortex is highly developed in all the Anthropoidea, and in some of the Platyrrhines (e.g. the Cebinae) its intrinsic laminar differentiation has proceeded further than in the Catarrhine monkeys, and even further than in the anthropoid apes and man. As already pointed out, in this respect the Cebinae make an approach to the specialized visual cortex of *Tarsius*.

The cerebellum in the marmosets, though relatively simple, shows a greater lateral expansion of the hemispheres than in lemurs of equivalent size, and the floccular lobes thus appear relatively smaller.

In most of the Old and New World monkeys the cerebral cortex becomes richly convoluted (figs. 124 and 125). Particularly conspicuous and distinctive of the convolutional pattern are the sulcus centralis which forms a boundary line between the motor cortex

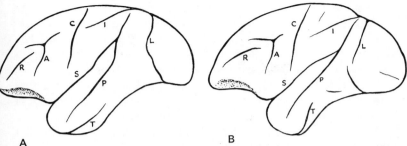

A   B

FIG. 124. Cerebral hemisphere of a Platyrrhine monkey, *Cebus* (A), and a Catarrhine monkey, *Macaca* (B), to show the close identity of sulcal patterns. *A*, arcuate sulcus; *C*, central sulcus; *I*, intraparietal sulcus; *L*, lunate sulcus; *P*, parallel sulcus; R, sulcus rectus; *S*, Sylvian sulcus; *T*, inferior temporal sulcus.

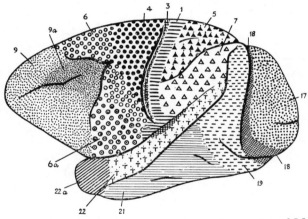

FIG. 125. Cyto-architectonic chart of the cerebral cortex of *Macaca*. See Key to Fig. 109, p. 237.

(area 4) and the general sensory cortex (areas 1 to 3), and the sulcus lunatus which limits anteriorly the visuo-sensory cortex on the lateral surface of the hemisphere. Because the appearance of this latter sulcus is so characteristic of monkeys and apes it is frequently called the 'simian sulcus'. Other sulci normally present in the monkey's brain are the intraparietal sulcus which separates the area parietalis (7) of the cortex from the area preparietalis (5), and the arcuate sulcus which intervenes between the area frontalis agranularis (6) and the area frontalis granularis (8). Thus on the dorso-lateral surface of the hemisphere the sulci in general are disposed more or less transversely as limiting sulci which develop at the junction of the transversely arranged cortical areas. As already noted, this patterned arrangement of sulci provides rather a striking contrast with the common lemuroid type of pattern in which the sulci tend to be disposed longitudinally, cutting across the cortical areas. The temporal lobe is marked by a long parallel (superior temporal) sulcus which forms the lower boundary of the superior temporal convolution and frequently joins the Sylvian sulcus at its posterior end. On the medial surface of the occipital lobe there is constantly a deep calcarine sulcus, producing an axial folding through the visual cortex.

The fissural pattern of the larger Platyrrhine monkeys is re-markably similar to that of the Catarrhine monkeys (fig. 124), and yet (judging from the constancy of certain differences in the skull and dentition, etc.) it is to be presumed that these two groups be-came segregated quite early in their evolutionary development, possibly even at the prosimian level. It is at least reasonable to sup-pose that the brain of the common ancestor of the Old and New World monkeys could hardly have been less primitive than that of the marmosets, and it may be inferred, therefore, that the close identity of the fissural pattern in Platyrrhine and Catarrhine brains is the result of evolutionary parallelism. Such a degree of parallelism in convolutional pattern is by no means exceptional in mammalian evolution—it has occurred, for example, in the Perissodactyla. It means, merely, that in brains having a common plan of construc-tion (based ultimately on a common heritage) the cortex during development tends to become folded in conformity with similar growth processes.

The thalamus in the Anthropoidea is considerably more ad-vanced in its development than in lower Primates, and the expan-sion of its dorsal elements has thrust back the posterior part to form a prominent rounded projection, the pulvinar. The pulvinar has thus come to be regarded as a characteristic feature of the brain of the higher Primates; in fact, however, the nuclear element of which it is mainly composed is quite large in lower Primates (including the tree-shrews), but it is not in them extruded into such a conspicuous elevation. The lateral geniculate nucleus of all Anthro-poidea is characterized by the sharp definition of its constituent laminae, and by the fact that, as already noted, they are everted to varying degrees so as to form a ventrally directed concavity or 'hilum' at which the fibres of the optic tract enter, and a dorsal convexity from which emerges the optic radiation (see fig. 113C). Since the processes of inversion and eversion of the geniculate nuclei in the Prosimii and Anthropoidea have developed in opposite directions, it may be suggested that the nuclei have been derived from a simple type (such as is seen in the brain of the tree-shrew) in which the laminae remain still unfolded. By itself, of course, these contrasting features do not compel the inference of a diphyletic origin for the Prosimii and Anthropoidea from such a primitive ancestry, but it does indicate that this is a reasonable possibility.

The cerebellum in the larger monkeys is of considerable size and very intricately fissured. The increase in size is mainly determined by the expansion of the lateral lobes, or cerebellar hemispheres, and the latter become elaborated by the development of secondary lobules. The floccular lobe remains distinct and, as in lower mammals generally, it is lodged in a deep excavation in petrous bone—the subarcuate fossa. Since the lateral lobes of the cerebellum have a two-way fibre connection with the neopallial cortex, and in functional co-operation with the latter play an important part in the control of the voluntary musculature, it will be realized that they tend to expand *pari passu* with the expansion of the neopallium. Thus, they attain their maximum elaboration in the human brain.

The size of the brain and the degree of complexity of its convolutional pattern are related *inter alia* to body weight. But while an increase of body weight by itself leads to an increase of absolute brain weight, it results in a decrease of relative brain weight. In order to get some indication of the relation of increase in brain size to factors other than body weight (e.g. degrees of intelligence), a cephalization coefficient has been devised which is expressed by the formula $k = \dfrac{E}{S^r}$, where E is the brain weight, S the body weight, and r a constant which has been calculated to be 0·65. By the application of this cephalization coefficient, it appears that the greater size of the brain in the large anthropoid apes is mainly due to the factor of body weight and thus is not an indication of enhanced intelligence. However, in the complexity of the convolutional pattern even the small gibbon shows an advance on the Catarrhine monkeys, particularly in the fissuration of the parietal lobe. In the large apes, and to the greatest extent in the gorilla, the convolutional pattern closely resembles—though in a somewhat simplified form—that of the human brain. The differences are mainly related to the relatively greater expansion of the association areas in the human brain. One of the results of the expansion of the parietal region of the cortex in *H. sapiens* is that the visual area becomes pushed back to such an extent on to the medial aspect of the cerebral hemisphere that only a small portion of it normally appears on the lateral aspect. Associated with this displacement of the visual cortex is the almost complete disappearance of the 'simian sulcus' except for a small and inconspicuous furrow which can usually be identified only with difficulty. In some individuals, however, a much more primitive condition is retained, and the 'simian sulcus' in such occasional specimens may be quite a conspicuous fissure (see fig. 126).

The basic convolutional pattern of the monkey brain is quite evident in that of the anthropoid apes, even though it may to some extent be obscured by the development of secondary sulci. In the human brain, the latter are far more numerous and, because of their high variability in different individuals, they tend still more to obscure the pattern of the main, or primary, sulci. For reference, it may be noted that, according to Schultz, in male gorillas the cranial capacity ranges from 473 c.c. to 752 c.c. [133]. The greater complexity of the convolutional pattern of the modern human brain, as compared with the apes, is related to its much larger size relative to the body weight; the cephalization coefficient is actually about three times that of a chimpanzee. This increase is presumably related in some way to human intelligence, but it is important to recognize that such a relation, so far as it exists, is evidently very indirect. The cranial capacity in modern man shows an astonishing range of variation, from less than 900 c.c. to about 2,250 c.c., and yet (contrary to popular assumption) there is no very obvious correlation in apparently normal individuals between cranial capacity and intellectual ability; at any rate, it has not been possible to establish any marked correlation by statistical studies. But it may well be that the modern human brain differs from that of the apes in more than gross volume, and there is indeed some evidence that such differences exist in the cell density of the cerebral cortex. For example, quantitative studies have led to the conclusion that, in the visual cortex, the grey/cell coefficient (i.e. the

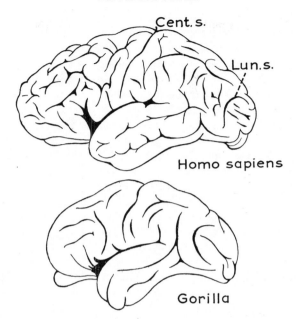

FIG. 126. Brain of a gorilla compared with an exceptionally small and primitive brain of *Homo sapiens*. (Adapted from G. Elliot Smith, *Essays on the Evolution of Man*, 1927.)

relation of the volume of grey matter to the volume of its contained nerve cells) is higher by about 50 per cent in man than it is in the chimpanzee, and it has been surmised that this reflects a greater richness of fibre connections in the cortex of the human brain [48]. In any case, however, it is true to say that, anatomically speaking, the differences between human and ape brains appear to be purely quantitative, for there is no known neomorphic element which distinguishes the former.

It is well to bear in mind these references to brain size and quantitative differences in cell density when considering the evidence provided by endocranial casts of fossil hominids. For example, the mean cranial capacity of the Middle Pleistocene species *H. erectus* was only about 1,000 c.c., that is to say, not much more than two-thirds the cranial capacity of *H. sapiens*, but it would not be legitimate to infer therefrom that individuals of this genus were correspondingly unintelligent. In fact, the archaeological evidence of their living sites in China makes it clear that they were skilful hunters, had a considerable facility in the fabrication of stone implements, and also knew how to make fire for culinary purposes. Yet, as the comparative illustrations in fig. 127 show, the brain of *H. erectus* in its absolute size, and also in its general proportions, occupies a place approximately intermediate between *H. sapiens* and the modern large apes.*

In the Early Pleistocene *Australopithecus* the cranial capacity was hardly bigger than that of a large gorilla; so far as is known from the available (mostly incomplete) specimens its maximum size probably did not greatly exceed 700 c.c. But since the limb and trunk skeleton indicates that the australopithecines were rather small in stature and, in some varieties, of slight build, it is probable that the *relative* size of the brain was somewhat greater. It is, of course, not possible to assess the level of intelligence from endocranial casts of *Australopithecus* for, apart from the questionable significance of the actual size of the brain, we have no means of knowing details of its intrinsic organization such as might be expressed in the grey/cell coefficient. Because of the small absolute size of the brain, it was at one time assumed by some anatomists that *Australopithecus* should be properly regarded as an 'ape' in the taxonomic sense, the argument being that 'hominids' are characterized by a large brain and therefore *Australopithecus* could not be a hominid. But this is to confuse the taxonomic term Hominidae (or its adjectival form, hominid) with *Homo*. There is a general consensus of opinion that the divergence of the Hominidae from the Pongidae as a separate line of evolution probably dates back to the Miocene period, perhaps about twenty-five million years ago, and almost certainly not later than the earliest part of the Pliocene, over ten million years ago. On the other hand, there is no evidence from the palaeontological record that the brain began to expand to modern dimensions before the latter part of the Early Pleistocene, i.e. less than a million years ago, and then it subsequently expanded very rapidly. It follows, therefore, that for a matter of at least several million years the earlier representatives of the hominid sequence of evolution continued to retain a brain of simian dimensions. Thus, while a large brain may be taken as a

---

* Attempts have been made to delineate the details of the convolutional pattern of *H. erectus* and other extinct types of hominid from endocranial casts, but the identification of individual sulci and gyri by such means is far too insecure to permit of positive statements or conclusive inferences.

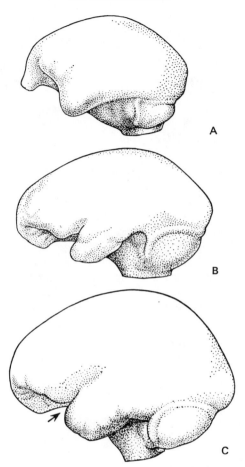

FIG. 127. Lateral view of the endocranial casts of (A) a male gorilla; (B) *Homo erectus*; and (C) *Homo sapiens*. The arrow points to the Sylvian notch ($\times \frac{1}{3}$). (From *The Fossil Evidence of Human Evolution*, Chicago Univ. Press, 1955.)

diagnostic character of the genus *Homo*, it may not be taken as a criterion of the Hominidae as a whole.*

The early evolutionary history of the brain of the Anthropoidea

---

* A close parallel is provided by the Equidae. In the earliest known representative of this family, the Eocene genus *Hyracotherium*, the brain was but a small fraction of the size of the brain of the modern *Equus*, and in the equid sequence of evolution the brain did not begin to expand towards modern dimensions before the end of the Miocene.

is but scantily known. Endocranial casts of *Mesopithecus* and *Liby-pithecus* show that in these Pliocene genera the brain was just as well developed as it is in the modern cercopithecoids, and the convolutional pattern of the cerebral cortex was also closely comparable. On the other hand, a partial endocranial cast of the extinct ape, *Dryopithecus* (*Proconsul*), is of particular interest because it indicates that the brain of these Early Miocene pongids was still primitive by comparison with the modern large apes. For example, the convolutional pattern is cercopithecoid rather than pongid in its simplicity, and also in the disposition of some of the sulci. Further, the petrous bone contains a large subarcuate fossa which must have accommodated a conspicuous floccular lobe of the cerebellum as in monkeys generally. This fossa has disappeared in the modern large apes, though it may be present in some species of gibbon. So far as this meagre evidence goes, it seems likely that the pongid type of brain had by no means completed its evolutionary development by the Early Miocene; such a conclusion conforms well with the evidence of the skull and limb bones that the Miocene apes still retained a number of anatomical characters indicative of a transition from a cercopithecoid phase of development.

7

The Evidence of
the Special Senses

AMONG the features wherein the Primates stand in contrast with other mammalian orders are the elaboration and refinement of those special sense organs and their neural centres in the brain which together provide the basis for a high degree of sensory discrimination. It is particularly by the enhancement and differentiation of the neural centres themselves that the Primates acquire the ability to define with much more precision the varying influences of their environment, and also the capacity for reacting with much greater accuracy thereto. Further, the discussion in the previous chapter will have made it evident that the progressive development and elaboration of these sensory centres have conditioned the evolutionary expansion of the brain as a whole.

The visual sense is the most informative of the discriminative senses, for it provides a means whereby objects may be recognized near at hand or in the far distance in regard to their position, form, texture and colour, and enables spatial properties to be defined with an accuracy which is hardly approached by the use of other sensory mechanisms. When the visual sense is developed in conjunction with the development of the tactile pads characteristic of the prehensile extremities in Primates generally, the combination of the two provides opportunities for exploring objects of the immediate environment, and for comprehending their significance, which is not possible in non-Primate mammals. In contrast to the visual sense, the sense of smell provides little information other than the quality and intensity of the odour of an object—it gives no information about its other qualities, or even (to more than a very slight degree) its position in relation to the observer. With the increasing perfection of the visual sense, therefore, and also because the sense of smell does not serve the immediate practical use in arboreal life for which it is required in terrestrial life, the olfactory organ and its neural centres undergo a progressive atrophy in the evolution of Primates.

THE OLFACTORY SENSE

In most lower mammals the external nares or nostrils are surrounded by an area of naked and moist glandular skin which constitutes the rhinarium. Possibly this moist surface aids the purely olfactory sense by enabling the animal to recognize the direction of air-currents which carry an odour from a distance. Primitively, the naked rhinarium is prolonged down from the nostrils over the median part of the upper lip to become continuous with a strong fold of mucous membrane which binds the lip firmly to the underlying gum. This condition is shown in fig. 128A. During embryological development a number of processes grow forward and downward from the head region and, by their ultimate fusion, construct the whole of the face. Thus, the median nasal process which occupies the midline forms the septum of the nose and the labial portion of the rhinarium, while on either side the maxillary processes (represented in fig. 128 by the stippled area) extend towards the midline to meet the margins of the median process; they form the cheek region and the hairy lateral parts of the upper

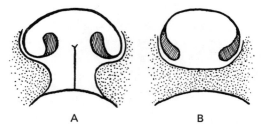

A   B

FIG. 128. Diagram illustrating the construction of the external nose and upper lip in (A), the tree-shrews and lemurs, and (B), *Tarsius* and the Anthropoidea. (Adapted from J. D. Boyd, *J. Anat.* 1932.) The maxillary components are represented by the stippled areas.

lip. In the tree-shrews and lemurs this primitive mammalian type of rhinarium persists. That is to say, the maxillary processes do not meet and fuse in the midline, and the median nasal process is therefore left exposed in the middle of the upper lip to form the labial part of the rhinarium. In some lemuroids (e.g. *Galago*) the maxillary processes make contact with each other, and in this case the labial portion of the rhinarium is in a deep groove and the whole of the superficial part of the upper lip is covered with hairy skin.

In *Tarsius* and in all the Anthropoidea the construction of the external nose is different (figs. 128B and 129). The labial part of the rhinarium is no longer present, for in the process of development the maxillary processes effect a fusion in front of the median nasal process so that the latter is completely buried below the surface. Thus the upper lip shows a continuous smooth surface uninterrupted by a median groove. With these changes are associated (1) a reduction of the elements of the upper lip formed from the median nasal process, including an attenuation of the fold of mucous membrane whereby the lip is attached to the gum, and (2) the muscularization of the whole of the upper lip across the midline, the muscular tissue having been carried to this position by the extension of the maxillary processes. These two features are

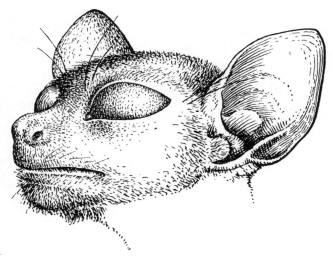

FIG. 129. Face of *Tarsius* to illustrate the pithecoid appearance of the nose and upper lip. Note also the large ears and the facial vibrissae. (After H. H. Woollard, *Proc. Zool. Soc.* 1925.)

related to the increased freedom and mobility of the upper lip which characterize all the higher Primates and which permit a degree of facial expression which is not possible in the lower Primates. Incidentally, it seems clear that the versatility of facial expression in the higher Primates plays a highly significant role as a means of communication between members of family or social groups.

In *Tarsius* and the Anthropoidea, the area of naked moist skin around the nostrils has entirely disappeared, presumably an outward manifestation of the retrogression of the olfactory apparatus as a whole in these Primates. Herein they show a strong contrast with the lemurs and tree-shrews, and the relatively primitive status of these latter is emphasized.

In the monkeys the disposition of the nostrils provides the basis of the distinction between 'Platyrrhine' and 'Catarrhine'. In the Platyrrhine monkeys the nostrils are relatively wide apart and separated by a broad septum, while in the Catarrhines the septum is relatively narrow (fig. 130). This distinction between the New and Old World monkeys is not a sharp one, however, for in some of the former (e.g. *Ateles*) the nostrils are fairly close together, while

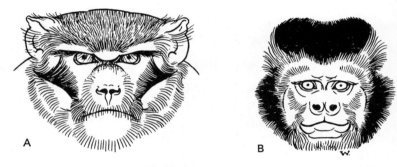

FIG. 130. (A) Catarrhine monkey (*Macaca*) and (B) Platyrrhine monkey (*Cebus*), showing the differences in the disposition of the nostrils. (By courtesy of the British Museum.)

in the latter (e.g. *Colobus*) they may be almost as wide apart as in a typical Platyrrhine. These superficial differences in the disposition of the nostrils are not obviously reflected in the nasal skeleton.

The differences in the external proportions of the snout region of the skull in the Primates are of course associated with corresponding differences in the dimensions of the nasal cavity. In the tree-shrews and lemurs the cavity is relatively large, and it is greatly complicated by the labyrinthine system of turbinal processes (conchae) which form a series of scroll-like structures projecting into the cavity from the lateral walls and roof. These processes are folded on themselves and, being covered on their surface by mucous membrane much of which is lined with olfactory sensory epithelium, they provide for a considerable extension of the latter. In general, therefore, there is a correlation between the degree of complexity of the turbinal system and the acuity of smell. A well-developed system may be regarded as a primitive mammalian character, and in lowly mammals the processes are relatively numerous (in some there may be as many as nine, e.g. *Orycteropus*).

In fig. 131 is shown the lateral wall of the right side of the nasal cavity in the pen-tailed tree-shrew. Most of the turbinal processes are laminar extensions of the ethmoid bone and are called ethmoturbinals. There are four of these, which are numbered from front to back. Along the roof is a process, the nasoturbinal, which extends

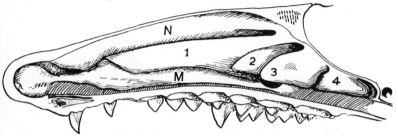

FIG. 131. Lateral wall of the nasal cavity of the pen-tailed tree-shrew. The ethmoturbinals are numbered 1-4. *N*, nasoturbinal; *M*, maxilloturbinal (×3). (*Proc. Zool. Soc.* 1934.)

along the under surface of the nasal bone, and adjacent to the floor of the nasal cavity is the maxilloturbinal which is a separate skeletal element having an independent centre of ossification. In many lower mammals (e.g. rodents, carnivores and ungulates), the maxilloturbinal is highly complex, being folded into an intricate maze of secondary convoluted scrolls. The schematic diagram in fig. 132 represents the appearance of an oblique section through the nasal cavity immediately below and parallel with the cribiform plate of the ethmoid which forms the roof of the cavity in its posterior part. This diagram shows the general arrangement of the nasoturbinal and ethmoturbinals in the tree-shrew. Laterally, and overhung by the main ethmoturbinals (or endoturbinals), are seen two small accessory scrolls—the ectoturbinals.

In the lemurs the turbinal system is similar to that of the tree-shrews; that is to say, besides the nasoturbinal and maxilloturbinal, there are four endoturbinals and two ectoturbinals. So far as is known, the only exception to this is the aberrant genus *Daubentonia*, which retains five endoturbinals and at least three ectoturbinals. The fact that the turbinal system is less reduced in *Daubentonia* than in the other lemuriforms is indicative of its more primitive status and is correlated with its relatively larger olfactory bulb.

The Lemuriformes and Lorisiformes show one major distinc-

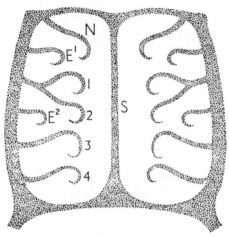

FIG. 132. Schematic diagram of a transverse section through the posterior part of the nasal cavity of a tree-shrew or lemur immediately below the cribiform plate of the ethmoid, to illustrate the components of the system of turbinal processes. S, nasal septum; N, nasoturbinal; E^1, E^2, ectoturbinals; 1, 2, 3, 4, endoturbinals.

tion in the arrangement of the turbinal system; in the latter the first ethmoturbinal is excessively large and extends down to overlap the maxilloturbinal. The functional significance of this feature is obscure, but it appears to be a lorisiform specialization contrasting with the more generalized condition in the lemuriforms.

In the higher Primates the turbinal system has undergone considerable reduction. In all the Anthropoidea (even in the more primitive types such as the marmoset) the diminution of the snout region, associated with the recession of the jaws, has led to a marked restriction of the nasal cavity (fig. 133). The maxilloturbinal is small and quite simple in its structure, and only two endoturbinals remain (corresponding to the superior and middle conchae of human anatomy). They also become reorientated so that they are disposed in a vertical rather than an antero-posterior

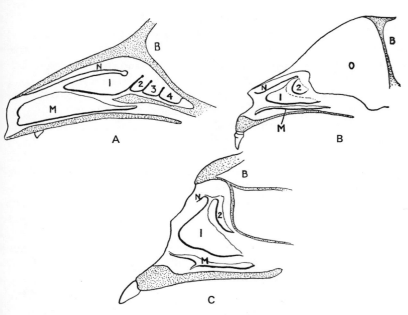

FIG. 133. Lateral view of the nasal cavity in (A) *Lemur*, (B) *Tarsius*, and (C) *Hapale*. *B*, brain cavity; *M*, maxilloturbinal; *N*, nasoturbinal; 1, 2, 3, 4, ethmoturbinals.

series. The nasoturbinal has shrunk to an inconspicuous ridge—the agger nasi—and only one ectoturbinal remains (as the bulla ethmoidalis) under cover of the first endoturbinal. That the ancestral stock from which the Anthropoidea were derived had a more elaborate turbinal system may be inferred from general considerations of comparative anatomy, and it is interesting in this connection to note that even in modern man one or two additional rudimentary turbinal processes may occasionally be found near the roof of the nasal cavity, while in the human embryo four endoturbinals can quite often be recognized.

In *Tarsius* the turbinal system is very similar to that of the Anthropoidea, for in addition to the small and attenuated maxilloturbinal there are only two endoturbinals. The nasoturbinal is slender in comparison with lemurs, but more conspicuous than the agger nasi of monkeys. The reduction of the turbinal system in *Tarsius* is clearly secondary to the restriction of the nasal cavity, but it seems evident that this restriction has come about in a manner different from the analogous changes in the Anthropoidea. Whereas in the latter it is the result of the retraction of the nasal cavity associated with the recession backwards of the jaws relative to the brain-case, in *Tarsius* it is due to the enormous development of the orbits, with the formation of a very thin interorbital septum, which has led to the obliteration of a large part of the nasal cavity by lateral compression. Hence the restricted nasal cavity is not placed below the front end of the brain-case as in monkeys—on the contrary, it has been displaced relatively far forward in front of it. These different relationships are shown in fig. 133B and c. It seems probable, therefore, that the close resemblance of the turbinal system in *Tarsius* and the Anthropoidea may be the expression of convergent evolution, the result of a limitation of space in the nasal cavity which has been conditioned by quite different factors.

So far as can be inferred from the size and external appearance of the nasal region of the skull, the olfactory apparatus of the fossil lemuroids of the Eocene was better developed than in their modern successors, and this inference is supported by the fact that the olfactory bulbs were relatively larger. By contrast, the retrogressive changes which have affected the olfactory apparatus in Primates generally seem to have proceeded more rapidly in the evolution of the tarsioids, for in some of the Eocene genera (particularly *Pseudoloris*) the nasal cavity was evidently as restricted as it is in the modern *Tarsius*, and in *Necrolemur* and *Tetonius* the olfactory bulbs were very small (see p. 253).

THE EYE

We have already emphasized that all the modern Primates are characterized by a high degree of elaboration of the visual centres of the brain, and that the increasing dominance of the visual sense played a significant part in the evolution of the group as a whole. The information supplied to the visual centres depends ultimately on the organization of the sensory membrane of the eye, the retina. This is a complex structure composed of the sensory elements (photoreceptors) which are sensitive to light, and layers of nerve cells and their processes through which the impulses initiated by the light stimuli are conveyed to the optic nerve and tract and so to the visual centres of the brain.

In all mammals except among those of nocturnal habits there are two kinds of photoreceptor in the retina, rods and cones. The rods function at low degrees of luminosity and thus have a high sensitivity, but they permit only low powers of discrimination. They come into action in what has been termed 'twilight' or 'scotopic' vision, and they are particularly responsive to movements of objects in the peripheral part of the field of vision. The cones, on the other hand, are sensitive to a higher intensity of light, and provide the basis for 'photopic' vision which allows a high degree of fine discrimination of spatial relationships and also an appreciation of colour and texture. In most mammals, cones predominate towards the centre of the retina and rods towards the periphery. In nocturnal mammals the cones tend to disappear entirely, leaving a pure 'rod retina'.

In the Anthropoidea generally, as in the human eye, the central area of the retina shows a local differentiation—the macula lutea (yellow spot), and here there are practically no rods at all. In the middle of the macula is a small depression, the fovea, where all, or most of, the layers of the retina are absent except a single layer of cones. The foveal region of the retina is also devoid of bloodvessels. As a result of this local excavation, the light is brought in direct access to the cones without previously passing through other layers of the retina and without interference from retinal bloodvessels. This, then, is the point in the retina of greatest visual acuity.

In lower Primates the retinal structure shows variations in relation to diurnal or nocturnal habits. In the tree-shrews, for example, *Tupaia* has a retina rich in cones in contrast to *Ptilocercus* (a crepuscular animal) in which the sensory elements appear to be entirely composed of rods. The nocturnal lemurs (e.g. *Loris*, *Nycticebus* and *Cheirogaleus*) have the rod type of retina, but in species of the genus *Lemur* (contrary to previous statements) there is quite a rich complement of cones—as indeed might be expected from the observation that these creatures are active during the daylight. All lemurs differ from the monkeys in having no true fovea, but the arrangement of the retinal bloodvessels suggests an 'incipient' avascularity of the central area, the vessels here being fewer and finer than in more peripheral regions [117]. It is interesting to note that a relatively avascular central area of similar appearance is also present in *Tupaia*. In the lemurs and tree-shrews, as in mammals generally,

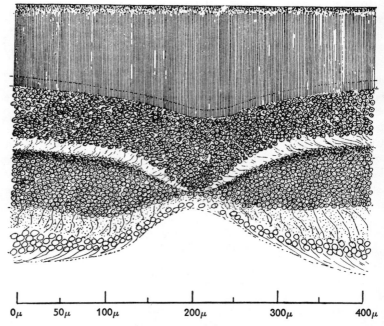

FIG. 134. Section through the retinal fovea of *Tarsius*. (After S. Polyak, *The Vertebrate Visual System*, Chicago University Press, 1957, fig. 512.)

the distribution of the photoreceptors shows a gradient of density which increases towards the central area of the retina, making possible a greater acuity of vision in this region. The development of a fovea in the Anthropoidea is a further refinement which greatly amplifies this acuity.

The differentiation of a retinal macula is not confined to the Anthropoidea for it is also quite well developed in *Tarsius* (although previously reported to be absent in this prosimian). Polyak [117] has reported that ophthalmoscopic examination shows a large yellow-pigmented macula with a small fovea, and he has also figured and described in detail the histological appearance of the fovea in section (fig. 134). This is an unexpected observation because the retina is of the nocturnal type, the photoreceptors being apparently entirely composed of rods. Since a fovea is assumed to be essentially a device in the higher Primates for accentuating the visual discriminatory functions served by cones, it is difficult to explain why there should be one at all in a rod retina. But a similar condition is found in one other Primate, the Platyrrhine monkey *Aotus*, which is also nocturnal and has a rod retina. It may perhaps be inferrred that in both these Primates the fovea is really a vestigial structure which has persisted, after the genus became adapted for nocturnal habits, from predecessors of diurnal habits with a cone retina and the normal type of fovea. Certainly this seems the most probable explanation for the fovea in *Aotus*, for it is the only genus among the Anthropoidea which has a pure rod retina and this is clearly a secondary adaptation. The same explanation may also apply to *Tarsius*, the assumption being that the ancestral tarsioids from which it was derived were diurnal creatures with a cone retina. On the other hand, in so far as the relative size of the orbits can be accepted as a reliable indication, it is likely that at least some of the fossil tarsioids of which the skull is known (e.g. *Pseudoloris* and *Tetonius*) were nocturnal. In any case it now seems clear that the foveal differentiation of the retina first appeared at the tarsioid level of Primate evolution, and is thus phylogenetically much more ancient than has commonly been supposed.

In the monkeys and apes the structure of the retina shows certain variations in histological details, but there is no evidence of a consistent, progressive refinement in the organization of its sensory elements from the 'lower' to the 'higher' levels of the Anthropoidea [176]. In the marmoset, however, the foveal excavation is not so complete as in the other monkeys; the outer nuclear layer of the retina (i.e. the layer of nuclei of the receptor cells) stretches across it so that the photoreceptors are not directly exposed to light stimuli. In this feature, the structure of the fovea is very similar to that of *Tarsius*. On the other hand, in some monkeys (e.g. *Cercocebus*) the differentiation of the macular region—assessed by the density of photoreceptors, the proportion of cones to rods and the perfection of the foveal excavation—even surpasses that of the human retina.

It is a characteristic of the Primates that the optical axes of the eye tend to become less divergent in the more advanced types, though among the prosimians the degree to which this has occurred shows considerable variation. In the Anthropoidea the eyes undergo a rotation during embryological development, so that they come to look directly forward with the optic axes parallel. This position of the eyes—which is reflected in the plane of the orbital apertures of the skull—provides an essential basis for stereoscopic vision. It allows an almost complete overlap of the fields of vision of both eyes, and enables a point in the visual field to be focused on corresponding points of both retinae simultaneously. Herein the Anthropoidea stand in contrast with many lower mammals in which the eyes are placed laterally with the optic axes diverging to the extent that the visual fields do not overlap to more than a slight degree; each eye thus receives a different image. The rotation forwards of the eyes is correlated with the recession of the snout region; hence in the lower Primates in which this recession has not proceeded so far as in the monkeys the plane of the orbital aperture is directed more or less laterally as well as forwards. The importance of the development of the macula and of full stereoscopic vision in the evolution of the higher Primates can hardly be overestimated. These visual factors were particularly emphasized by Elliot Smith in one of his essays on the evolution of man [27], where he remarks that the development of the macula lutea made possible a fuller appreciation of the details, the texture and the colour of objects seen, enabling the eyes to follow the outlines of objects and appreciate better their exact size, shape, and position in space, while stereoscopic vision not only enhances still further these discriminatory functions but facilitates the increasing perfection of tactile skill and manual dexterity by conferring more accurate control on the movements of the hand and fingers. It is worth while emphasizing one of the implications of Elliot Smith's studies, that the progressive advancement in the later evolution of the Primates is conditioned, not by the perfection of the visual apparatus alone, but by this in association with the retention of prehensile digits and the development on the latter of tactile pads of high sensitivity.

Broadly speaking, the retinal structure in the Catarrhine monkeys and apes is quite similar to that of the human eye, except for minor variations in different genera. As we have noted, also, it shows no gradation of increasing differentiation in an ascending series of the Anthropoidea. Yet there can be no doubt that man is capable of powers of visual discrimination which vastly exceed those of other Primates. This serves to emphasize the general principle that the progressive refinement of functions of sensory discrimination in general, which is one of the outstanding features of the evolution of the Primates, is predominantly effected not (as far as we have evidence) by an increasing elaboration and differentiation of the peripheral receptor organs themselves, but by the sensory centres of the central nervous system which make possible a more complete analysis of gradations of stimuli to which the receptors themseves are *already* differentially susceptible, and which also permit a more effective and accurate synthesis of the sensory impulses at the higher functional levels of the brain. The preliminary analysis at lower levels is determined by sorting mechanisms whereby nerve fibres conveying information of different spatial or qualitative import are sorted and distributed to localized groups or layers of nerve cells, and the final synthesis is effected by complex fibre systems which interconnect different sensory areas at the cortical level. In this connexion, it is interesting to note that *Tupaia*, as in Primates generally, is capable of a considerable degree of colour discrimination.

Stereoscopic vision depends not only on the position of the eyes, it also depends on the sorting of optic nerve fibres whereby those conveying impulses related to corresponding points in the visual fields of the two eyes are carried to the same region of the brain. Hence, in man and the higher Primates the optic nerve fibres from the two eyes do not undergo a complete, or almost complete, decussation at the optic chiasma as they commonly do in lower vertebrates; a large proportion (about 40 per cent) remain uncrossed and become relayed to the ipsilateral cerebral hemisphere where they end in close relation to the crossed fibres which come from corresponding sectors of the opposite retina. In other words the incomplete decussation at the optic chiasma makes possible the representation of functionally equivalent parts of the retina of both eyes in the same part of the brain—an essential requirement for stereoscopic vision. The proportion of uncrossed optic fibres in most of the lower Primates is not certainly known from quantitative data. Only a very small proportion remain uncrossed in *Tupaia*, and an inspection of stained sections of the brain of *Tarsius* indicates that in this genus they are comparatively few. On the other hand, in some of the lemurs the proportion is relatively high— perhaps as high as in the monkeys.

The retinal fibres are finally conveyed by way of the optic tract to the lower visual centres of the brain, of which the lateral geniculate nucleus is particularly important because it serves to sort the retinal impulses and relay them to the visual cortex. We have already made reference to the laminated structure of the nucleus in Primates (p. 244). It is also a characteristic of some non-Primate mammals; in the cat, for example, it has three layers of which the outer two receive optic fibres from the contralateral eye, and the middle one from the ipsilateral eye. In all but some of the lower Primates the lamination is more complex, and in many genera there are six layers of which three receive crossed fibres and three uncrossed. The *degree* of laminar differentiation, as assessed by the sharpness of their definition and their individual compactness, is on the whole more highly developed in the Catarrhine monkeys and anthropoid apes. In some of the monkeys, indeed, the structural differentiation even exceeds that of the human geniculate nucleus, and in some specimens of the macaque monkey the laminae in the central region of the nucleus may be multiplied to eight. In the Platyrrhine monkeys the lamination is variable; in some genera there are six laminae, in others only four. In *Tarsius* no more than three laminae are present, as in some non-Primate mammals, a simplicity which seems to be out of conformity with the foveal differentiation of the retina and the unusually elaborate structure of the visual cortex. Among the lemurs the lamination of the geniculate nucleus, though complex by reason of the wide variation in the size of the cells which constitute the several laminae, is less sharply defined than in the Anthropoidea. The number of laminae in lemurs is also variable; in some genera there are six as in the monkeys, but it is interesting to note that they are arranged in a different serial order in their relationship to the two eyes. For whereas in all the Anthropoidea (so far as is known) the arrangement is such that the second, third and fifth laminae receive ipsilateral retinal impulses and the first, fourth and sixth contralateral, in those lemurs with six laminae the corresponding relationship is second, third and fourth on the one side, and first, fifth and sixth on the other. It is evident, therefore, that the intrinsic organization of the lemuroid geniculate nucleus differs from that of the Anthropoidea in more than its character of 'inversion' (see p. 244). In the tree-shrews the serial order of the cell laminae is again different, for while in them there are five sharply demarcated layers and a sixth rather indistinct one, contralateral optic nerve fibres end in the first, third, fourth and fifth, and ipsilateral in the second and sixth.

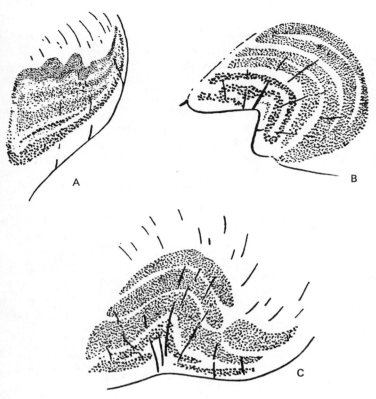

FIG. 135. The appearance in transverse section of the lateral geniculate nucleus of (A) *Lemur*; (B) *Cercopithecus*; (C) man, showing the arrangement of the cell laminae.

The functional implication of the six-layered type of geniculate nucleus, three being related to each eye, is not known. It is perhaps significant that the six layers are only present in that part of the nucleus related to central vision—that is, to the central area of the retina which serves a high degree of visual acuity and also full colour vision. The proposition has naturally been advanced, therefore, that the triple representation of the central area of each retina in the geniculate nucleus may provide the anatomical substratum for the trichromatic hypothesis of colour vision, but no direct experimental evidence has yet been obtained to substantiate this interesting conjecture.

We have remarked that the structural differentiation of the retina and the lower visual centres in man does not surpass that of some of the cercopithecoid monkeys or anthropoid apes, and the same may be said of the visual cortex. Although, as we have already recorded (p. 260), some quantitative data have been reported which indicate that there is a difference in the cell density of the human visual cortex as compared with that of the anthropoid apes, and that this difference may be related to richer fibre interconnections in the former, it seems probable that the exceedingly acute powers of visual discrimination in man are predominantly the result of the expansion of the cerebral cortex as a whole—particularly of those areas and fibre systems which interrelate the visual cortex with other sensory receptive areas through the intervening association areas, and with cortical regions such as those of the temporal lobe which are believed *inter alia* to serve functions connected with visual memorization. It is by such elaborations of cortical organization that man is enabled to make fuller use of the variety of retinal impulses conveyed to the visual cortex by way of the optic nerves and tracts and the geniculate nuclei, and to attain a capacity for complex processes of visualization of which the rudiments have been progressively developed during the long course of evolution of his Primate precursors.

THE EAR

The structure and complexity of the essential part of the apparatus of hearing, that is to say the cochlea of the inner ear and the spiral organ of Corti which it contains, show little or no variations in the different genera of Primates which indicate significant differences in auditory acuity (though there may be differences in the length of the cochlea which are related to the range of pitch over which sounds are audible). It might be supposed, from the circumstance that in the higher Primates the external ear is relatively small and appears to have undergone retrogressive changes, that this indicates a lesser degree of auditory acuity than that possessed by lower mammals with large mobile ears. To a certain extent this is no doubt true, but it needs to be recognized that the dimensions of the external ear are not only related to the sense of hearing. The size is in part determined by the use of thin, membranous ears in mammals generally as a temperature regulating mechanism, for

being richly supplied with bloodvessels, they can by vascular dila-
tion in hot environments facilitate the loss of body heat. There is
also some evidence, experimental as well as observational, that in
some animals a direct relationship exists between the surface area
of the external ear and the mean environmental temperature. In
other words, a large ear is not by itself an index of auditory acuity.
It seems probable, for example, that the contrasting size of the
ears in the galagos and lorises, members of the same family, is
related to different requirements for temperature regulation. Both
these groups are nocturnal and inhabit a similar environment of
tropical forest, but the slow-moving lorises have smaller ears while
the actively moving galagos, living presumably at a higher meta-
bolic level, have conspicuously large ears. The contrast in the size

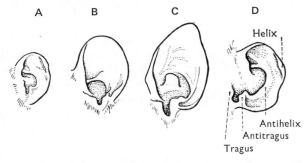

A   B   C   D

FIG. 136. External ear of (A) *Tupaia*; (B) *Cheirogaleus*; (C) *Tarsius*;
(D) *Cebus*. (Approximately natural size.)

of the ears in the two subfamilies of the tree-shrews evidently has
a different significance, for they not only inhabit the same geographical territory but they are both active arboreal creatures. But
Ptilocercus (the only genus of the subfamily Ptilocercinae) is nocturnal in habit while the Tupaiinae are diurnal. It is likely, therefore, that the large mobile ear of *Ptilocercus*, contrasted with the
unusually small ear in the Tupaiinae, actually is related to the
need for a greater auditory acuity when active at night. The external ear of *Tupaia* (fig. 136A) appears to have undergone retrogressive changes leading to a general appearance which bears a
remarkably close resemblance to that of the higher Primates. It is
rounded and fixed closely to the side of the head so that it has very
little mobility. The peripheral margin is incurled to form a marginal ridge corresponding to the helix in the ear of monkeys, apes,

and man. The small flap called the tragus overlaps the auditory aperture in front. Opposite it is the antitragus from which a strong fold curves upwards and then forwards to form the antihelix. Above, the antihelix divides into two terminal ridges, the superior and inferior crura.

In many lemurs the external ear retains a simple construction of the type well illustrated in *Cheirogaleus* (fig. 136B). Here the tragus and antitragus are small and unobtrusive, while the antihelix is continued forward above the auditory aperture as a single horizontal ridge sometimes termed the supratragus. In other genera such as *Microcebus* and *Galago* the ear, besides being much larger, is ribbed into a series of curved ridges and grooves and can be folded down on itself by muscular action. It also shows a minor specialization in the expansion of the supratragus to form a flap-like fold.

In *Tarsius* the external ear is the most specialized of the modern Primates (fig. 136C). It is conspicuously large, the free part is marked by ridges and grooves as in *Galago* and is also capable of folding, the tragus is a prominent eminence, and the supratragus forms a well-defined flap which is larger than that of *Galago* and resembles the equivalent structure in bats. That the auditory sense is highly developed in *Tarsius* may be inferred from the fact that the terminal nuclei of the auditory nerves (the cochlear nuclei) and their connections in the brain are exceptionally large.

In all the Anthropoidea the external ear is relatively small and shows but little variation in the different groups. It is rounded in outline, and its margin is always folded in to some extent above; indeed, in some Platyrrhines (e.g. *Alouatta*) the entire free edge may be inrolled as it commonly is in the human ear. The supratragus is almost always a simple ridge. Among the Old World monkeys part of the upper free margin of the ear may in macaques and baboons be produced into a definite point, a primitive mammalian feature which is usually absent in the Platyrrhines. The inrolling of the margin in the human ear brings about an inversion of this pointed region; but the point itself is often to be recognized as a small eminence (Darwin's tubercle) on the inner edge of the helix. The mobility of the ear in the higher Primates is much restricted, but it is interesting to note that even in man there are quite a number of intrinsic and extrinsic muscles which provide very forcible evidence for the derivation of the Hominidae from an ancestral type in which the ear was freely mobile, as it is in most lower Primates and in mammals generally.

Since the small ossicles of the middle ear, the malleus, incus and stapes, sheltered as they are in the bony chamber of the tympanic cavity, serve in all mammals the same mechanical function of conducting vibrations of the drum to the cochlea, it might be supposed that they would show little variation in their morphology. It might further be supposed that such differences as they do show would be likely to provide valuable evidence of morphological affinity for the very reason that they could hardly be directly attributed to different environmental requirements. In fact, however, we know very little of the mechanical factors which determine that certain morphological differences in the ossicles may be related to the need for detecting particular qualities of sound; sounds of a kind which are unimportant for one genus may be exceedingly important for another.

Our knowledge of the comparative anatomy of the ossicles of the ear in mammals is almost entirely based on a monograph published in 1879 by Doran. According to his observations [22], in *Tupaia* the malleus (in being neckless and devoid of a laminar process) 'differs from that of any other insectivorous mammal' and quite closely resembles that of some of the lower Primates—e.g. certain lemurs and marmosets. This statement is of particular interest because at the time it was made the taxonomic status of the tree-shrews as primitive Primates related to the lemurs was not yet recognized. In the lemurs the ossicles are similar to those of some of the Platyrrhine monkeys, and in this respect *Tarsius* resembles the lemurs. The Catarrhine monkeys differ from the anthropoid apes and man, and approximate to the Platyrrhines, in the straight and little divergent crura of the stapes, while, as regards the Hominoidea, the ossicles of the Pongidae (particularly the stapes) indicate a closer affinity with man than with the monkeys.

TACTILE SENSATION

The skin is everywhere plentifully supplied by plexuses of sensory nerves and is sprinkled with sensory receptors which respond to tactile, pressure, thermal and pain stimuli. But there are certain regions where tactile sensation is particularly acute. In primitive mammals these are the regions to which are attached specialized tactile hairs, or vibrissae. In the tree-shrews and lemuriform lemurs the full complement of vibrissae characteristic of placental mammals in general is retained. Thus on the face there is a mystacial group over the maxillary region, a supraorbital group, a genal group over the zygomatic part of the cheek, a mental group near the point of the chin, an interramal group on the under surface of the lower jaw, and a carpal group situated on the forearm just proximal to the wrist and usually towards the inner side. These vibrissae are implanted by large bulbous roots in small cutaneous tubercles which are richly innervated. The facial vibrissae derive their nerve supply from the fifth cranial (trigeminal) nerve, and it is for this reason that this nerve and the cranial foramina through which its branches pass are so large proportionately in primitive mammals. By making contact with objects in the immediate environment, the sensory hairs convey information of their proximity and are thus of particular advantage for nocturnal animals.

It may be said that, in the evolution of the Primates, the more primitive tactile organs represented by the vibrissae have been gradually replaced by the development of the more delicately informative tactile pads on the terminal phalanges of the digits. These tactile pads were acquired as a secondary result of the transformation of sharp claws into flattened nails, a transformation which (it seems) was primarily related to the need for a more efficient pliability in the grasping functions required for arboreal acrobatics. In all the higher Primates the terminal digital pads are densely strewn with sensory corpuscles and pressure receptors which constitute a tactile discriminatory mechanism of great precision.

While the lemuriforms possess the full set of vibrissae typical of primitive mammals, in the lorisiforms they are much reduced, and on the face they may be represented by little more than the mystacial group; the carpal vibrissae are usually absent. In *Tarsius* the vibrissae are also reduced, though less so than in the lorisiform lemurs; carpal vibrissae have been described in the young animal, but they are absent (or perhaps only difficult to define) in the adult.

Mystacial and supraorbital vibrissae still persist in the New World monkeys, and in the marmosets there is also a carpal papilla to which two or three vibrissae are attached. In the Old World monkeys the facial vibrissae are still less conspicuous, and in the anthropoid apes and man they are no longer present. But it is interesting to note that even in the human foetus a small and transient cutaneous papilla may occasionally be found in the carpal region, bearing witness to the original presence in human ancestry of the carpal vibrissae which characterize the lower Primates.

8

The Evidence of
the Digestive System

IN attempting to assess evolutionary relationships by the comparative study of visceral structures which do not directly or indirectly affect skeletal anatomy, it is well to recognize certain obvious limitations. Such a study can of course only provide information about the terminal products of evolution which survive to-day. It is certainly the case that some groups of modern Primates show distinctive features in their visceral anatomy, and these must evidently be regarded as the end-results of evolutionary trends which have distinguished these groups from other groups. As such, they provide quite important evidence for the consideration of phylogenetic relationships. But since we are not able to demonstrate these trends objectively by reference to the fossil record, we have no means of knowing at what stage in evolutionary diversification they first began to manifest themselves and whether, therefore, they were present in the *earlier* representatives of each group. But, in spite of these limitations, the comparative anatomy of living types does contribute valuable collateral evidence of an indirect nature which is found to conform quite well with the more direct palaeontological evidence of the course of Primate evolution.

THE TONGUE

Since the epithelial covering of the tongue contains scattered in it the peripheral receptors of taste sensation—the taste buds—it might appropriately have been considered in the previous chapter among the organs of special sense. But the tongue also serves other important functions such as those associated with deglutition and with the toilet of the teeth, and these functions may be reflected conspicuously in its structure.

As in mammals generally, the tongue in all Primates is covered on its upper surface by a variety of papillae which give to it a roughened texture. The most numerous and smallest of these are the conical papillae which are closely packed to form a fine velvety pile. They serve a purely mechanical function in the trituration of the food, and in some types of mammal, e.g. carnivores, they are strongly developed as recurved thorny processes which aid in grasping food which has been brought into the mouth. Sprinkled among the conical papillae, and usually more numerous towards the tip and sides of the tongue, are the rounded, knob-like fungiform papillae, most of which contain a number of taste buds. The greatest density of taste buds, however, is concentrated around the large vallate papillae which are situated at the back of the tongue. In the smaller Primates there are commonly three of these vallate papillae disposed in a triangular arrangement (fig. 137), but their number is partly related to the size of the tongue and there may

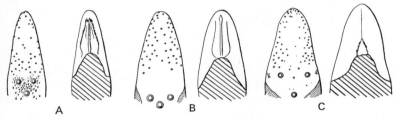

FIG. 137. Upper and lower aspects of the tongue of (A) *Galago* (× 2), (B) *Tarsius* (× 2), and (C), *Callitrix* (× 1½). The fungiform papillae are represented by the small circles, and the vallate papillae by the larger concentric circles below. The outline of the sublingua is indicated on the under surface of the tongue.

be as many as nine in some of the larger genera. It is a characteristic feature of the tongue of the lemurs that its dorsal surface behind the vallate papillae (i.e. the pharyngeal surface) is thickly covered with thorny conical papillae; they are most obtrusive in the lemuriform lemurs but they are also present in the lorisiforms, extending down towards the epiglottis. They represent a specialization peculiar to the lemurs among the Primates, which recalls strongly the appearance of this part of the tongue in carnivores (fig. 138).

On the under surface of the tongue in tree-shrews and lemurs, attached along its median axis but free at its tip and lateral margins, is a remarkably well developed sublingua. This structure has been regarded by some authorities as an essentially primitive formation, the morphological equivalent of the tongue of submammalian vertebrates, and it is commonly present in marsupials and also in certain

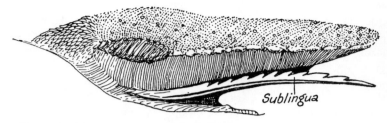

FIG. 138. Tongue of *Lemur*, showing the well developed sublingua, and the large conical papillae covering the pharyngeal surface. (After F. Wood-Jones, *J. Anat.* 1918.)

other lower mammals. In the lemurs, however, it has attained a high degree of specialization, reaching towards the tip of the tongue and forming a thin flexible plate of a horny consistency with a median axial ridge, the lytta, on its under surface. The apex of the sublingua is always markedly denticulated, and the lateral margins are also finely serrated. This pronounced development of the sublingua with its apical denticles and its free mobility is apparently associated with the peculiar specialization of the lower incisors and canines. It has already been pointed out that these fine, styliform teeth are disposed in a closely set pectinate arrangement and are, in fact, actually used for combing the fur. Observations on the living animal have made it clear that they are kept clean by the denticulated sublingua, which thus functions as a toothbrush. In the aye-aye the typical lemuroid modification of the front teeth is not present; in this animal, therefore, the sublingua is not denticulated, but the tip of the lytta is produced into a hook-like structure which is evidently adapted for cleaning the interval between the rodent-like lower incisors.

In the tree-shrews (particularly some of the Tupaiinae) the sublingua also extends towards the tip of the tongue and has a serrated margin, but it is not so conspicuously developed as it is in lemurs generally (fig. 139).

In the tongue of *Tarsius* the conical papillae are quite fine and form rather elongated hair-like processes, and there are no large thorny papillae on the pharyngeal surface as in the lemurs. The sublingua is of simple form, somewhat fleshy in consistency, adherent at the tip and only slightly free at its lateral margins, and it is not serrated. The median lytta is well marked and ends anteriorly in a free point which is slightly bulbous (fig. 137). In its main features the tarsioid sublingua approximates to the generalized structure found in marsupials and shows none of the specializations characteristic of the lemurs.

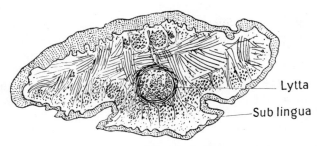

FIG. 139. Transverse section through the tongue of the pen-tailed tree-shrew, showing the sublingua and the lytta (×15). (*Proc. Zool. Soc.* 1934.)

The tongue of the marmoset (which probably represents the most primitive expression of this organ in the Anthropoidea) is similar in general form to that of *Tarsius*. It shows the same elementary arrangement of three vallate papillae and the absence of any evident conical papillae on the pharyngeal surface. On the under surface is a small triangular sublingua of soft consistency with finely crenulated margins which are free to a limited extent, but it does not extend forwards to the tip of the tongue. In the other Platyrrhine monkeys the sublingua is represented by no more than a vestigial fold of mucous membrane—the plica fimbriata—as is also the case in the anthropoid apes and man. In the Catarrhine monkeys, the plica fimbriata is often absent altogether.

THE ALIMENTARY TRACT

The alimentary tract in the Primates is on the whole rather generalized in its construction, showing little tendency to any extreme variation. It might be anticipated that the variations which do occur in the alimentary system would bear so intimate a relation to the nature of the diet that they would be of little or no value for assessing morphological affinities between one group and another. But in a monograph on the comparative anatomy of the intestinal tract in different orders of mammals Chalmers Mitchell [90, 91] has noted that, in spite of the similarity of the diet in herbivorous marsupials and placental mammals, there is a marked difference in their gut pattern, and also that, for example, the terrestrial Carnivora display a pattern of intestinal tract essentially similar even though almost every kind of diet is found amongst them—purely carnivorous, piscivorous, omnivorous or frugivorous. The unexpected conclusion is thus reached that 'however great may be adaptive resemblances, the inherited element dominates the structure' of the alimentary tract. Within the limits of the Primates the variations which appear to have some taxonomic significance mainly affect the lower end of the alimentary tract—the large intestine or colon—a part of the gut which is not so much concerned with active processes of digestion as with functions of excretion.

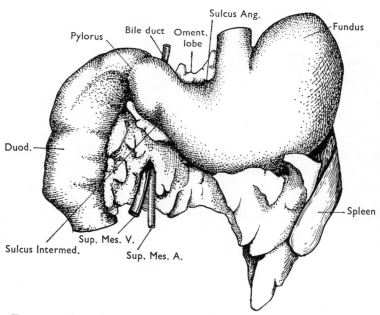

FIG. 140. Anterior view of the stomach, the duodenum and related structures in the pen-tailed tree-shrew (×4). (*Proc. Zool. Soc.* 1934.)

The stomach preserves a primitive simplicity in most of the Primates. Indeed, except for proportional differences, its form in the tree-shrews is reproduced quite closely in man. The stomach may be subdivided (but not sharply) into the main part or body, the fundus and the pyloric part (fig. 140). The last named terminates at the pylorus which marks the junction of the stomach with the first part of the small intestine, the duodenum. The Old World monkeys of the subfamily Colobinae are the only Primates with an unusually specialized stomach; in this group it is proportionately very large and also elaborately sacculated, modifications which may be associated with peculiarities of their vegetarian diet,

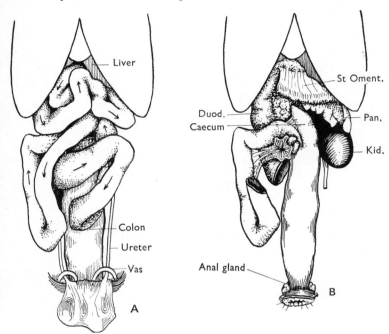

FIG. 141. Dissection of the abdominal viscera of the pen-tailed tree-shrew, with (A), the small intestine in position, and (B), the greater part of the small intestine removed ($\times 1\frac{1}{2}$). (*Proc. Zool. Soc.* 1934.)

and perhaps also to be correlated with the absence or reduction of cheek pouches in these monkeys.

The small intestine of the Primates is fairly uniform in its general conformation and disposition but, partly because of differences in peritoneal relationships, its component parts—duodenum, jejunum and ileum—are not so distinctly demarcated in the lower Primates as they are in the higher. In the primitive mammalian arrangement of the small intestine, its whole length, including the duodenum, is suspended freely from the posterior abdominal wall by a median dorsal mesentery, and it descends in a series of mobile coils from the pyloric end of the stomach to the large intestine. This simple arrangement is found in the tree-shrews (fig. 141), and it also persists in some of the lemurs, in *Tarsius*, and even occasionally in the Platyrrhine monkeys (e.g. the squirrel monkey). In other Primates (as, indeed, in many non-Primate mammals) the lower end of the duodenal part of the small intestine becomes loosely anchored down to the posterior abdominal wall by a band of peritoneum called the cavo-duodenal ligament. This fixation of the duodenum appears to be one of the factors which conditions, or is associated with, a rotation of the intestine so that the junction of the small with the large intestine, the ileocolic junction, comes to

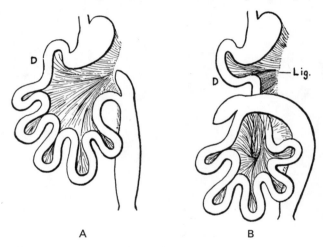

FIG. 142. Diagram to illustrate the fixation of the duodenal part of the small intestine, *D*, by the cavo-duodenal ligament, *Lig.*, which occurs in most Primates. In (A), the primitive mammalian disposition of the intestine and mesentery is depicted. In (B), the intestinal tract has undergone a rotation in association with the fixation of the duodenum.

be placed in front of the commencement of the small intestine (fig. 142). The rotation ultimately leads to a complication of the mesentery of the proximal part of the colon, which is no longer disposed as a simple, straight fold of peritoneum. With the exception of the Hominoidea, the duodenum in most Primates is suspended by a definite mesentery, or mesoduodenum, so that it is

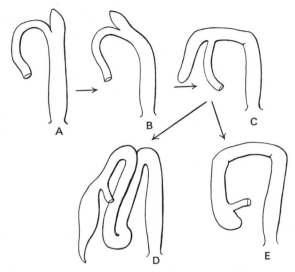

FIG. 143. Diagram to illustrate the changes which occur in the evolutionary development of the large intestine. (A) the condition in *Ptilocercus*; (B) *Tupaia*; (C) *Tarsius*; (D) the lemurs; (E) Anthropoidea.

freely movable. In the anthropoid apes and man, this part of the small intestine becomes secondarily retroperitoneal during foetal development, and as a result it is bound down and firmly fixed to the posterior abdominal wall throughout almost its entire length.

The colon shows certain modifications in different groups of Primates which are relevant to discussions on their interrelationships. The most primitive type of mammalian colon consists of a simple straight tube of relatively wide calibre which extends down directly to the anal canal and is suspended from the posterior abdominal wall by a continuous dorsal mesentery (fig. 143A). At its commencement, where it is joined by the small intestine, is a cone-shaped diverticulum called the caecum. Such a type of colon is retained unmodified in the pen-tailed tree-shrew. In the genus *Tupaia* the upper end of the colon is inclined to the right, thus producing an incipient demarcation between a transverse colon and a descending colon (fig. 143B). This displacement of the ileocolic junction is usually carried further in *Tarsius*, and in this genus quite a well-defined transverse colon may be present (fig. 143C). In the Anthropoidea the ileocolic region with the caecum typically migrates down towards the right iliac region, and there now appears a division of the colon into ascending, transverse and descending components (fig. 143E). A significant exception is found in the squirrel monkey (*Saimiri*) in which the large intestine is very short and only slightly curved to the right at its upper end, presenting an appearance very similar to that of *Tarsius*.

In all the modern lemurs the colon shows a conspicuous specialization which contrasts quite strongly with the simpler condition of other Primates. In this group the colon is characteristically elongated and bent sharply on itself to form a long, narrow loop (fig. 143D). This colic loop (ansa coli) may become elaborately twisted into a complicated spiral formation (e.g. in *Galago*, *Loris*, *Indri* and *Perodicticus*). In the development of the colic loop, whether in its simple or elaborated form, the lemurs parallel the condition found in certain lower mammals such as ungulates. It is difficult to ascribe this modification to special requirements of the diet, for it is found well developed in the lorises, galagos and the lemuriform lemurs (including *Daubentonia*) among which the diet is by no means uniform. Only in some of the smaller lemuriforms, *Cheirogaleus* and *Microcebus*, is the ansa coli not present; these genera preserve a more primitive condition not unlike that of *Tarsius* except that the descending colon is not quite so straight.

The characteristic lemuroid tendency, manifested in almost all the modern types, for the ansiform elaboration of the large intestine suggests that, as a group, they are rather far removed in their taxonomic affinities from *Tarsius* and the higher Primates; in this respect the evidence of the colon pattern conforms with that of the other aberrant features of the group which we have noted in previous chapters.

A caecum is found in all Primates and in almost all groups of mammals;* it may be regarded, therefore, as a typical mammalian feature which was presumably present in the ancestry of the Primates. In the tree-shrews it is of simple form. In the lemurs it is relatively large and often markedly dilated. It may be sacculated (*Perodicticus* and *Galago*) or simple and globular (*Cheirogaleus*). In some genera (*Lemur* and *Daubentonia*) the tip of the caecum terminates in a well-differentiated conical appendix which in mac-

* A significant exception is the order Insectivora in which, except for the aberrant group of elephant-shrews (Macroscelidoidea), the caecum is absent. It is also absent in many edentates and sporadically in some of the genera of other mammalian orders.

roscopic appearance, and also in histological structure, quite closely resembles the vermiform appendix of the anthropoid apes and man. However, there is no indication of such a structure in a number of other lemurs (e.g. *Loris* and *Cheirogaleus*).

In *Tarsius* the caecum is simple in shape but unusually long—almost equalling in length the whole of the rest of the colon. It has no sacculations, nor does it show any terminal differentiation equivalent to the appendix.

In the New World monkeys the caecum is conical, non-sacculated, and relatively voluminous, tapering gradually to a pointed extremity which may be elongated and thus simulate an appendix (fig. 144A). But it does not contain local accumulations of lymphoid tissue concentrated to the extent which is characteristic of a true vermiform appendix; it is to be regarded, rather, as the pointed end of the caecal pouch proper. In the Old World monkeys the caecum is generally smaller and terminates in a blunt, rounded extremity (fig. 144B). Its wall is elaborated by the formation of sacculi and on its surface are seen narrow bands of muscle (taeniae coli), which extend down from the longitudinal muscle coat of the ascending part of the colon. The vermiform appendix is completely absent in these monkeys.

In the anthropoid apes and man the caecum closely resembles that of the Catarrhine monkeys in its relative size and shape, and in the presence of well-defined taeniae coli and sacculations. It differs markedly, however, in having attached to it a narrow, tubular appendix. The functional significance of the vermiform appendix is still obscure, but there seems no adequate reason for supposing (as has been argued) that it is a primitive mammalian trait which has been retained in the Hominoidea and lost in the Catarrhine monkeys, or that it is a 'degenerate' structure of no functional value. It is evidently a specialized formation which, among the Anthropoidea, is peculiar to the anthropoid apes and man.

THE LIVER

The comparative anatomy of the solid viscera associated with the alimentary tract provides only equivocal evidence for the interrelationships of the Primates, for they show very considerable variability in their form even in closely allied genera of each group. It is well recognized by the human anatomist, also, that the form of

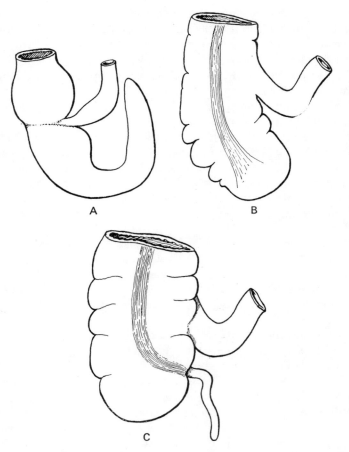

FIG. 144. Ileocaecal region of (A) *Callicebus*; (B) *Macaca*; and
(c) man.

these viscera may be modified and distorted by transient changes
in the hollow viscera which surround them and that their precise
contour, therefore, can vary even from one individual to another.
However, a brief reference may be made to the liver, for the lobar
divisions of this organ show certain features of interest.

In practically all Primates except the Hominoidea the liver is
suspended from the diaphragm and the abdominal wall by a mesen-
teric fold of the peritoneum. In the anthropoid apes and man it is

directly attached to the under surface of the diaphragm over an
area uncovered by peritoneum, the 'bare area' of the liver. In the
Colobinae, the abnormally large size of the stomach results in a
somewhat analogous condition, for the liver becomes tightly compressed against the diaphragm and adherent to it. Generally speaking, the lobulation of the liver is more complex in the lower than
the higher Primates, so that in a graded series the lobes appear to
become progressively reduced in number. In the tree-shrews there
are three main lobes, the right lateral, central and left lateral (fig.
145). Of these the right lateral lobe is the smallest and the central
lobe (which is incompletely subdivided into right and left lobules)

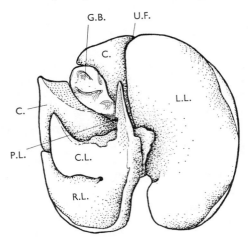

FIG. 145. Under surface of the liver of the pen-tailed tree-shrew
(×3). (*Proc. Zool. Soc.* 1934.) *G.B.*, Gall bladder; *U.F.* Umbilical
fissure; *C.* Central lobe; *C.L.* Caudate lobe; *P.L.* Papillary lobule;
L.L. Left lateral lobe; *R.L.* Right lateral lobe.

is the largest. The right lateral lobe has projecting down from
its medial part a conspicuous caudate lobe, and from the base of
the latter extends a pointed process, the papillary lobule. In the
lemurs the central lobe is relatively smaller, and its papillary process is considerably reduced. The liver of *Tarsius* is very similar
to that of the lemurs, but while it shows a somewhat more advanced character in the less distinct lobulation and in the smaller
size of the caudate lobe, the primitive papillary lobule is actually
better developed.

In the Anthropoidea the right lateral lobe comes to dominate the other lobes by its relative increase in size, and in the anthropoid apes and man (particularly the latter) it is the largest of them all. The central lobe tends to lose its individuality and partially fuses with the right and left lateral lobes. The caudate lobes and papillary lobules are still further reduced in size, and the latter loses the distinctive 'papillary' form which is characteristic of the lower Primates. Among the anthropoid apes the gorilla is exceptional in its hepatic anatomy, for in the complexity of its lobulation it resembles some of the lower Primates; the significance of this aberrant feature is unknown.

It has been remarked by Straus [157] that 'the four (modern) anthropoid apes and man possess in common a number of visceral characters that clearly pronounce their affinities'. He has also confirmed the statement of previous observers that the configuration of the liver in the gibbon more closely resembles that of man than any of the large apes. This might be taken to substantiate in some degree the thesis that the evolutionary sequence of the Hominidae diverged from that of the Pongidae at the hylobatine stage of their evolutionary development, and that in this phylogenetic sense the Hominidae are more closely related to the Hylobatinae than to the large anthropoid apes. But, because of the seemingly capricious vagaries of hepatic morphology in the Primates, which may show quite wide variation in closely related types, evidence of this sort naturally requires very cautious consideration.

9

The Evidence of
the Reproductive System

CONSIDERING the general uniformity of the main functions of the reproductive system, it might be supposed that its structural components would be likely to show little variation even among animals of widely differing habitat. In fact, they do present certain differences of a generic type, more especially in the superficial details of the external genitalia, though it is by no means obvious in all cases what the functional significance of these differences may be. But apart from such variations which fluctuate from one genus to another, it is possible to detect, in proceeding from the lower to the higher Primates, a gradational series of modifications which reflect in one way or another a progressive efficiency in the reproductive processes characteristic of the order as a whole—particularly those processes directly concerned with gestation.

THE MALE REPRODUCTIVE ORGANS

The main features of the male reproductive organs of the Primates are seen in their most elementary and generalized form in the tree-shrews. In fig. 146 is shown a dissection of the urethral tract and related structures of *Ptilocercus*. At the base of the bladder is the prostate gland, a compact and non-lobulated structure encircling the commencement of the urethra, which in shape and histological appearance is very similar indeed to the human prostate. Behind this gland is the prostatic utricle (uterus masculinus), a piriform sac homologous with the uterus of the female and, like it, derived embryologically from the paramesonephric ducts. On either side of the utricle are seen the vas deferens (duct of the testicle), the vesicular diverticulum, and the vesicular gland. The function of the vesicular gland is unknown, and it does not appear to have any well-defined homologue as a separate element in the higher Primates. Probably the gland becomes incorporated with the diverticulum in a common structure corresponding to the

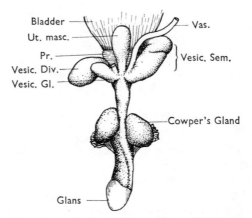

Bladder — Vas.
Ut. masc.
Pr.
Vesic. Div.
Vesic. Gl. — Vesic. Sem.
— Cowper's Gland
Glans —

FIG. 146. Dissection of the male urethra of *Ptilocercus* viewed from behind (×3). (*Proc. Zool. Soc.* 1934.)

vesicula seminalis of human anatomy. The vesicular diverticulum, which is essentially a receptacle for the temporary storage of spermatozoa, opens into the urethra by a duct, the common ejaculatory duct, which it shares in common with the termination of the vas deferens. Alongside the urethral canal lower down are the bulbourethral (Cowper's) glands. The penis is relatively short and stout, its tip being formed by an oval, almond-shaped glans. The glans penis is demarcated behind by a well-defined groove, the coronal sulcus. Unlike most Primates (and, indeed, unlike most placental mammals) the penis does not contain a bony skeleton, the os penis or baculum.

In fig. 147 is seen from the side a dissection of the male external genitalia of the tree-shrew, showing the scrotal sac which in the mature animal is occupied by the testis. It will be noted that the scrotal sac lies at the side of the body of the penis, that is, in a parapenial position, as it also does in certain lemurs (e.g. *Perodicticus*). To some extent this may be regarded as a sort of transitional stage between the prepenial position of the scrotum characteristic of the marsupials and the postpenial position of most Primates.*

It is characteristic of almost all the Primates that at sexual maturity the testis descends permanently to occupy the scrotal sac.

---

* In some species of gibbon the scrotal sacs may be parapenial, or even prepenial; this apparently primitive condition must surely be a secondary adaptation in these particular species.

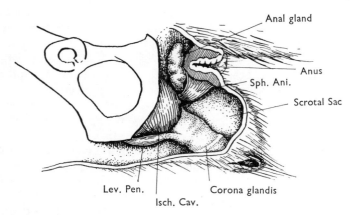

FIG. 147. Lateral view of the scrotal sac, penis and associated structures in *Ptilocercus* (×4). (*Proc. Zool. Soc.* 1934.)

Exceptions are found in some of the lemurs (*Loris*, *Perodicticus*, and probably *Daubentonia*) in which the descent is seasonal as it commonly is in many lower mammals. There is some doubt whether descent is permanent in the marmosets, but apart from this group it appears to be the case in all the Anthropoidea. In some of the latter the testes take up their final position in the scrotum some little time after birth. In the gibbons they descend during the juvenile period of life, but in the large anthropoid apes and man they have normally completed their descent by the time of birth. In the tree-shrews, the testes are permanently retained in the scrotum in *Tupaia*, but in the more primitive *Ptilocercus* descent is intermittent. Within the scrotum the testes are contained within a pouch of peritoneum, the saccus vaginalis, which during embryological development is extruded through the inguinal region of the abdominal wall from the general peritoneal cavity. The connection between the two is the processus vaginalis which remains an open canal in the lower Primates, including most Platyrrhine and Catarrhine monkeys, but this normally becomes completely obliterated in the Hominoidea so that the saccus vaginalis in the latter is an entirely closed cavity.*

The morphological details of the penis show wide intergeneric variations in the Primates, but in general the lemurs display a

* A patent processus vaginalis has been reported to persist normally in the orang-utan.

common tendency to specializations wherein they stand in contrast with other groups. While the epithelial covering of the glans penis is only moderately rugose or marked by a fine papillary pattern in some genera (e.g. *Nycticebus, Cheirogaleus* and *Daubentonia*), in many others it is beset with small or large recurved spicules (grappling spurs) disposed in various degrees of complexity and quite similar to the spinous development on the penis of many lower mammals (e.g. rodents). In *Galago* the spicules may be elaborated into a bidentate or tridentate form. In all the lemurs a bony skeleton is found in the penis, but this baculum varies considerably in its relative length and contour, and to some degree in its relationship to the urethral canal. In the Indriidae and in *Cheirogaleus* it forks distally, while in most of the Lemuridae, and in the Lorisiformes and *Daubentonia*, it terminates in a single median thickening. In *Tarsius* the penis is relatively simple in its constructional details, but the glans is elongated and slender, being covered by a smooth or but slightly granular epithelium; the lips of the urethral orifice form rather prominent flap-like folds, and there is no baculum.

In the Anthropoidea the penis is also relatively simple in its morphology. In the Platyrrhine monkeys the glans is of a regular ovate or hemispherical form, but in *Ateles* it is covered by very small cornified denticles. A baculum is commonly present, but it has been reported to be lacking in *Lagothrix*. In all the Catarrhine monkeys (with possible exceptions) and in all the anthropoid apes there is a small baculum in the glans extending back for a short distance into the body of the penis. There is no baculum in the human penis. The glans in the Old World monkeys is usually rounded or oval, often expanded into a rather conspicuous knob, and the margin of the glans at the coronal sulcus is commonly marked by quite deep notches. In the anthropoid apes the glans penis in the gibbon, orang and chimpanzee is slender and pointed, showing no expansion like the monkeys. In the gorilla, on the contrary, it forms a conical expansion with a well-defined margin. In this respect, the gorilla approaches more closely to the monkeys, and also to man, than do the other apes.

The accessory organs of the male genital system offer little evidence relevant to the problem of Primate interrelationships. The seminal vesicles vary considerably in size, and are said to be absent altogether in *Daubentonia*. The prostate gland is always a compact structure of conical form, in some cases completely encircling the urethra as in man, but usually folded round its dorsal aspect only; in monkeys (and to a lesser extent in the anthropoid apes) it may be superficially subdivided by a transverse groove into cranial and caudal lobes. The prostatic utricle is absent in at least some of the prosimians (e.g. the genus *Lemur*), and in *Tarsius* it is present in the form of a minute diverticulum of the urethra which has virtually no lumen. In all the Anthropoidea (including man) the utricle is present as a small and inconspicuous diverticulum, completely embedded in the substance of the prostate gland. In some genera of the monkeys the ramifications of its lining epithelium give a much more glandular appearance than in others.

THE FEMALE REPRODUCTIVE ORGANS

The female external genitalia of the Primates, like those of the male, show many variations which differ from genus to genus, and even from species to species, so that they are of little value for assessing the taxonomic affinities of the major groups. The appearance of the vulva in the pen-tailed tree-shrew is shown in fig. 148.

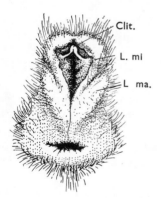

FIG. 148. External female genitalia of *Ptilocercus* (×4).

At the anterior margin of the vulval orifice is a small clitoris (the homologue of the glans penis of the male). From it a fold of mucous membrane extends on either side to become continuous with labium minus, which forms the immediate boundary of the vulval orifice. More laterally is an ill-defined elevation of hairy skin which represents the labium majus of human anatomy (the homologue of the scrotal area of the male). The variations of these structures in the Primates mainly relate to the relative length of the clitoris and the relative prominence of the labia. In the lemurs, for example, the clitoris may be very long and pendulous, and in some genera (e.g. *Galago*) it may be supported by a baculum. The labial folds are not very well marked in the lorisiforms, but in the lemuriforms the labium majus may be quite conspicuous and is covered by a rugose skin having a glandular structure. A curious distinction between these two lemuroid infraorders is found in the relation of the urethral canal to the clitoris. In the lemuriforms the usual (and more primitive) condition common to Primates and most other mammals is found, the urethral orifice being situated at the base of the clitoris though in some cases (e.g. *Lemur*) reaching towards the tip of the latter. In the lorisiforms the clitoris is traversed by the urethral canal which opens at its tip, thus simulating the male arrangement.*

In *Tarsius* the clitoris, which is bifid at its tip, is quite small and concealed from view by prominent labia minora. In the New World monkeys it may be very short and concealed (*Callitrix*, *Callicebus*) or long and pendulous (*Saimiri*, *Ateles*), and in some genera, e.g. *Cebus*, there may be a small baculum embedded in its tip. The vulval orifice is bounded by labia minora which are commonly well-defined, and in some genera folds of skin evidently corresponding to the labia majora are also present. But both these folds are highly variable in their development, and there is often some doubt how far, in monkeys and apes, they can be strictly homologized with the labia majora and minora of human anatomy [170]. In the Old World monkeys the female genitalia also show wide variations, particularly in the relative development of the clitoris, and in most cases there are no obvious skin folds which correspond to the labia majora.

In the anthropoid apes the labia minora are commonly well developed, but the labia majora, which are present and distinct in the foetal and juvenile stages of development, undergo retrogression subsequently, and in the adult they may be very inconspicuous or even absent. In this respect the human female may be said to retain permanently a developmental condition which in the apes is

---

* A similar type of perforate clitoris is found in the Hyaenidae and some rodents.

only temporary. The clitoris in the anthropoid apes is relatively prominent.

In the course of its embryological development the uterus and oviducts are derived from paired ducts, the paramesonephric or Müllerian ducts. In the monotremes and marsupials, and also in many rodents, these ducts remain ununited throughout their extent and thus lead to the formation of two separate uteri. In most placental mammals, however, the paramesonephric ducts fuse to a variable extent at their lower ends to form a single body, or corpus uteri, with bilateral extensions called the uterine horns. The latter are continuous with the oviducts along which ova are transported to the uterine cavity.

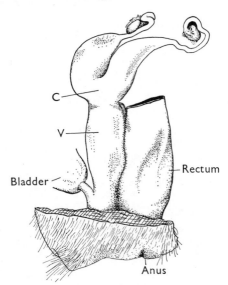

FIG. 149.   Internal genital organs of the female in *Ptilocercus* (×3).
C, Corpus uteri; V, vagina.

In the tree-shrews the body of the uterus is relatively quite small, and the long uterine horns taper into the oviducts (fig. 149). In lemurs the uterus is also bicornuate in all genera, but the body of the uterus tends to be relatively larger. This tendency is carried somewhat further in *Tarsius* in which the length of the cornua is not much more than half the length of the corpus uteri. In the Anthropoidea the uterine parts of the paramesonephric ducts become completely fused to produce a uterus simplex, that is to say, a uterus with a single undivided corpus and no lateral horns (fig. 150). In some of the monkeys, e.g. *Callitrix*, the body of the uterus may show a low median groove and its lateral angles may extend out towards the oviducts, suggesting their derivation from the bicornuate type of uterus. It is also the case that, even in the human female, a bicornuate uterus may occur anomalously as the result of incomplete fusion of the paramesonephric ducts during embryological development.

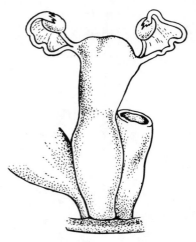

FIG. 150.   Internal female genital organs of *Cercopithecus* (× 1).

In the Anthropoidea the mature females (with a few possible exceptions) show a sexual cycle with an approximate monthly rhythm, and they are capable of breeding throughout the year. The cycle is accompanied by changes in the uterine mucous membrane and vaginal secretions, and in the apes and Catarrhine monkeys, as in the human female, by menstrual bleeding. In the New World monkeys, menstruation appears to be unusual, but it has been observed in certain genera, e.g. *Cebus* and *Ateles*, though usually rather slight and often irregular. In *Tarsius* there is a rhythmic cycle affecting the vaginal secretions lasting about a month, and definite menstrual bleeding of scanty extent has been observed in occasional cycles in captive animals. It is a marked characteristic of many Catarrhine monkeys of the subfamily Cercopithecinae that the external genitals and surrounding skin become conspicuously turgid and swollen at the phase of the menstrual cycle which precedes ovulation, and in some species the swelling reaches grotesque proportions. This rhythmic activity of the sexual skin is represented in *Tarsius* by a moderate congestion of the genital region, but in the Platyrrhine monkeys such changes are slight or absent. Among the anthropoid apes, very obvious swelling of the sexual skin occurs in the chimpanzee, but in the other genera there is no more than a slight turgescence of the vulva at the corresponding phase of the menstrual cycle. If cyclical swelling of the sexual skin characterized human ancestry, all traces of this phenomenon have disappeared in *H. sapiens*.

In the lower Primates breeding is commonly restricted to limited seasonal periods, as in lower mammals generally. A slight discharge of menstrual blood has been reported to occur in treeshrews and certain lemurs during the period of coming into heat (pro-oestrus), but these observations lack adequate confirmation. According to the observations of Sprankel [154], in a state of captivity *Tupaia* may breed throughout the year, with a gestation period of 43 to 46 days.

THE PLACENTA

Among the comparative studies which may provide significant evidence for the interpretation of phylogenetic problems is that concerned with the processes whereby the embryo is implanted and nourished in the uterus, and more particularly with the structure and mode of formation of the placenta. It is to be noted, however, that the validity of this evidence is more limited than some have supposed, for placental development and structure may show quite wide variations within the limit of a single mammalian order. The subject is highly specialized and technical so that it is not possible to treat it adequately here, but because of its possible relevance to taxonomy we may consider briefly some of the evidence which it has supplied. For this purpose, we rely heavily on the standard monograph of the late Professor J. P. Hill on developmental processes in the Primates [50], and in order to clarify the meaning of terms which may be unfamiliar to those who have no acquaintance with embryology a highly schematic diagram of a developing embryo and its membranes, broadly representative of one of the higher Primates, is shown in fig. 151.

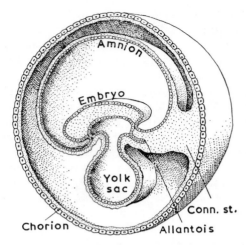

FIG. 151. Schematic representation of an embryo of a higher
Primate in the earliest stages of its development.

When the developing ovum reaches the uterine cavity, it throws
out certain membranes—the foetal membranes—which directly or
indirectly establish contact with the mucous membrane lining the
uterus. The outermost membrane is the chorion, which completely
invests the chorionic cavity containing the embryo. The chorion is
covered by a layer of epithelium, the trophoblast, and is lined on
its inner aspect by embryonic connective tissue, or mesoderm,
which comes to be richly supplied by foetal blood vessels. The
trophoblastic epithelium undergoes differentiation into two well-
marked layers of which the more superficial forms a continuous
stratum of nucleated protoplasm, or syncytium, which is not divided up into separate cells. It is the trophoblast which, by its in-
vasive activity, burrows into the uterine wall and leads to the em-
bedding of the developing ovum there. The surface of the chorion
is rapidly complicated by the development of villous processes
which cover its surface, and these chorionic villi penetrate deeply
into the uterine mucosa. Dorsal to the embryo there is formed the
amniotic cavity the wall of which is a membrane called the amnion.
This cavity expands until it completely envelops the embryo, and
the amniotic fluid which it contains has an important protective
function. In the lower Primates, as in mammals generally, the amniotic sac is formed by folds which rise up around the margin of the

developing embryo and coalesce above it. The gut cavity is in communication with the yolk sac in the early embryo, and although this
is of far greater importance for oviparous vertebrates, it still serves
a minor nutritive function in mammalian embryos. The caudal end
of the embryo in higher Primates is attached to the chorion by a
mass of mesoderm called the connecting stalk, and it is through the
latter that blood vessels extend from the embryo itself to the chorion
and the chorionic villi where they are brought into close relationship
with the maternal blood vessels in the uterine wall. From the caudal
end of the embryonic gut there extends back into the connecting
stalk a diverticulum called the allantois. In birds and reptiles the
allantois expands to form an extensive membrane which lies in
close apposition to the shell of the egg, and, being richly supplied
with blood vessels, it serves as an embryonic 'lung'. In mammalian
embryos the allantoic diverticulum, which in most mammals is
quite extensive, appears to serve a somewhat analogous function,
for by its outgrowth it helps to convey blood vessels from the
embryo for the vascularization of the chorion. With this brief, and
necessarily over-simplified, outline of the foetal membranes and
associated parts, we may now draw attention to some of the
different types of placenta which have been defined on the basis of
their intrinsic structure.

   In the simplest type of placenta the chorionic villi, covered by
trophoblastic epithelium, merely interlock with corresponding
crypts formed in the epithelial lining of the uterine cavity. Such a
placenta is termed *epitheliochorial*, and even when it is fully functional the maternal blood remains separated from the foetal tissues
of the chorion by the endothelial lining of the uterine blood vessels,
the connective tissue of the uterine mucous membrane, and also
the uterine epithelium. A more advanced and functionally efficient
type of placenta is the *syndesmochorial* in which the uterine mucous
membrane becomes partly denuded of its epithelial covering by
the erosive action of the trophoblast so that this barrier between
the maternal and foetal blood is removed. This type of placenta
is found in some of the ungulates. A further stage of elaboration is
represented by the *endotheliochorial* placenta in which the trophoblastic epithelium comes into direct contact with the vascular endothelium of the uterine capillaries (as seen in Carnivora, certain
bats, and some of the Insectivora). Lastly the most efficient
placenta from the point of view of the intimacy of the relation

between foetal and maternal circulations is the *haemochorial* type which is found in the higher Primates, most Insectivora and some of the rodents. In this the trophoblastic syncytium erodes its way right into the blood vessels of the uterine wall, opening up large blood spaces filled with maternal blood with which it is in immediate contact without even the intervention of the endothelial lining of the maternal vessels. In the case of the epitheliochorial placenta, in which there is no intimate merging of the chorionic tissues with those of the uterus, the placenta becomes detached from the uterine wall at birth without bringing away with it any of the uterine mucous membrane. Hence this type of placenta is said to be non-deciduate. In the other types of placenta, particularly the haemochorial, parturition involves the separation of a good deal of the uterine tissue which is shed as part of the after-birth. Such a placenta is termed deciduate.

In the evolutionary development of the placental structure in the Primates, and of the processes whereby the embryo becomes implanted in the uterus, J. P. Hill distinguished four stages or grades—lemuroid, tarsioid, pithecoid and anthropoid (= hominoid). These stages are taken to represent progressvie structural elaborations which appear to reflect a progressive functional efficiency, but at the same time (as might be expected) each group of the modern Primates shows certain divergent specializations peculiar to itself which are unlikely to have characterized the ancestry of other groups.

The placentation of all the lemurs which have been studied is of the epitheliochorial type, and it also shows a primitive character in the fact that the chorionic villi are distributed diffusely and in most cases uniformly over the surface of the chorion. In both these features the placenta is very similar to that of some of the ungulates. The wall of the amniotic sac is formed by the coalescence of folds, and the allantois arises as a relatively large flask-shaped and vesicular diverticulum which extends out to the chorion. Taken in conjunction with each other, all these characters appear to indicate that the placentation of the lemurs is fundamentally primitive. But it has also been argued that it has been derived from a more complicated type by a process of secondary simplification. The evidence for such an interpretation is, of course, indirect, and partly rests on the assumption that, since the placenta of such primitive mammals as many of the insectivores is haemochorial, the latter is more likely to represent the basal type from which the placentation of the Primates was derived. On the other hand, if (as commonly assumed) the epitheliochorial type is the less efficient functionally, it is difficult to accept the suggestion that it could be developed as an evolutionary reversion from the haemochorial type. Thus, unless more convincing evidence to the contrary is forthcoming, it seems more reasonable to assume that the primitive characters of the lemuroid placenta are genuinely primitive. It is interesting to note, however, that some features in the development of the lemuroid placenta do foreshadow to a slight but significant extent features which come to be distinctive of the higher Primates. These are the rapid and early differentiation of the chorion as a whole, and its precocious vascularization from blood vessels which are conveyed to it by the allantoic outgrowth. In *Galago*, there are still more significant approaches to the higher Primates, for in this genus there is a differentiated, localized area of the diffuse placenta which is distinguishable by its thickened character and its histological structure, and which, moreover, is in contact with an area of the uterine wall partly denuded of its epithelium; this has been interpreted as a primordial stage in the evolution of the localized, discoidal and haemochorial placenta which distinguishes *Tarsius* and the Anthropoidea.

In *Tarsius* the placenta is of the haemochorial, deciduate type and quite closely resembles that of the monkeys. It is a circumscribed discoidal structure and thus contrasts strongly with the diffuse placenta of lemurs. In its development it approximates to the Anthropoidea in the very early differentiation of the chorion and the manner in which this becomes vascularized. The allantois is not vesicular as in lemurs and lower mammals; instead, it grows out as a slender strand into a solid connecting stalk of mesodermal tissue which is developed very rapidly. This connecting stalk is similar to that of all the higher Primates (including man), and probably results from a precocious proliferation of the mesodermal covering of the reduced allantoic diverticulum. It provides a favourable matrix for the rapid proliferation of blood vessels and, as a result, the vascularization of the chorion is still further accelerated—a process which is clearly of considerable advantage for the efficient nutrition of the embryo in its earliest developmental stages. In certain details of its minute structure the placenta of *Tarsius* is unique among the Primates, and on these grounds Hill considered that it is too specialized to be representative of the actual forerunner of the type of placenta found in the Anthropoidea, but he accepted the probability that the latter arose from some ramification of the tarsioid stock in which these structural peculiarities (which are, it seems, by no means of a fundamental nature and doubtless merely generic specializations) were not present. Although the placenta of *Tarsius* does show a remarkable approximation to that of the higher Primates, it also presents one or two features in which it is more primitive, and reminiscent, so to speak, of the generalized mammalian condition, such as the mode of formation of the amniotic sac from folds which undergo secondary fusion.

The strong contrast which *Tarsius* shows in its mature placenta with that of the lemurs appears to be at variance with the implications of other anatomical evidence, and of the palaeontological record, that they are related at least to the extent that they are appropriately placed in the same suborder, Prosimii. But a similar contrast is seen between the lemurs and the tree-shrews. It has been reported that the placenta of *Tupaia* in its mature condition is of the haemochorial type though in earlier stages of development it is endotheliochorial [89]. On the other hand, according to J. P. Hill [51] the vascular endothelium of the uterine capillaries does persist to maturity but in such an attenuated form as to be difficult to detect. Even so, the placenta shows a significant advance over that of the insectivores in the degree to which the villous processes of the trophoblast are developed. Moreover, the allantois is much reduced, and the placenta is double and discoidal as it is in most monkeys. This demonstration of Primate characteristics in the tupaioid placenta is in accord with the numerous Primate features which we have noted in the total morphology of the tree-shrews. It is to be noted, however, that the tupaioid placentation, like that of other groups of Primates, shows certain specializations peculiar to itself.

In the Anthropoidea the early developmental processes concerned in the establishment of the placenta are hastened and abbreviated still further. A very definite advance over the tarsioid condition is manifested in the mode of formation of the amniotic sac, for this no longer arises from folds but makes its appearance by cavitation, as it were spontaneously, in the dorsal part of the mass of cells from which the embryo subsequently develops (in more precise terms, from the ectodermal part of the inner cell mass). The chorion is formed more precociously than in *Tarsius* and, as in the latter, the vesicular type of allantois is replaced by a massive connecting stalk in which foetal blood vessels appear at a very early stage. In the Platyrrhine and Catarrhine monkeys, with extremely few exceptions, two separate discoidal placentae of an elaborated haemchorial type are found. Each group shows certain differences which are mainly related to the profuseness of the trophoblastic proliferation, and in this respect the Catarrhine monkeys are the more progressive. By comparison 'the development of the Platyrrhine placenta is a slow and cumbrous process, involving so much time that it only reaches a condition of what we may call structural efficiency at quite a late period in gestation', while 'the Catarrhine, on the other hand, by abbreviation and acceleration of the developmental processes, has speeded up the development of its villous placenta in the most remarkable way' so that 'it is capable of carrying on its full functions immediately the foetal circulation is established, and it does so not only much earlier, but in what would seem to be a much more efficient manner (judging from structural relations) than is the case in the Platyrrhines' [50].

The placentation in man and the anthropoid apes (so far as the latter is known) is very similar to that of the Catarrhine monkeys in which, indeed, its distinctive characters are already clearly foreshadowed. The early embryo elaborates its placenta in a still more precocious manner and gains a most intimate relation with the maternal tissues by very rapidly burrowing its way by trophoblastic activity into the wall of the uterus. Thus it no longer undergoes its initial development in the lumen of the uterus, as happens in all the lower Primates, for while it is still quite minute it becomes deeply submerged within the substance of the uterine wall. Unlike the monkeys, also, only one discoidal placenta is found in the hominoids.

The progressive differentiation and elaboration of placental mechanisms, which evidently promote a more and more efficient means of securing the optimum conditions for the developing embryo, comprise one of the most outstanding features of Primate evolution. Correlated with this, evidently, is the reduction in the number of young produced at a birth. In general, plural gestation is to be regarded as a primitive mammalian feature, and it is characteristic of almost all Primates that they give birth to only one at a time. In the tree-shrews there are normally two, one developing in each uterine horn. In the lemurs twins are not uncommon in *Loris, Galago, Nycticebus* and the Cheirogaleinae, but otherwise they are exceptional. In the Anthropoidea the primitive status of the Callitrichidae is emphasized by the fact that this is the only group which normally produces twins (and sometimes triplets). In all other higher Primates, multiple births occur only as occasional anomalies.

10

The Evolutionary Radiations
of the Primates

I N the foregoing chapters we have considered the various anatomical systems of the Primates with the intent to see what inferences
may be drawn from them severally in regard to the origin and interrelationships of the main subsidiary groups of the order. It will
already be apparent from this evidence that some of these groups
probably became segregated at a very early time in the evolutionary
history of the order as a whole, for this seems the most reasonable
explanation of certain of the structural contrasts which they show.
It may be, also, that they underwent some degree of parallel development after their initial divergence from a common parent
stem. In the present chapter it is proposed to collate and synthesize
(and to some extent recapitulate) the evidence of the various anatomical systems in order to give a more comprehensive picture of
the phylogenetic history of the Primates so far as the available data
allow this to be done.

Palaeontological studies make it clear that living mammals represent but a very few of the end-products of diverse evolutionary
trends. These few have survived throughout the geological ages
as the result of some particular structural and functional advantage
which they have achieved, and through which they have secured
a more or less complete harmony with the environment in which
they find themselves. But the vast proportion of mammalian
genera and species have become extinct, and these are to be regarded as so many evolutionary experiments which have ultimately met with no success.

As already emphasized, it is necessary to take into full consideration both Recent and fossil types in order to arrive at a conception
of the evolutionary trends which have characterized the direction
of phylogenetic development, and in this way to apprehend the
precise relationship of one group with another. For the distinctive
features of the Primates or of their subdivisions are not to be de-

fined merely by a consideration of the anatomical characters of the few terminal products of evolution which happen to have survived to the present time, but rather by the dominating evolutionary tendencies which have characterized these groups since their initial differentiation from a basal and generalized ancestral stock. Only by this comprehensive approach can we assess the possibilities or probabilities of the direct derivation of one group from another. Thus, for example, if a survey of a whole series indicates that there has been a prevailing and progressive tendency in that series towards the reduction and loss of a primitive character X, it may be deemed unlikely that the series would have given rise, during the later phases of its evolution, to another series in which the character X is retained or evinces a tendency to progressive specialization. In assessing possibilities and probabilities of this kind, it is of course necessary to take account of the complexities of the selective influences involved, and also of the genetic complexities upon which the phenotypic characters are based. As we have seen (Chap. I), positive reversals in evolution in the sense that characters once lost are redeveloped in precisely their original form must surely be very exceptional. It has to be admitted, however, that only a reasonably complete palaeontological documentation can give a final answer to questions of this kind in individual instances, and only such a documentation can determine how far evolutionary parallelism is responsible for some of the structural resemblances in different groups of Primates. Unfortunately, the fossil record of the Primates is still far from complete, particularly in the case of the earlier phases of prosimian evolution; until it is more complete their interrelationships can only be inferred by assessing possibilities and probabilities in the light of the available evidence.

The data presented in the previous chapters provide considerable support for the proposition that the Primates are properly to be subdivided into two main suborders, Prosimii and Anthropoidea. This classification, which is now generally accepted, supersedes one formerly held by several authorities in which the tarsioids were placed in a separate suborder between the lemurs and the Anthropoidea. This earlier mode of classification was thought to be justified because the modern tarsier not only lacks some of the specializations which are distinctive of the lemurs, it also shows a number of advanced characters in which it appears to approach the Anthropoidea. Indeed, some comparative anatomists had gone so far as to link the tarsioids with the Anthropoidea rather than with the lemurs in their scheme of classification. But the accumulation of fossil evidence has shown how difficult it may be to distinguish clearly between the early lemuroids of Palaeocene and Eocene times and the early tarsioids, and it also seems probable that some of the advanced characters of the tarsioids are not strictly homologous in their total morphological pattern with those in which they resemble the higher Primates. Taking all the evidence into account, the inclusion of the tarsioids with the lemurs in the suborder Prosimii, but their separation within the suborder as a separate infraorder Tarsiiformes, seems most nearly to express their true phylogenetic relationships. It may be, of course, that the accession of fossil material will eventually demonstrate that one or other of the alternative classifications of the subdivisions of the Primates which are advanced from time to time is more in accord with these relationships. But until or unless this proves to be the case, it is desirable to conform to the generally accepted classification and its taxonomic terms, if only to avoid nomenclatural confusion and misunderstanding.

THE AFFINITIES OF THE TREE-SHREWS

For almost a century attention has been drawn by comparative anatomists to certain characters in which the Tupaioidea resemble the higher Primates and in particular the lemurs, but it was only in later years that these resemblances were considered to be significant enough to place the groups within the order Primates. Their assignment at one time to the very heterogeneous order Insectivora (which includes hedgehogs, moles, ground-shrews, elephant shrews, etc.) was mainly based on the negative evidence of primitive characters which they retain in common—that is to say, what we have called 'characters of common inheritance' (see p. 26). But 'the characters of independent acquisition' which the tree-shrews have developed distinguish them sharply from all the present-day Insectivora and make it evident that, in spite of the primitive aspect of their anatomy taken as a whole, they have followed to a considerable extent some of the trends of evolution distinctive of the Primates. These trends are manifested most clearly in the construction of the skull, the advanced degree of differentiation of the visual mechanisms of the eye and brain, the reduction of the olfactory apparatus, the cusp patterns of the teeth (particularly in

Ptilocercus), the free range of movement of the hallux, the serrated sublingua of the tongue, certain details of the placentation and so forth. It is thus evident that the resemblances to the Primates are by no means all of a negative kind (though even these have their value as negative evidence). Naturally, the possibility presents itself that the Primate features of the tree-shrews may have been independently acquired as the result of convergent evolution dependent on similar habits of life, but it is difficult to explain on this basis the complexity of the resemblances in structural minutiae. The argument for convergence might be valid for some of the more general features, but it can hardly account for the detailed similarities. A brief recapitulation of some of the anatomical evidence will serve to emphasize this point. For example, the intricate mosaic of bony elements which enter into the formation of the orbito-temporal region of the skull, and the construction of the auditory region and cranial base (including the disposition of the various vascular foramina here), comprise a total morphological pattern which is in many respects similar to that found in the lemuriform skull and which finds no exact parallel in any other group of placental mammals. Again, the elaboration of the visual apparatus is not simply expressed by a large optic nerve and an extension of the area of the visual cortex; it is reflected in a differentiated retinal structure which includes an incipient avascularity of the central area such as is found in lemurs, in an elaborate and sharply defined lamination of the lateral geniculate nucleus, in a visual cortex which in its histological complexity approaches very closely that of the higher Primates, and in a marked degree of colour discrimination. Further, it has been recently shown that in *Tupaia glis* the retina has a spectral sensitivity curve similar to higher Primates and different from that of squirrels with a similar arboreal habitat [181]. If we add to these features the numerous other details to which we have called attention, such as those of the turbinal system of the nasal cavity, the ossicles of the middle ear, and the musculature of the limbs, the Primate affinities of the anatomy of the Tupaioidea seem to be reasonably well assured. Their relationship to the lemurs in particular is emphasized by two further points of considerable interest. One of these (which we have already mentioned) is the distinctly lemuroid habit of using the procumbent lower incisors as a 'dental comb'. The other concerns the observation that the parasites which infest closely related animals commonly show an equivalent degree of taxonomic relationship, and it is thus of some significance that a genus of ectoparasites *Docophthirus*, which has been recorded from tree-shrews, is also found in lemurs and it has been remarked by Hopkins that this lends support to the placing of the tupaioids among the lemuriforms [55].

There is also, of course, the palaeontological evidence to take into account. The Plesiadapidae of Palaeocene and Lower Eocene antiquity show resemblances to the Tupaioidea which led some authorities to accord them a close taxonomic association even before it was generally accepted that the Tupaioidea should be included in the order Primates. The fact that intensive studies (particularly of the dentition) have now firmly established the Primate affinities of the plesiadapids serves still further to reinforce the argument for the Primate affinities of the tree-shrews. The morphological gap which separates the modern tree-shrews from the modern lemurs may appear somewhat abrupt, but it must again be stressed that the former still retain a remarkably primitive assemblage of characters, while the latter have acquired a number of highly specialized characters.

In recent years the taxonomic position of the tree-shrews has once more become the subject of animated discussion, and varying opinions have been expressed on this problem. Increasing emphasis has been based on arguments that the morphological similarities with the Primates are the result either of the possession in common of primitive mammalian characters or of convergent evolution based on a similar arboreal mode of life. It has been stated, for example, that the extension forwards of the palatine bone to make contact with the lachrymal bone on the inner wall of the orbit is really a primitive condition to be found in some insectivora. Whether it is indeed primitive, however, is open to question, and in any case the pattern of sutural contacts made by the palatine in *Tupaia* differs from that of any known insectivore in reproducing with some degree of precision the pattern that is highly characteristic of the Malagasy lemurs. It is claimed, also, that in relation to its arboreal habitat the brain of the nocturnal marsupial phalanger, *Trichosurus*, shows certain resemblances to the tree-shrew brain; such resemblances, however, are of doubtful significance, and in regard to the complexity and differentiation of the visual cortex by no means close. Apart from this, the tupaiids are of course not marsupials; they are eutherian mammals and it is with the orders of the latter that they should be compared. The reproductive behaviour of tree-shrews studied in captivity has been shown to be very primitive [86]. The female gives birth to the young (usually two) in a separate nest of leaves previously constructed for them and only suckles them once every forty-eight hours, and sexual maturity is reached in four months—a very rapid rate of maturation for Primates in general. It has been argued that this type of behaviour and growth is so primitive as to exclude the tree-shrews from the Primates altogether. But primitive characters by themselves need not eliminate them from this order; it has always been well recognized that in many ways they certainly are very primitive and it is to be presumed that the basal primitive stock of Eocene or Palaeocene times must have included such primitive characters as they still retain.

Features peculiar to the tupaiids have also been invoked as supposedly demonstrating their non-primate status. These have been mentioned in preceding chapters, but, by way of example, we may here refer back to three of them. The placenta has been stated to be of an endotheliochorial type, in contrast to the haemochorial type of higher Primates. But the placenta of lemurs is even further removed morphologically from the higher Primates, for in them it is epitheliochorial in structure. The second peculiar factor relates to the lower visual centre of the brain, the lateral geniculate nucleus. In the tree-shrews, as already noted, this nucleus has a remarkably well-defined cellular lamination similar in degree to Primates and in strong contrast to insectivores. However, it has been shown experimentally that the serial order of the cell layers receiving respectively crossed and uncrossed optic fibres from the retina is not the same as that of the higher Primates. But this is also the case (albeit in a different degree) with the lemurs as compared with the higher Primates, though the former are still included in this order. Lastly, experimental observations have shown that the main motor tract from the brain, the pyramidal tract, extends down the spinal cord in its dorsal column, and not in the lateral column as it does in both the Primates and the Insectivora. Herein the tree-shrews resemble rodents, but within this order there may be variations (e.g. the porcupine) in which part of the tract descends in the lateral column. Thus the taxonomic significance of this feature is somewhat uncertain.

Even those who would exclude the tree-shrews from the Primates have expressed the view that if the former may more properly be regarded as leptictid-like insectivores, at any rate among living mammals they are the closest Primate relatives. This conclusion receives support from serological reactions in which the tree-shrews do show affinities with the Primates and, unexpectedly, according to some authorities slightly closer affinities with the higher Primates than with the lemurs. But, at the same time, they also show some serological affinities with insectivores. In a study by Hafleigh and Williams [46] it was concluded that 'the high correspondence of *Tupaia* to HSA (i.e. human serum albumin) which is within the range of other prosimians and twice the value of the insectivore (hedgehog) included in the series provides further support to the view that the tree-shrew is a true Primate'. A still more remarkable blood reaction has been shown by Murray Johnson in applying blood protein electrophoretic studies to problems of mammalian taxonomy, for whereas haemoglobin migration in Insectivora is cathodal in direction, in *Tupaia* (as in Primates) it is anodal [64].

Taking into account the different views about the Tupaioidea, these seem to reflect two major alternative interpretations (as expressed by Dr G. G. Simpson [150]—either they are the most insectivore-like Primates, or they are the most Primate-like insectivores. From the taxonomic point of view it is perhaps not of great importance which of these two concepts is the more correct. To solve the dilemma it has even been suggested that the tree-shrews should be placed apart in a completely separate order of mammals, all to themselves. We have here adopted the view that, on the balance of the total evidence available, they should still be regarded as extremely primitive members of the order Primates that branched off very early indeed from the basal primate stock in Palaeocene times, or even earlier, and that (as would be expected) they have developed some aberrant specializations peculiar to themselves. Other Primates, of course, have developed surprisingly aberrant specializations peculiar to themselves, for example the modern aye-aye and the fossil group of the Carpolestidae, but they are not for this reason excluded from the order. And if the Primate-like structural, functional and serological traits of the tree-shrews are all to be explained on the basis of the retention in common of primitive features together with the development in common of convergent features, it is surely very remarkable indeed that the patterned complexity of all these characters should so closely replicate that of the Primates.

In retrospect, then, there now seems little doubt that the reluctance of taxonomists in the past to recognize the essentially Primate affinities of the tree-shrews may have been determined by the fact that in many respects these little animals are still very primitive and generalized, and thus contrast rather strongly with the advanced characters which are taken as diagnostic of some of the other groups of Primates. But it is a remarkable fact that the order Primates is distinguished from other existing mammalian orders in just this way—that it includes representatives, still living, which reflect (in a modified form) almost all levels or grades of its evolutionary history from the most primitive to the most advanced. T. H. Huxley, in one of his essays in *Man's Place in Nature* (1863), noted this in his observation that 'Perhaps no order of mammals presents us with so extraordinary a series of gradations as this— leading us insensibly from the crown and summit of the animal creation down to creatures, from which there is but a step, as it seems, to the lowest, smallest, and least intelligent of the placental mammals.'

It is interesting to consider why, of all the existing mammalian orders, the Primates alone should be able to present such a remarkable *échelle des êtres* in the representatives of the order which are still extant. The explanation is to be found in the fact that by their arboreal mode of life, sustained from the time when the basal stock of generalized placental mammals first began to undergo diversification, each successive grade in the evolutionary series from the less advanced to the more advanced has found its own ecological niche among the branches. For example, the smallest and more primitive types, by confining their activities mainly to the more attenuated branches of the tree-tops, lead a secluded life within the protection of foliage and have thus become effectively segregated from the larger types. Because of the nature of their immediate environment, also, many of them have tended to develop nocturnal and insectivorous habits. The medium-sized Primates have found their natural environment mainly among the larger boughs and, by developing their capacity for running and leaping along the latter, are able to extend their range very considerably in rapidly moving from tree to tree across wide stretches of forest.

The development of habitual brachiating habits by the anthropoid apes, again, has allowed them to benefit from a wider range of environmental possibilities, particularly in the case of the larger types which swing among the larger boughs, or (as a secondary result of their size) even move through the undergrowth along the ground. It is probably true to say, also, that by avoiding extreme specializations, and thus preserving a considerable functional plasticity, the arboreal Primates have acquired a degree of adaptability which must have played a significant part in the conservation of the more primitive groups. In other mammalian orders, broadly speaking, each successive grade of evolutionary advance has replaced, and led to the extinction of, antecedent grades as the result of progressive specialization and competition within the same general type of environment. But in the Primates each successive grade has developed a new ecological domain, leaving behind representatives of antecedent grades (more or less modified for their local habitat of course) in occupation of the particular arboreal environment for which they had already become adapted. It may be said, indeed, that the trees of African and Asiatic forests still retain in rough outline a stratified population of Primates which represents the successive grades of the evolutionary tree of this order. And the lowliest representatives among the living Primates are the tree-shrews.

We may now picture the emergence of the Primates in the Palaeocene (or more probably in the latter part of the preceding Cretaceous Period) in the form of small arboreal creatures very similar to the modern tree-shrews, alert, active and agile among the branches, relying more and more on the discriminative potentialities of the visual sense, and less and less on the more limited scope of the olfactory sense. The protection of the foliage would presumably have conferred an advantage in the struggle for existence not available to terrestrial types, and this was undoubtedly one of the main factors which permitted the development of an increasing complexity of general organization without the need for side-tracking specializations on which the terrestrial mammals had perforce to depend for their continued existence. It is the increasing complexity of *general* organization (particularly in the higher functional levels of the brain) which gives to the Primates their distinctive capacity for a wider range of adjustments to any environmental change.

LEMURS

The existing lemurs can be divided into two somewhat contrasting groups, differing in the structural composition of the tympanic and orbito-temporal regions of the skull, the arrangement of the vascular foramina in the cranial base, the external genitalia (of the female), certain features of the nasal cavity, etc. It is this contrast in morphological details which justifies the separation of the two groups in the infraorders Lemuriformes and Lorisiformes. Yet they possess a number of distinctive characters in common, and (as we have noted, for example, in the composition of the orbital wall) certain of the features characteristic of one group may as an exception be found to be present in individual genera of the other.

Of the two groups there has in the past been a tendency to regard the lemuriforms as the more divergently specialized, mainly because the position of the bony ectotympanic ring *inside* the tympanic bulla was assumed to be the result of an aberrant development. But we have seen good reason to suppose that this disposition of the ectotympanic is, after all, a distinctive character of primitive Primates in general. It is found in all the Eocene lemuroids of which the skull is known and also (but only to a very limited extent) in *Necrolemur*, it is present in the most primitive of the Primates still extant—the tree-shrews, and from the developmental point of view it is primitive in the sense that the ectotympanic bone retains its essentially embryonic ring shape. If this interpretation is correct, it need not be assumed that the lemuriforms and the lorisiforms are each the end-products of widely diverging lines of evolution emanating from a common ancestral stock far more primitive than either, for it may well be that the Lorisiformes were secondarily derived from the Lemuriformes, that is to say after the latter had acquired the pattern of morphological characters by which, as an infraorder, they are taxonomically defined. The palaeontological evidence, incomplete as it is, supports such an inference, for all the Eocene lemurs so far known appear to come within the definition of the Lemuriformes on the basis of their skull structure, and the earliest lorisiforms so far known are the Miocene genus *Progalago* and the Pliocene genus *Indraloris*.

There is another reason for inferring that the Lemuriformes and Lorisiformes have a nearer phylogenetic relationship than is often supposed, and that is their possession of some quite peculiar specializations in common—for example in the dentition, the details of the carpal skeleton and the characteristic loop of the colon. The evidence of the dentition is particularly impressive, for both infraorders show an identity of detail in the reduction of the upper incisors (associated with a retrogression of the premaxilla), and in the formation of an elaborate dental comb by the styliform lower incisors and lower canine (associated with the development of a caniniform front lower premolar tooth). In the Eocene lemuriforms of the family Adapidae this dental specialization was not present; thus, if the Lorisiformes had become segregated as an independent group *before* the evolutionary stage represented by the adapids, it would be necessary to postulate that this extreme type of dental specialization had been developed separately in the two groups. In view of the elaborate nature of the specialization, however, and taking into account the probability that it arose as an adaptive response to some quite special needs of a unique nature, an independent development seems very unlikely. As we have seen, the dentition of the Tupaioidea gives an indication of the initial stages in the evolution of the lemuroid dental comb, for in these primitive Primates the incisors are procumbent and styliform and actually serve the function of a comb for cleansing the fur. In the later stages of lemuroid evolution the lower canines also became procumbent and styliform and were assimilated into the comb. In some of the tupaioid genera (e.g. *Anathana*) a procumbency of the canine is already evident and this clearly adumbrates the more extreme lemuroid condition.

It is by no means clear what factors were responsible for the elaboration of the dental comb as it is seen in the modern lemurs, for there appears to be nothing peculiar in the texture of their fur which would demand such a specialized structure. Possibly the tree-shrews provide a clue to this problem, following the observation by Höfer [53] that they employ their procumbent incisors for combing the long stiff hairs of their bushy tails. It may be, then, that in the early stages of the evolution of the modern lemurs before they had fully acquired the characteristic prehensility of their limbs, a bushy tail of the tupaioid type was used as a temporary expedient for its balancing functions in arboreal activities. Another possibility which has been suggested is that the common ancestral stock from which the lemuriforms and lorisiforms were derived

inhabited a region where they had to contend with severe infesta-
tions of ectoparasites, to the extent that their survival depended
on the elaboration of a fine dental comb to protect themselves
against them. These are, of course, no more than speculations,
and as such they have value only so far as they offer possible or
plausible explanations for phenomena which otherwise appear to
be inexplicable. Of the other structural specializations common
to the two main groups of modern lemurs, the serrated sublingua
is presumably a secondary corollary of the dental specialization.
On the other hand, the functional significance of features such as
the colic loop and the peculiarities of the carpal pattern is far less
obvious, though, like the dental comb, they may be taken to imply
a derivation from a common ancestral stock in which these
specializations had already become to some degree manifest.

From inferences based on the comparative anatomy of Recent
types and from the palaeontological evidence now available, it is
possible in broad outline to draw a picture of lemuroid evolution.
The earliest prosimians must have been very similar in their basic
anatomy to the modern tree-shrews, and the fossil remains of the
Plesiadapidae make it clear that tree-shrew-like creatures were
already in existence in Palaeocene times. This is not to say that any
of the known plesiadapid genera were directly ancestral to the
modern lemurs—they had probably become too far specialized in
their dentition to have played such a role. But it may well be that
some more generalized type of plesiadapid was ancestral. That
the Lemuriformes were well established as a taxonomic group by
the Middle Eocene is made evident by the various genera of the
family Adapidae. In their skull structure and their limb skeleton
(so far as the latter is known) the Adapidae conform very closely
to the modern lemuriforms while at the same time preserving a
number of features in some respects more primitive. The dentition had not yet acquired any of the distinctive characters of the
Recent types, for the lower incisors were not procumbent, the
lower canines were unspecialized, and the four premolars of the
generalized mammalian dentition were still retained. The brain of
the Eocene lemuriforms was also exceedingly primitive by comparison with their modern representatives, but it showed a significant feature in the precocious enlargement of the temporal lobe
of the cerebrum, an advance which was distinctive of these early
Primates among all other groups of Eocene mammals and which

clearly foreshadows the elaboration of this region of the cerebral cortex in the higher Primates.

It is uncertain in what geographical region the Adapidae first emerged as a taxonomic unit, but during the Eocene they spread widely over both hemispheres and became differentiated to form two subfamilies, the Adapinae in Europe and the Notharctinae in the American continent. The European subfamily, which was on the whole the more generalized, is represented by the genera *Pronycticebus* and *Adapis*; of these the former is the more primitive, particularly in the simple construction of the premolars. In *Adapis* the last premolar had undergone some degree of molarization, an elaboration that has been regarded as a specialization unlikely to occur in the ancestral line of those genera of modern lemurs in which the tooth preserves a simple, bicuspid character. In fact, however, it does not necessarily exclude *Adapis* from such a phylogenetic relationship, for the secondary simplification of the premolar by the loss of a cusp would involve no more than a negative reversal of evolution and, as we have seen (p. 8), negative reversals are quite a common phenomenon in phylogenesis.

The American Notharctinae, though very similar indeed in their skeletal and dental morphology to the European Adapinae, evidently underwent some degree of independent and parallel evolution after their segregation in the New World, for they are distinguished by the fact that in the development of quadritubercular from tritubercular upper molars the fourth cusp (hypocone) arose by a fission of the protocone and not, like the true hypocone in most Primates, as a new upgrowth from the cingulum. The earlier notharctine genus *Pelycodus*, in a succession of species, demonstrates rather clearly this derivation of the pseudohypocone which in the later genus *Notharctus* became fully developed. While it seems reasonable to assume that the Adapinae (whether the genus *Adapis* or some other generic group of this subfamily) included the ancestral stock which gave rise to the modern lemurs, it is commonly held that the Notharctinae comprised a divergent line which eventually became extinct. But the possibility has also been suggested that they provided the basis for the subsequent evolutionary development of the New World monkeys, a suggestion which conforms with the observation that in some of the Platyrrhine genera (e.g. *Callicebus*) the fourth cusp of the upper molar is a pseudohypocone of the notharctine type. Unfortunately, however, there is no fossil record of the Platyrrhine monkeys until the Miocene, by which time they had become differentiated to the extent that they already closely resembled existing types.

Following the Eocene, the palaeontological history of the Old World lemurs also shows a serious gap, for the next record of their evolutionary progress comes from the Early Miocene of East Africa and this provides good evidence that by then the Lorisiformes had become clearly differentiated. Their primary origin is obscure, but if it is now accepted that the characteristic features of the adapid skull (including the construction of the tympanic region) were not in fact aberrant specializations but basic features of the primitive Primate skull, then the derivation of the lorisiforms from the Eocene lemuriforms offers no theoretical difficulty. The Early Miocene genus *Progalago* is very closely related, and probably ancestral, to the modern African genus *Galago*. In certain details of the skull, dentition and endocranial cast it was somewhat more primitive than the latter, but the characteristic features of the Galaginae were already well established. In particular, the procumbency of the lower incisors and canines (associated with retrogressive changes affecting the upper incisors) was almost as extreme as it is in modern types. Thus between the latter part of the Eocene and the beginning of the Miocene (even though this was a matter of at least ten million years) it seems that the diversification and modernization of the lemuroid sequence of evolution proceeded rather rapidly. Meanwhile, other groups of lorisiforms, representing the ancestral precursors of the Asiatic lorisiforms, must have migrated into the southern regions of Asia, but the fossil record presents only one fragmentary specimen of these, the Pliocene genus *Indraloris*. It may be presumed also, that the precursors of the lemuriform lemurs which are now confined to Madagascar wandered south from the northern hemisphere (where their earliest fossil remains have been discovered) and in some manner in mid-Tertiary times (probably by natural raft transportation—for there is no good evidence that a land-bridge existed) reached their present habitat from the African mainland; none of their representatives have survived elsewhere. Isolated in Madagascar from the competition with many other groups of mammals which the lorisiform lemurs had to face in Africa and Asia, the lemuriform lemurs later underwent the astonishing diversification which culminated in a number of aberrant genera such as *Mega-*

ladapis, Palaeopropithecus and *Archaeolemur*, types which survived into the Pleistocene but which are now extinct. Among the most remarkable results of this late evolutionary efflorescence of the Lemuriformes was the emergence of monkey-like types such as *Archaeolemur*, in which the general features of the skull and dentition paralleled so closely those of true monkeys that on their first discovery they were actually mistaken for the latter. This parallelism has a certain taxonomic relevance, for the fact that some of the early lemuroids developed along similar lines demonstrates that they possessed a genetic constitution with evolutionary potentialities equivalent in some respects to those of the precursors of the Anthropoidea. This betokens a closeness of relationship between the lemurs and the higher Primates which (in addition to other valid considerations) seems effectively to dispose of an argument which has occasionally been advanced—that the lemurs are so divergent in their morphology from the Anthropoidea that they should be excluded altogether from the order Primates.

With one exception, all the existing genera of the Lemuriformes are clearly the end-products of an evolutionary trend culminating in such morphological characters as the retrogression of the upper incisors, the reduction of the premaxilla, the forward extension of the orbital plate of the palatine bone, the retention of the tympanic annulus in its embryonic form, the elaboration of the dental comb, and so forth. The exception is the aye-aye (*Daubentonia*), which poses a particularly puzzling problem because, in spite of the fact that it shows undoubted affinities with the lemuriforms in a number of characters—for example, the construction of the tympanic region of the skull, the elongation of the fourth digit of the pes with the enlargement of the hallux which bears a flat nail, and the presence of the ansa coli—it appears to have followed a very different evolutionary trend in the hypertrophy of the incisors (which have assumed a rodent form), the persistence of a large premaxilla and of a fronto-maxillary contact in the orbital wall, the retention of sharp claws on all the digits of the manus and pes except the hallux, and the development of a convolutional pattern of the brain which in several respects is quite unlike that of the lemurs. Certain of these characters of *Daubentonia* are apparently quite primitive characters, and it is mainly for this reason that the genus is presumed by some authorities to be the end-product of a side-line of lemuriform evolution which separated from the main line at a very ancient geological date. The discovery of the early Primates of the Palaeocene and Eocene assigned to the families Plesiadapidae and Apatemyidae, with hypertrophied incisors showing some resemblance to those of *Daubentonia*, seemed to support this conception, and, as we have seen, it has been suggested that some of these extinct types may have been ancestral to the modern genus. But it has to be admitted that there is no palaeontological evidence of intermediate types to substantiate such a thesis. It is perhaps more likely, therefore, that *Daubentonia* is really an aberrant off-shoot of the lemuriform stock which arose comparatively late—after the latter had become isolated in Madagascar. In a sense, this implies a partial reversal of the prevailing evolutionary trend which has characterized the Lemuriformes as a whole, if, as we suppose, they had already achieved the specializations which they have in common with the Lorisiformes (e.g. the dental comb) before they reached Madagascar. On the other hand, so far as the upper incisors are concerned, we know that a reversal of this kind did occur in *Archaeolemur*, for in this Pleistocene lemuriform these teeth are large and spatulate and thus very different from the reduced and peg-like upper incisors which are typical of the Madagascar lemurs.

In order to summarize the main anatomical features by which the lemuroids as a whole are distinguished from other taxonomic groups of the Primates, the prevailing evolutionary tendencies which have characterized their phylogenetic development may be summarized as follows:

(*a*) In the *skull*—progressive reduction of the premaxilla (except in *Daubentonia*)—retention of the primitive annular type of ectotympanic enclosed within the bulla (except in the Lorisiformes)—divergent modifications of the entocarotid artery which either undergoes atrophy (Lemuriformes) or takes an unusual course through the foramen lacerum (Lorisiformes)—participation of an extension of the palatine (Lemuriformes) or the ethmoid (Lorisiformes) in the medial wall of the orbit.

(*b*) In the *dentition*—reduction of the upper incisors (except in *Daubentonia*, Plesiadapidae, Apatemyidae, and *Archaeolemur*) with procumbency of stiliform lower incisors and associated modifications of the lower canine to form a 'dental comb', and caniniform specialization of P_2—frequent tendency for molarization of the last premolars.

(c) In the *limbs*—progressive development of a 'biramous' type of manus and pes—reduction of the second digit of the manus—in the carpus compression of the capitate, reduction of the lunate and displacement of the os centrale—in the tarsus a tendency towards an elongation of the calcaneus and navicular, which reaches its maximum expression in the Galaginae—compression of the meso-cuneiform—relative elongation of the fourth digits of the manus and pes—retention of a modified claw on the second pedal digit.

(d) In the *brain*—development of axial rather than limiting sulci on the dorso-lateral surface of the cerebrum—inversion of the lateral geniculate nucleus—precocious development of the temporal lobe early in the phylogenetic series.

(e) In the *special sense organs*—development in many genera of a rod retina of nocturnal type—retention of the primitive number of endoturbinals in the nasal cavity—retention of a naked rhinarium, with a median sulcus in the upper lip—elaboration of the external ear.

(f) In the *digestive system*—specialization of a serrated sublingua and marked development of thorny papillae on the pharyngeal part of the tongue—elaboration of the colon pattern to form an 'ansa coli'.

(g) In the *reproductive system*—elaboration of the penis with the development of 'grappling spurs' and a baculum—perforation of the clitoris by the urethra (in the Lorisiformes)—retention of a primitive epitheliochorial placenta of a diffuse, non-deciduate type, with a vesicular allantois and the formation of the amniotic sac from folds which secondarily coalesce.

TARSIOIDS

Our survey of the various anatomical systems of the only surviving tarsioid genus, *Tarsius*, has shown that they comprise a remarkable combination of primitive and specialized characters, and also a mixture of characters some of which are typically lemuroid while others appear to approximate closely to those distinctive of monkeys (particularly the Platyrrhines). The primitive characters may be instanced by the simple tribosphenic molars, the digital formula of the manus, the unelaborated gut pattern and the relatively smooth cerebral cortex—the specialized characters by the enormous eyes and orbital cavities and the modifications of the hindlimbs for saltatory progression—the lemuroid characters by the well-defined sublingua, the digital formula of the pes, the presence of toilet digits on the pes (though there are two of these and not one as in the lemurs), the large mobile ears, and a lateral geniculate nucleus of the inverted type—and the simian characters by the details of the rhinarium, the abbreviated snout, the restriction of the nasal cavities, the partial separation of the orbital cavity from the temporal region by an expanded alisphenoid, the basal position of the foramen magnum, the differentiation of a macula in the retina, and the enlarged occipital lobe of the brain with a highly complex visual cortex. It is not surprising, therefore, that conflicting views have been advanced regarding the taxonomic position of the tarsioids within the order Primates. Some authorities have been impressed by the lemuroid affinities (particularly the resemblances with some of the Lorisiformes in skull structure and the saltatory modifications of the hindlimb) and have associated them closely with the lemurs in their scheme of classification. Others have been impressed with the simian characters to the extent that they have separated them widely from the lemurs and placed them in a common taxonomic group (Haplorhini) with the Anthropoidea [52]. Yet others have placed them in a separate suborder occupying an intermediate position between the lemurs and the higher Primates, and the proposition has also been advanced that the Anthropoidea are derived from a tarsioid ancestry. As we have already indicated, the accretion of palaeontological material has produced evidence which is indirectly relevant to this problem. In the first place, it has emphasized the difficulty of determining with certainty (at any rate by the dentition) whether some of the earliest prosimians are related to the primitive lemurs or to the primitive tarsioids because of the mixture of characters which they show. It thus appears likely that the two groups are the descendants of a common ancestral stock, and this indicates a degree of phylogenetic relationship which is appropriately reflected in the classification which associates the tarsioids with the lemurs in the common suborder Prosimii. Second, it is clear from the fossil record that the tarsioids became differentiated and achieved their characteristic infraordinal specializations extremely early—certainly by the Lower Eocene and possibly even during the Palaeocene. For example, the Eocene types *Necrolemur, Nannopithex, Hemiacodon* and (to a lesser extent) *Teilhardina* show the elongation of the calcaneus, and in the case of the first two genera the tibio-fibular fusion, while the gross enlargement of the orbital cavities was already fully developed in *Pseudoloris*. Further, the endocranial casts of *Tetonius* and *Necrolemur* show that in these early times even the brain had approximated closely to the proportions distinctive of the modern tarsier. The European genus *Pseudoloris* is of particular interest because, though a smaller creature, its skull and dentition are almost a replica of those of *Tarsius* (except that, *inter alia*, in its lower dentition it has a vestigial first premolar and an enlargement of the most anterior tooth). It may thus be truly said that *Tarsius* is a 'living fossil', in the sense that it is but a slightly modified successor of a tarsioid genus which existed in Eocene times.

Clearly, if the tarsioids as a taxonomic group had already developed their peculiar specializations to such an advanced degree in the early part of the Eocene, it is difficult to suppose that they could have provided a basis for the subsequent evolution of the Anthropoidea in which such specializations are absent. It is thus only possible to maintain the thesis of a tarsioid ancestry for the higher Primates if it is supposed that the latter took their origin (presumably in Palaeocene times) from a basal pro-tarsioid stock before the characteristic tarsioid specializations had become established. But, on the morphological criteria by which the infraorder Tarsiiformes is usually defined, it is doubtful whether such a basal stock could properly be termed tarsioid. As we have already suggested, it may be that some of the advanced, or simian, characters of the skull which have been attributed to the modern *Tarsius* are illusory, the result of fortuitous resemblances. For example, the reduction of the snout is to a great extent only apparent, its posterior part being overlapped and concealed by the expanded orbits; the restriction of the nasal cavity is also dependent upon the enlargement of the specialized orbits and seems to have come about in a manner quite different from the analogous changes in the monkeys; the reduction of the olfactory parts of the brain may be a secondary consequence of these peculiar modifications of the skull; the broad and rounded shape of the brain as a whole is perhaps attributable to the distortion and antero-posterior compression of the brain-case produced by the large orbits; the displacement of the foramen magnum basally and the participation of the alisphenoid in the posterior wall of the orbit is possibly also secondary to the enormous development of the eyes. Thus, these features (or at any rate some of them) may not be morphologically equivalent in the strict sense to the corresponding features in the Anthropoidea.

But, apart from these examples of possible convergence, there remain a number of features which, taken together, seem clearly to betoken a real affinity between *Tarsius* and the higher Primates, particularly in the differentiation of a retinal macula, the intrinsic structure of the visual cortex, the construction of the rhinarium, the haemochorial placenta, and so forth. It is, of course, possible that some of these morphological resemblances have developed in the tarsioid sequence of evolution independently of the Anthropoidea, but even so they are at least indicative of a community of origin from an ancestral stock with common potentialities for evolutionary development in similar directions. In other words, it may be inferred with reasonable assurance that they were initially derived from prosimian ancestors which were quite closely related.

It is not feasible to draw more than a very fragmentary picture of the evolutionary radiations of the tarsioids, for the reason that the fossil remains of many of the known genera are themselves very fragmentary. The evidence for the tarsioid nature of some of the types is based only on the dentition, but the fact that in a few genera this diagnosis is confirmed by the evidence of the skull, endocranial cast and limb bones allows us to place some confidence in this criterion of dental morphology. The place of origin of the Tarsioidea is uncertain. Seeing that early forerunners of the various subdivisions of the Primates are so richly represented in the Old World, it has been supposed that the group emerged from the basal Primate stock somewhere in the Eastern Hemisphere. On the other hand, fossil tarsioids were already in existence in North America in the Middle Palaeocene, and the earliest known European tarsioid (*Teilhardina*) is of later date, Lower Eocene. Whatever may be the case, it is certain that almost at the beginning of the Tertiary epoch primitive tarsioids had spread widely into both the Old and New World. They may have undergone an independent evolution in the European and American continents, or possibly successive waves migrated to both regions from some Asiatic source.* Only one subfamily is known to have been com-

* A fossil prosimian from Asia, *Hoanghonius*, found in deposits of Oligocene or late Eocene date in China, has been assumed to be a tarsioid. But this type is represented by no more than a few mandibular fragments with part of the lower dentition [172].

mon to the Old and New Worlds, the Omomyinae. So far as the dentition is concerned, these are among the most generalized of the fossil tarsioids. In the Old World they are represented only by the European genus *Teilhardina*; in North America they flourished over a lengthy period extending from the Upper Palaeocene to at least the Upper Eocene or Lower Oligocene, the latest of this series including the genera *Chumashius* and *Macrotarsius*.

Apart from the Omomyinae, the American tarsioids tended towards rather extreme specializations in the dentition, manifested in a gross hypertrophy of the front teeth and, in some genera, an enlargement and lateral compression of the last premolars. At the same time the molars generally retained their primitive tribosphenic form. The specialization of the last premolars reached its culmination in the genus *Carpolestes*, in which the teeth developed a serrated, shearing edge. This exaggerated modification of the dentition has led some authorities to regard it as a familial distinction, and to recognize this by creating a new tarsioid family, Carpolestidae. Others have questioned whether the Carpolestidae really are tarsioids, or even Primates. But the Primate status of the group seems well assured by the character of the molars, and in fact *Carpolestes* represents no more than the extreme of a trend which is also evident in other related tarsioids. For example, as we have seen (p. 107), the hypertrophy and compression of the last premolar is quite obtrusive in *Tetonius*, whose tarsioid status would be difficult to contend, and several of the American genera provide a fairly closely graded morphological series leading up to the extreme condition of *Carpolestes*.

In the European region of the Old World it is evident that, as in the New World, the Eocene tarsioids diverged in their evolutionary development along a number of different lines. Some, such as the genus *Periconodon*, developed rather unusual features in the cusp pattern of the molar teeth in which they show some resemblance to certain Platyrrhine monkeys. Others showed a marked propensity for developing quadritubercular molars which in some respects approach the molars characteristic of the higher Primates. These genera include *Necrolemur* and *Microchoerus* which are grouped together in the subfamily Necrolemurinae. Apart from its molars, *Necrolemur* shows in its skull certain traits which have also been regarded as advanced, or 'pithecoid', characters, such as the disposition of the entocarotid foramen and the development of a tubular auditory meatus. But the latter, as we have noted, is not wholly externalized as it is in the higher Primates—part of it still retains a position within the tympanic bulla. Both *Necrolemur* and *Microchoerus* are also specialized to the extent that the front tooth of the lower dentition (probably the canine) is markedly hypertrophied. The genus *Caenopithecus* perhaps belongs to the same group, but the precise status of this fossil is uncertain. By some authorities it is associated with the Adapidae, but the characters of the upper dentition (in particular, the abrupt contrast between the premolars and molars) suggest tarsioid affinities. The fragmentary remains of the skull indicate that the orbits are not so large as in *Tarsius*, the zygomatic bone is strongly built, there is a pronounced recession of the snout region, and the mandibular symphysis is synostosed. In some respects, indeed, the skull of *Caenopithecus* certainly seems to have been more pithecoid in appearance than that of other Eocene prosimians.

Except for *Pseudoloris* (which may have been ancestral to the modern *Tarsius*) and the more generalized representatives of the Omomyinae, most of the known tarsioid genera manifest a degree of specialization in one direction or another which makes it unlikely that they themselves provided the basis for the subsequent development of any of the modern Primates, though some of them may have been quite closely related to such ancestors. The question naturally arises why so large a proportion of fossil prosimians showed pronounced specializations, and why so few manifested the more generalized morphology which might be supposed to characterize the ancestral precursors of the Primate groups which exist to-day. The explanation depends on the fact that in the course of evolutionary diversification those types which become successfully stabilized by adaptive specializations for a particular environment are likely to endure for a considerable period of time, at any rate so long as the environment remains unchanged. Their total population over such a time period is thus likely to be much greater than that of the generalized types which are transitional from lower to higher levels of evolutionary advance, for, in so far as they are transitional in type, these tend to be transitory in time (particularly in periods of rapid evolutionary diversification). It follows, therefore, that the probability of preservation as fossils (always a slender probability) is much less for the transitional types simply because in their case the total population in time and space from which a random individual may by chance be perpetuated in fossilized form is smaller. Yet another factor needs to be re-emphasized in this connection. Most of the Recent Primates are arboreal, forest-living types of tropical habitat, and those that have acquired terrestrial habits have only done so secondarily. It may be presumed, therefore, that the main course of the evolutionary history of the Early Primates was pursued in forested territory. But if, even under the most favourable circumstances, the chances of an individual being preserved as a fossil are slender, they are very much more remote in the environment of humid, equatorial forests. It may be, then, that many of the prosimian fossils so far discovered were marginal, aberrant types, adapted for some other kind of terrain which would render fossilization less improbable.

The great diversity of prosimian types which spread widely over the world during the Palaeocene and Eocene makes it clear that it was a period of intense evolutionary activity for the basal Primates. To-day the Primates comprise a relatively limited number of sub-orders, infraorders, and families. In those early times the groupings were certainly more complex, for many of the initial evolutionary radiations were abortive experiments and terminated in extinction. Thus, attempts to classify them in terms of the major taxonomic groups of the surviving Primates are likely to involve an element of arbitrariness. But it is desirable as a matter of practical convenience to classify them in accordance with these well-recognized categories so far as this can be done without stretching the definitions of the latter to the point of extravagant compromise, even at the risk of over-simplifying their taxonomy—at any rate as a provisional device pending further amplification of the fossil record. The multiplication of major taxonomic groups to accommodate separately all the varied types of extinct Primates may ultimately be justified when more complete palaeontological evidence accrues; to multiply them prematurely may lead to unnecessary confusion.

We may summarize the distinguishing characters of the infraorder Tarsiiformes by listing their main evolutionary tendencies as follows:

(a) In the *dentition*—reduction of the incisors usually to two or one (the lower incisors disappeared altogether in some of the fossil genera such as *Tetonius* and *Necrolemur*)—enlargement to varying degrees of the canine—retention of simple premolars and in several genera an early loss of the first premolar—marked specialization of P_4^4 in some of the American tarsioids—molars either retaining the primitive tribosphenic form (American tarsioids, *Pseudoloris*, *Tarsius* etc.), or becoming quadritubercular (*Necrolemur*) or even multitubercular (*Microchoerus*).

(*b*) In the *skull*—progressive enlargement of the orbits accompanied by a marked compression of the nasal cavities—participation of the ethmoid in the medial wall of the orbit and of the alisphenoid in the posterior wall—displacement of the foramen magnum to the basal aspect of the skull—the ectotympanic produced into a long meatus by ossification of the annular membrane and subsequently becoming wholly extruded from within the bulla—entrance of entocarotid artery through the bulla wall.

(*c*) In the *limbs*—the retention of a primitive structure in the forelimb, including the digital formula and the disposition of the carpal elements—progressive and early specialization of the hindlimb for saltation, associated with tibio-fibular fusion and extreme lengthening of the calcaneus and navicular—retention of modified claws on the second and third pedal digits.

(*d*) In the *brain*—a broadening of the cerebrum as a whole with expansion and pronounced structural differentiation of the visual cortex and the formation of a prominent occipital lobe—inversion of the lateral geniculate nucleus—retention of primitive features in the cerebral commissures and the cerebellum.

(*e*) In the *special sense organs*—development of a nocturnal type of retina with the differentiation of a macula—reduction of endoturbinals following a constriction of the nasal cavities by the enlarged orbits—disappearance of a naked rhinarium—elaboration of the external ear.

(*f*) In the *digestive system*—retention of a sublingua of simple form and of a primitive colon pattern.

(*g*) In the *reproductive system*—retention of primitive features in the external genitalia and in the bicornuate form of the uterus—development of a haemochorial discoidal placenta with precocious vascularization of the chorion and a rudimentary allantois.

MONKEYS AND APES

That the fossil record of the earliest stages of the phylogeny of the Anthropoidea is still meagre is presumably due to the fact

(which we have already stressed) that the main lines of their evolutionary development were followed by arboreal creatures whose
natural habitat was in forested regions. Only by a happy combination of fortuitous circumstances is it likely that individuals in such
an environment are likely to be preserved as fossils which eventually become available for study. Interesting examples of such
chance circumstances have been found in the Fayum desert region
of Egypt, the Siwalik Hills of India, and in the richly fossiliferous
deposits of lacustrine origin which occur in the region of Lake
Victoria in East Africa. The East African deposits, which are of Early
Miocene date, contain numerous remains of arboreal Primates—
lemurs (*Progalago*), cercopithecoid monkeys (*Mesopithecus*), and
primitive anthropoid apes, e.g. *Pliopithecus* (*Limnopithecus*) and
Dryopithecus (*Proconsul*). The geological evidence (as well as that
of the fossil fauna and flora) indicates that in Miocene times the
local environment there consisted of forests of the 'gallery' type
extending along the banks of rivers which flowed into a lake occupying what is now the eastern side of Lake Victoria. It seems
evident that the remains of some of the creatures inhabiting these
forested valleys were carried down by the rapidly flowing rivers
and then quickly covered over by deposits of lacustrine silt. But
such conditions favourable to the fossilization of forest-living
creatures are probably unusual—at least, deposits of a similar
nature have only rarely been found. The fossiliferous deposits in
Egypt were probably laid down under somewhat similar circumstances, for the Fayum area was evidently richly forested in Oligocene times and intersected with rivers. It is in these deposits that
some of the earliest fossil Anthropoidea have been found, e.g. *Oligopithecus* and *Aegyptopithecus*, ranging in antiquity up to 30
million years ago [140]. From the Miocene site in East Africa the
earliest remains of undoubted cercopithecoid monkeys have been
recovered. Unfortunately these remains are limited to fragments of
jaws with portions of the dentition. The molars had already developed the bilophodonty which is distinctive of the modern cercopithecoids; indeed, the dentition (so far as it is known) is hardly
distinguishable from modern colobine monkeys except, possibly,
on the basis of small metrical differences. It is evident, therefore,
that the sub-order Cercopithecoidea was already well differentiated
by the early part of the Miocene. The discovery of other remains of
Mesopithecus in Lower Pliocene deposits of southern Europe, and

of the allied genus *Dolichopithecus* as well as the modern genus *Macaca* in Pliocene deposits of later date in France, shows, further, that by the end of the Miocene the suborder had begun to undergo diversification into a number of types and to spread widely over the Old World. But of the immediate precursors of *Mesopithecus* we have only dubious evidence. A number of jaws with teeth from the Oligocene of Egypt (*Parapithecus* and *Apidium*) have been interpreted as primitive cercopithecoids in which the bilophodonty of the molars had not as yet become fully established, but this interpretation has been seriously questioned (see p. 117). The Parapithecidae, and also *Amphipithecus* from the Eocene of southern Asia, both of which retained the primitive dental formula that still characterizes the New World monkeys, i.e. with three premolars, may be basal representatives of the cercopithecoid stock preced ng the stage of bilophodont specialization, but they may equally well be (and have commonly been assumed to be) extremely primitive hominoids having affinities with the Hylobatinae. A satisfactory statement on the taxonomic status and phylogenetic affinities of these early types must await the discovery of more complete remains.

The evolutionary origin of the Platyrrhine monkeys is even more obscure, for the earliest fossil record of this group, from the Miocene of Colombia and Patagonia, has brought to light genera (*Homunculus*, *Cebupithecia* and *Neosaimiri*) which are quite closely similar to Recent ceboid genera. We have noted the possibility that the Platyrrhines developed independently of the Old World monkeys from the notharctine group of lemuriforms, a possibility which seems feasible when account is taken of the fact that opportunities for faunal interchange of tropical arboreal types between the Old and New World were probably very unfavourable from the Eocene onwards. On the other hand, the fact that the blood serum proteins indicate a closer affinity between the New World and the Old World Anthropoidea than between either of them and the modern lemurs suggests that they had a common origin from a non-lemuroid Primate stock. On the basis of the comparative morphology of the different infraorders of the Prosimii, it is reasonable to assume that such an ancestral stock is likely to have been more closely related to the Tarsiiformes than to the others. Here, again, the problem of the phylogenetic relationship of the New World monkeys can only be determined when the palaeontological record becomes more completely documented.

It is possible with some reason to argue the thesis that the whole group of cercopithecoid monkeys represents a specialized sideline of evolutionary development which branched off from the main line of hominoid evolution towards the end of the Oligocene, and which followed an aberrant course in its dental specialization and in other features such as conspicuous ischial callosities, cheek pouches, and so forth. In one sense such an interpretation of the fossil record might be valid so far as it applies to the *later* representatives of the Cercopithecoidea (i.e. the end-products of cercopithecoid evolution), but in another sense it would also be true to say that the anthropoid apes passed through a cercopithecoid phase of evolution because in many respects the Cercopithecoidea are morphologically the more primitive and generalized. So far as dental morphology is concerned, at any rate, the fossil record of the Hominoidea can be traced back into the Lower Oligocene (that is, before the earliest evidence of fully differentiated cercopithecoids), and even into the Eocene if the identification of *Amphipithecus* as a primitive hominoid is accepted. That the mandibles and lower dentition of the Parapithecidae belong to a primitive Oligocene hominoid has been questioned on the grounds that, although the molars quite closely resemble those of the early fossil apes *Propliopithecus* and *Pliopithecus*, their cusp pattern is of a generalized type not uncommonly found in other early mammals such as the condylarths. But a consideration of the dentition as a whole—particularly the premolar series taken in conjunction with the molar characters— makes it certain that the Parapithecidae were Primates, and among the latter it finds its closest affinity with the Hominoidea. On the other hand, the conformation of the mandible and the characters of the front teeth also suggest an affinity with the tarsioids, and it is this evidence which has led to the supposition that the Hominoidea were directly derived from a basal stock of Primates closely allied to the tarsioids. While there remains some uncertainty about the taxonomic status of the Parapithecidae, there is less doubt about *Propliopithecus*, also of Oligocene antiquity. The dentition of this latter genus, though more primitive, resembles so nearly that of the Oligocene and Miocene hylobatine genera that their relationship seems well assured.

It may be accepted, then, that small anthropoid apes with a hominoid, and in some genera a gibbon-like, dentition had come into existence by the early part of the Oligocene period. Hominoid evolution evidently proceeded with some rapidity at that time, for by the Early Miocene East Africa was populated by a great diversity of apes of all sizes. Among these are the primitive gibbon-like creatures, *Pliopithecus*, whose probable relationship to *Propliopithecus* we have just noted. The limb-bones of *Pliopithecus* have demonstrated that it was by no means so specialized as the modern gibbon, even though in some of their morphological features the limbs show incipient changes which, so to speak, adumbrate the extreme modifications characteristic of Recent hylobatines. It seems probable that the East African species of *Pliopithecus* provided the basis for the subsequent evolution of fossil types of the same genus which, in later Miocene and in Pliocene times, spread widely over Europe and Asia, and those that reached the Far East in the Pliocene presumably gave rise to the modern gibbons. The remains of Pleistocene gibbons of modern type have been found in Java, and also in Szechwan in China [54]. The larger apes of the Early Miocene of East Africa (probably the descendants of smaller and more primitive pongids related to the *Propliopithecus-Pliopithecus* group) are represented by the genus *Dryopithecus* (*Proconsul*), of which, as already noted, there were three species ranging in size from that of a pygmy chimpanzee to that of a large gorilla.

The genus *Dryopithecus* is characterized by a dentition which is fundamentally of the pongid type, approximating quite closely to the dentition of the modern large anthropoid apes except that the incisors are not hypertrophied to the same extent and the cusp pattern of the molar teeth is more simple. The great variety of the Early Miocene apes in East Africa suggests that during that period the evolutionary diversification of the Pongidae was proceeding rather rapidly, and it may well be that this region constituted a centre of dispersal from which they gradually extended their territory into other parts of the world. By the Middle Miocene and during the subsequent Pliocene, many genera of apes spread into the Mediterranean area, Central Europe, and as far east as India. In Europe they were mainly represented by *Dryopithecus* and in India by a variety of types in one of which (*Ramapithecus*) the dentition shows some interesting resemblances to that of primitive hominids. The diversification of pongid genera in the later Miocene and Early Pliocene of India makes it likely that this region provided a secondary centre of dispersal, and it was probably from this centre that the ancestors of the modern orang-utan spread eastwards to find the present habitat which the latter occupies in Malaysia. Such an inference receives some support from the fact that one of the Middle Miocene genera from the region of India, *Palaeosimia*, shows in its molar pattern striking similarities with the orang.

Undoubtedly the most important discovery which has emerged from a study of the fossil remains of *Dryopithecus* is that, although the dentition is definitely of the pongid type and the skeleton presents certain pongid characters in many details of the skull and limb bones, the genus still preserved primitive features corresponding to a cercopithecoid phase of evolution. Their morphological details make it clear that these Early Miocene apes had certainly not achieved the degree of specialization found in the large apes of to-day. In particular, they had not developed the extreme type of brachiation characteristic of the modern apes; they were evidently capable of running and leaping among the branches and scampering along the ground like the modern cercopithecoid monkeys, and (at least in the smaller species) they must have been lightly built and agile creatures. It seems, then, that while the Cercopithecoidea had practically completed their evolutionary differentiation by the early part of the Miocene period, the Pongidae were still in process of developing their own distinctive specializations. In the details of the cusp pattern of the molars, *Dryopithecus* shows certain features which might be regarded as minor specializations, but in spite of these it seems likely enough that in East Africa the genus provided the ancestral stock from which the modern chimpanzee and gorilla were eventually derived. The remarkable variety of contemporaneous types in the East African Miocene clearly indicates that they must have occupied different ecological habitats, and it may be assumed that their diet was mainly vegetarian. But it is interesting to note that the ground-living baboons of today, and even chimpanzees, have been observed in the wild occasionally to capture and eat small animals. This sporadic carnivorous propensity may have been developed to a greater extent by some of the early fossil apes where vegetable matter was scarce, for example in savannah country. Particularly may this have been the case with those types that in their dental morphology appear to adumbrate the evolution in later times of the early hominids which, it now seems well assured, hunted animals for food. That the Dryopithecinae, and in particular the Late

AOM Z

Miocene and Early Pliocene genus of this subfamily, *Ramapithecus*, with a geographical range extending from India to East Africa, may have provided the ancestral basis for the later evolution of the Hominidae is also a perfectly feasible proposition, but before commenting further on this thesis it is convenient to summarize the prevailing evolutionary tendencies by which the suborder Anthropoidea as a whole has been distinguished from other groups of Primates. They were as follows:

(*a*) In the *skull*—progressive reduction of the snout region accompanied by a restriction of the nasal cavities—flexion of the basicranial axis and recession of the facial skeleton below the front part of the brain-case—great expansion of the brain-case—complete rotation forwards of the orbital apertures—relative enlargement of the entocarotid artery—participation of the ethmoid in the medial wall of the orbit—separation of the orbit from the temporal fossa by the expanded alisphenoid—ectotympanic forming a tubular auditory meatus accompanied by the disappearance of a prominent bulla (except in the Platyrrhines)—expansion of the frontal bones—displacement of the foramen magnum on to the basal aspect of the skull.

(*b*) In the *dentition*—incisors spatulate in form and consistently $\frac{2}{2}$ — canines often greatly enlarged, particularly in the male sex, but may become secondarily reduced (as in the Hominidae)—premolars bicuspid and finally reduced to two—quadritubercular molars.

(*c*) In the *limbs*—retention of primitive mammalian features in both fore- and hindlimb, including the disposition of the carpal and tarsal elements and the digital formula—transformation of claws into flattened nails on all the digits (except in the Callitrichidae)—progressive elongation of the forelimbs and retrogression of the pollex in the more arboreal types.

(*d*) In the *brain*—great expansion of the cerebral hemispheres which become richly convoluted—characteristic development of the central sulcus, the lunate sulcus, and other limiting sulci—very marked reduction of olfactory centres and a corresponding expansion and differentiation of the visual centres—elaboration of the cerebellum—eversion of the lateral geniculate nucleus.

(*e*) In the *special sense organs*—differentiation of a macula at the central point of the retina—reduction of the external ear—dis-

appearance of a naked rhinarium—marked reduction of the turbinal processes of the nasal cavity.

(*f*) In the *digestive system*—retention of primitive features in the
tongue—differentiation of the colon into ascending, transverse,
and descending portions—development of a vermiform appendix
in the higher types (but absent in all monkeys).

(*g*) In the *reproductive system*—retention of primitive features
in the external genitalia, except for a small baculum in the monkeys and apes—corpus uteri simplex—haemochorial, deciduate
placenta—very precocious vascularization of the chorion and spon-
taneous development of the amniotic sac.

THE EMERGENCE OF THE HOMINIDAE

It is now well recognized that the anthropoid ape family (Pongidae), and the family which includes *H. sapiens* as well as other
genera and species now extinct (Hominidae), show in their anatomical structure much closer affinities with each other than either
does with other groups of Primates. It is primarily for this reason
that they are grouped together in the same superfamily Hominoidea
and are thus contrasted with the other superfamilies of the suborder Anthropoidea. This classification carries the implication that,
at some time in the geological past, the Hominidae and the Pongidae arose as divergent evolutionary lines which originated in a common ancestral stock. There is a general agreement on this thesis,
but there is a considerable division of opinion regarding the geological period when the lines commenced their divergence. Probably most would agree that, by analogy with what is known of
rates of evolutionary differentiation and diversification in other
groups of mammals, the Hominidae and Pongidae became segregated as independent evolutionary sequences during the Miocene
or perhaps in the very early part of the Pliocene. But some authorities have supposed that the segregation occurred much earlier—
in the Oligocene or even in the Eocene period. Such extreme views
may perhaps have been dictated by too superficial an impression
of the contrasts between pongids and hominids, an impression
which is likely to obtrude itself if comparisons are limited to
modern man and modern apes. If, for example, *H. sapiens* is directly
(and exclusively) compared with, say, a gorilla, the morphological
differences are of course very striking—particularly in the size of the
brain, the development of the jaws and teeth, the proportionate

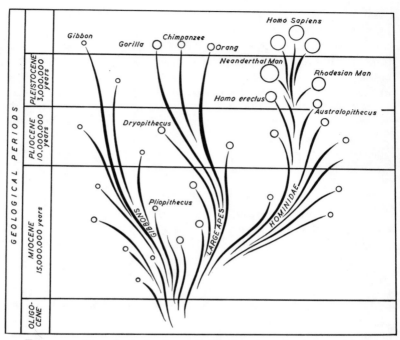

FIG. 152. Genealogical schema indicating the relationships of the Hominidae and the anthropoid apes so far as these can be inferred from the fossil evidence at present available. The circles are intended to represent approximate differences in the relative size of the brain. Note that the brain evidently did not begin to undergo the expansion characteristic of modern man until the Early or Middle Pleistocene.

dimensions of the limbs and so forth. But if extinct hominids of primitive type (with small brains, massive jaws, and large, powerful teeth) are compared with the more generalized pongids of former days (with a less specialized dentition and with limbs not showing the extreme disproportions of their modern successors), the anatomical differences between the two groups appear much less abrupt. As we have already emphasized, it is important in this connection to recognize that the familial term Hominidae and its adjectival form hominid are by no means limited to the genus *Homo.* The latter is the terminal product of a long evolutionary sequence and, as far as we know at present, only came into existence towards the early part of the Middle Pleistocene. Further, it was only with the emergence of the species *H. sapiens* that the brain expanded to the dimensions characteristic of modern man. In the more ancient hominid species *H. erectus*, of the early part of the Middle Pleistocene, the mean brain volume (as estimated by the cranial capacity) was considerably less than that of *H. sapiens*, and in the Early Pleistocene *Australopithecus* it could hardly have been much larger than in the modern gorilla. When we compare all the known types of hominid, living and extinct, with all the known types of pongid, living and extinct, it appears that the main anatomical differences which were developed in the earlier stages of their evolution are to be found in the morphological details of the dentition, and in the structural adaptations in the one group for arboreal brachiation and in the other for erect bipedalism.

The question naturally arises whether any of the known Miocene or Pliocene genera of anthropoid apes might have provided the basis for the subsequent evolutionary development of the Hominidae. Most of these fossils have been regarded as bearing a not very distant relationship to the Recent anthropoid apes, but it is also the case that some of them exhibit generalized features which might equally well indicate a relationship to the progenitors of the Hominidae. Such a proposition has met with two main objections, neither of which can be considered valid. One of these is based on the assumption that the large, pointed canines characteristic of the Pongidae are specialized features and therefore could not be antecedent to the smaller and apparently more simple canines characteristic of the hominid dentition. But, as we have noted, there is good evidence, partly based on the morphological features of the modern human canine and partly on that of fossil hominids, that the former represents a regression from a more powerfully constructed canine and is thus the result of a secondary simplification. The other objection seems to have been based on the tacit assumption that because the dentition of the fossil apes is typically pongid therefore the creatures were like the modern anthropoid apes in other respects—including the extreme modifications of the limbs for specialized habits of brachiation. Such an assumption we now know to be erroneous; the structure and proportions of the limbs of the Early Miocene apes were remarkably generalized and, indeed, of just such a type as might have become modified in the course of evolution in the direction, either of the brachiating limbs of the modern Pongidae, or of the limbs of the bipedal

Hominidae. On purely morphological considerations, then, there is no theoretical impossibility in the proposition that some group of the known Miocene apes may have been ancestral to the Hominidae; indeed, it may even be said to be a likely proposition. The time factor can hardly be said to raise a difficulty in the way of its acceptance, for if we suppose the hominid sequence of evolution to have become segregated from the pongid sequence in the Middle Miocene, this would allow at least twenty million years for the progressive development of all those hominid features which finally reached their fullest expression in *H. sapiens*.

Although we have fairly abundant fossil evidence of anthropoid apes throughout the Miocene and Pliocene periods, no indubitable record of the Hominidae is known before the beginning of the Pleistocene. As we have noted, it has been suggested that the Lower Pliocene ape *Oreopithecus* is a very early representative of the Hominidae, but the peculiarities of its dentition and the proportions of its limbs (evidently related to a specialized degree of brachiation) do not accord with such an interpretation. The earliest known, unequivocal, hominid is *Australopithecus*. a genus which was in existence in the Early Pleistocene. One of the outstanding features of this genus is the small size of the brain, and it was no doubt because of this feature that its hominid status was at first rather hotly contested. But all the important morphological features of the dentition are fundamentally of the hominid type (and very different indeed from the dentition of all known pongid genera, Recent or extinct), the skull shows a combination of features which, in spite of its general ape-like proportions, approximates to the hominid type of skull, and the postcranial skeleton, in particular the pelvis, establishes beyond question that it was adapted for an erect, bipedal gait of the hominid type. In other words, as we have previously emphasized, *Australopithecus* had reached an advanced stage of development in the direction of evolution which has characterized the Hominidae, and divergent from the direction followed by the pongid sequence of evolution. Incidentally, the small size of the brain in this genus has finally settled the question (previously debated on the speculative basis of indirect evidence) whether in hominid evolution the enlargement of the brain preceded the achievement of an upright posture, or vice versa. The evidence of *Australopithecus* in this regard is entirely consonant with that of the hominid species *H. erectus* in which though the limb structure (so far as it is known) seems to be in-distinguishable from that of modern man, the brain was consider-ably smaller. While the *Australopithecus* phase of hominid evolution is represented by abundant fossil remains from South Africa, there is now definite evidence that creatures of a similar type occupied more northern regions of the African continent, and perhaps (but the evidence for this is not very secure) even to spread the eastern regions of Asia.

The remarkable discoveries of australopithecine remains in South and East Africa have led some palaeontologists to recognize among them several different genera, and this multiplication of generic types has no doubt complicated unnecessarily the dis-cussions about their status in hominid phylogeny. Thus, the literature on these extinct hominids contains references to genera under the names of *Australopithecus*, *Plesianthropus Paranthropus*, *Zinjanthropus* and *Telanthropus*. We have here taken the more conservative view that all these types represent so many species or local varieties of a single genus, *Australopithecus*, for the morpho-logical evidence does not appear to warrant more distant relation-ships. In this connection, it seems proper to remark that palaeo-anthropologists should be urged to refrain from creating any new species or genus on the basis of a fossil specimen (particularly if it is fragmentary) *unless it can be demonstrated with reasonable assurance that the skeletal and dental characters of the specimen de-viate from those already known to an extent at least equivalent to the differences between recognized species or genera in Recent representa-tives of the same or allied groups.* It needs to be recognized, also, that no new fossil species or genus should be created by their discoverer without a formal diagnosis of the type, for it is only by reference to formal diagnoses that other students can judge whether a newly created species or genus is 'true' or not. *Australo-pithecus* and '*Plesianthropus*' are so similar in their morphological details that even a specific separation is hardly justified. On the other hand, the type called '*Paranthropus*' shows some distinctive traits in its larger size, certain characters of the dentition, and the exaggerated size of the jaws which has led to the development of powerful temporal muscles and, to accommodate their attachment on the skull, a low sagittal crest on the cranial roof. Such differ-ences, however, appear to call for no more than a specific distinc-tion at the most. The three 'genera' just mentioned have been based on numerous fossil remains found at different sites in South Africa. *Telanthropus*, discovered at one of these sites, has been distinguished as a separate genus on the basis of its smaller molar teeth and minor details of the maxillary region of the skull, but the remains of this type are so fragmentary that, at least for the present, it may with more propriety be regarded as an extreme variant of *Australopithecus*. That the australopithecine population displayed a considerable variability is of course not in doubt, but such a degree of morphological variation is to be expected in rapidly evolving types taxonomically representative of a single genus. In 1959, an almost complete Australopithecine skull was found in basal deposits of the Olduvai gorge in Tanganyika [71] and, in spite of the fact that it is patently so similar in its characters to the type which had been designated '*Paranthropus*', it was distinguished by yet another generic name, *Zinjanthropus*. Still more recently a mandible and some limb bones were found at a lower level in the Olduvai gorge. Details of these new specimens have yet to be fully analysed, but it can be said from the available reports that the jaw and lower dentition, except for rather trivial differences in the over-all proportions of the premolar teeth whereby they deviate from the South African specimens of *Australopithecus* (so far known), confirm its relationship to this genus; indeed they do not appear to merit even a specific distinction from some of the South African fossils. It seems probable that the East African australopithecines antedated those of South Africa. Indeed, according to the potassium-argon datings of Evernden and Curtis [74], the antiquity of the Olduvai specimens is of the order of 1·75 million years and the lower end of a humerus found in Early Pleistocene deposits near Lake Rudolf, and attributed to *Australopithecus*, has similarly been dated to 2·5 million years. At any rate, so far as can be determined on the evidence at present available, the *Australopithecus* phase of hominid evolution certainly extended over the early part of the Pleistocene—a matter of half a million to two or three million years ago, and (in South Africa) into the Middle Pleistocene.

The succeeding *H. erectus* phase probably lasted during the next two or three hundred thousand years, and it is certain that the two phases overlapped in time to some extent in different parts of the world. Like *Australopithecus*, *H. erectus* was characterized by massive jaws and large teeth, though both these features were somewhat less pronounced. But the brain had increased considerably in volume, with concomitant changes in the brain-case, and the limbs were evidently more perfectly adapted for an upright posture and gait. The representatives of *H. erectus* which inhabited China in the Middle Pleistocene were advanced in culture to the extent that they fabricated crude stone implements and had learnt the use of fire for domestic purposes. If the terms 'man' and 'human' are by convention applied to those later hominids which had acquired the intellectual ability not only to *use* tools but to *make* them, they are certainly applicable to *H. erectus*. It has now been demonstrated that they are also applicable to *Australopithecus*, for crude stone implements, called 'pebble tools', have been found both at the South African and the East African sites in the same deposits containing australopithecine remains and in close association with them (and no more advanced type of hominid has yet been detected with certainty in these deposits).

For many years, *H. erectus* was known only from fossil remains found in the Far East. There have since been found in Algeria fossil jaws and a parietal bone which so closely resemble those of the Chinese variety of *H. erectus* that they are really not distinguishable. In spite of this, these remains have been given a new name, *Atlanthropus*, an appellation which is unfortunate not only because it seems unjustified, but also because it introduces an unnecessary complication and may obscure their real significance for human evolution. These Algerian remains were found in association with stone implements referable to the Early Acheulian phase of Palaeolithic culture, and there is reason to suppose that they date from the early part of the second interglacial period, probably more or less contemporaneously with the later Javanese fossils but antedating Swanscombe man (p. 358). Another possible representative of *H. erectus* is 'Heidelberg man', known only from a massive lower jaw discovered in 1908, and probably of somewhat greater antiquity than the Algerian fossils. But the dentition of this type shows certain features which have led some authorities to think that its relation to *H. erectus* may not be very close; obviously, a decision on this question can only be reached when further fossil material of Heidelberg man is available for comparative study. Lastly the discovery in upper levels of the Olduvai gorge of a skull of primitive appearance associated with stone implements assigned to the Chellean phase of palaeolithic culture has raised the possibility that *H. erectus*—or a type closely related to it —extended its geographical distribution to East Africa.

EMERGENCE OF HOMO SAPIENS

In discussing the emergence of the species *H. sapiens*, which took place during the Pleistocene, it may be recalled that this geological period can be sub-divided on a relative chronological basis by reference to its climatic fluctuations. The latter half of the period (or perhaps rather more) was marked by a series of very cold cycles when much of the temperate regions of the earth became covered by glaciers and ice sheets. There were four of these major glaciations, separated by three interglacial periods during each of which the climate became much warmer. The last glaciation reached its initial climax probably about 50,000 years ago, and the first glaciation is reckoned to have occurred about half a million years ago.

To-day the family Hominidae is represented by the genus *Homo*, and by only one species *H. sapiens*—that is to say, modern mankind. The geographical varieties of this species, or 'races' as they are commonly termed, show considerable differences in superficial features such as skin colour, hair texture, and so forth, but they are much less easily distinguished by their skeletal characters. Nevertheless, the latter do show differences, and it is well to recognize the extent of these skeletal variations in modern man when the problem arises whether a fossil human skeleton is that of *H. sapiens*, or of some different and extinct species of *Homo*. For it has happened from time to time by failing to recognize the wide range of individual and racial variability in our own species that some authorities have claimed that the human remains which they have discovered are those of a hitherto unknown species, and have even christened them with a new specific name. This has had the unfortunate effect of confusing and distorting the perspective of the latter-day prehistory of mankind (so far as it has been revealed by fossil remains) to a quite ridiculous degree.

More careful and systematic comparisons have now made it clear that *H. sapiens* has a quite respectable antiquity. For example, during the time when the Palaeolithic (Old Stone Age) culture known as the Magdalenian flourished in Western Europe, the local population was composed of people who, judging from their skull and skeleton, were similar in physical characters to modern Europeans. The Magdalenians had developed a quite rich culture, and they were responsible for some of the most beautiful examples of cave paintings and sculptures such as those found in the famous caves at Lascaux in France. They lived during the latter phases of the last glaciation of the Ice Age in what is sometimes called the Reindeer Age (for the reason that reindeer herds occupied Europe in large numbers at that time). Now, pieces of charcoal left by the Magdalenians in one of the Lascaux caves have been analysed for their content of radiocative carbon, and this has given an antiquity of about 15,000 years. We can go back further into the Aurignacian period which immediately preceded the Magdalenian, and still we find from a number of fossilized remains that the local population was apparently not distinguishable from modern populations of *H. sapiens*. A radioactive carbon dating of 27,000 years has been reported for the period of the Aurignacian culture, and if the species *H. sapiens* was already fully differentiated at that time the final stages of its evolutionary emergence must have occurred still earlier. We know also from fossil evidence that the species had spread widely over the earth many thousands of years ago. For example, *H. sapiens* had certainly reached Australia, and even North America, at least 10,000 years ago, and at Florisbad in South Africa there was found in 1933 a human skull, also not to be distinguished from the *H. sapiens* type, whose antiquity has been estimated indirectly by the radioactive carbon method to have been at least 40,000 years.

The question now arises—is there any concrete evidence from the fossil record that *Homo sapiens* was actually in existence before the last glaciation of the Ice Age? Preceding the Aurignacian period in Europe there was a prolonged cultural period of the Palaeolithic termed the Mousterian, which can be conveniently divided into an Early Mousterian phase and a Late Mousterian phase (though this is really an over-simplification of the cultural sequences which followed, and partly overlapped, each other during those times). The Early Mousterian covered the latter part of the last interglacial period and extended into the onset of the last glacial period, while the Late Mousterian coincided with the climax of the first part of the last glaciation. A fair number of fossil remains of Early Mousterian man have been found in Central Europe and also in Palestine, so that we know a good deal about their cranial and dental anatomy. A striking character is

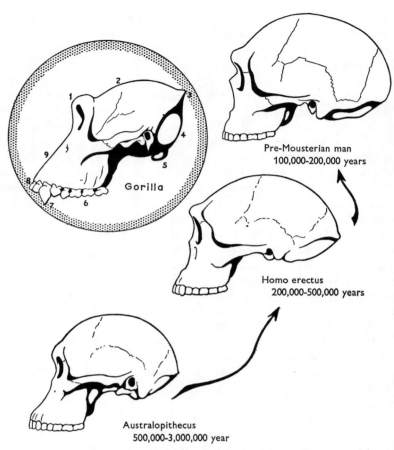

FIG. 153. This diagram illustrates the appearance of the skull in a series of fossil hominid types arranged in their temporal sequence. The antiquity of each type has been estimated by methods of relative dating and is to be regarded as no more than approximate within very broad limits, It is also to be noted that some of these types overlapped in time in different geographical regions. The representation of the genus *Australopithecus* is based on a skull found at Sterkfontein in South Africa, *Homo erectus* on a skull cap and portions of jaws found in Java, pre-Mousterian man on the Steinheim skull, the generalized type of Neanderthal man on one of the skulls found at Mount Carmel in Palestine, and the specialized type on a skull found at Monte Circeo in Italy. Inset on the left is shown, for comparison and contrast, the skull of an adult female gorilla; the numerals indicate a few of the fundamental characters which, *taken in combination*, comprise a total morphological pattern distinguishing

354

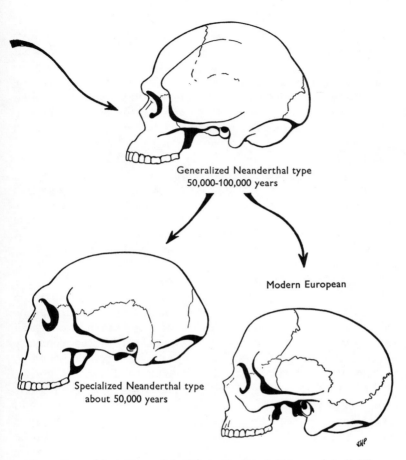

Generalized Neanderthal type
50,000-100,000 years

Modern European

Specialized Neanderthal type
about 50,000 years

the anthropoid ape type of skull from the hominid type of skull. The characters indicated are: (1) The forward projection of the brow ridges well beyond the front end of the brain-case. (2) The low level of the cranial roof in relation to the upper border of the orbital aperture. (3) The high position of the external occipital protuberance. (4) The steep slope and great extent of the nuchal area of the occiput for the attachment of the neck muscles. (5) The relatively backward position of the occipital condyles. (6) Except in advanced stages of attrition, the teeth are not worn down to a flat, even surface. (7) The canines form conical, projecting, and sharply pointed 'tusks'. (8) The large size of the incisor teeth. (9) The massive upper jaw. Note, also, that in the hominid type of skull a pyramidal mastoid process of quite distinctive pattern is consistently present.

All the skulls have been drawn to the same scale.

(Proc. Amer. Phil. Soc. 103, 1959.)

their wide variability, for while many individuals show primitive features such as strongly developed brow ridges, a somewhat retreating forehead, prominent jaws, and a feebly formed or absent chin eminence, others are very similar in their skull structure to the more primitive races of modern mankind. Further, the limb bones (so far as they are known) appear to be of quite modern type. Opinions vary on the question whether in some of their skull characters the Early Mousterians exceed the limits of variation found in *H. sapiens*. Even if this is the case, it is not altogether clear that they exceed the limits to the extent that they can properly be assigned to another species altogether. Probably it is wise to defer a decision on this point for the present—if they are not *H. sapiens* in the strict sense, they represent the immediate precursors of modern *H. sapiens* and may be conveniently designated as such. But their great variability is of particular interest from another point of view.

In later Mousterian times, characterized archaeologically by the full development of the typical stone-tool industry to which the term 'Mousterian' is properly attached, there existed in Europe and neighbouring regions the distinctive type of man now so well known as 'Neanderthal man'. The latter may be conveniently referred to as the specialized Neanderthal type in contradistinction to the generalized type that predominated in the Early Mousterian. The outstanding features of the specialized type are the massive brow ridges, retreating forehead, large projecting jaws, absence of a chin eminence, and certain peculiarities of the occipital region and base of the skull. Some of the limb bones, also, are unusual in the thickness and curvature of their shafts and the relative size of their articular extremities. In curious contrast, the size of the brain of Neanderthal man—as indicated by the cranial capacity— was surprisingly large, on the average even larger than that of modern *H. sapiens*. A sufficient number of Neanderthal skulls have been collected to permit of their study by statistical analysis; this has not only demonstrated their homogeneity as a local European population, it has also shown that in a number of dimensions (and proportional indices constructed therefrom) they lie outside the known range of variation of modern *H. sapiens*. All these facts led to the assumption, maintained by some anthropologists (but now not generally accepted), that the specialized Neanderthals constituted a distinct species, *Homo neanderthalensis*. At one time it was generally supposed that this extinct type was directly ancestral to modern man. But, as already mentioned, we now know that it was preceded by earlier types, and these, though showing a number of primitive features, were much more akin morphologically to modern *H. sapiens*. Moreover, the fossil and archaeological record makes it clear that at the end of the Mousterian period Neanderthal man disappeared from Europe quite abruptly, to be replaced by a population of the modern *H. sapiens* type. Presumably the latter spread into Europe from a neighbouring area, perhaps the Middle East, and by replacement led to the extinction of the specialized Neanderthal type.

Let us now return to the Early Mousterian populations and reconsider them in the light of their wide range of variability. At one end of this range are individuals which approach so closely to primitive races of *H. sapiens* that it is difficult to decide whether they can be taxonomically separated from this species. In other words, it seems that this degree of variability could readily have provided the raw materials, so to speak, for the evolutionary diversification of what were evidently two terminal types—the specialized Neanderthals which became extinct, and the modern forms of *Homo sapiens*. Such a conclusion fits in quite well with the evidence relating to the temporal sequence. Some authorities, it may be noted, have been tempted to interpret the variability of Early Mousterians in terms of the coexistence of genetically different types, some of which may have interbred. But this is to complicate the picture unnecessarily and somewhat arbitrarily. The high variability of the Early Mousterians—considered as a single general population composed of regional variants—was probably related to their dispersal over Europe in small hunting communities, their exposure to changing climatic extremes, and the increasing intergroup competition for the means of survival. Such circumstances are particularly favourable for the diversifying action of the selective processes in evolution. It is interesting to note, by the way, that the first Neanderthal skull to be studied—discovered in 1856 in a cave situated in a valley known as Neanderthal (near Düsseldorf)—was discussed at some length by T. H. Huxley in 1863; he expressed the opinion that it was 'the most pithecoid of human crania yet discovered', but he concluded that, at the most, it demonstrated 'the existence of a man whose skull may be said to revert somewhat towards the pithecoid type'. In this statement Huxley showed a remarkable insight, for the accumulation of evidence since his time has certainly made it clear that some of the 'pithecoid' features of Neanderthal man were the results of a secondary retrogression.

As far as is known at present, the 'specialized Neanderthals', in the strict sense in which this term is commonly used, were limited in their geographical distribution to Europe and certain adjacent areas. But there lived at about the same time, or perhaps rather later, very similar types of prehistoric man in other parts of the world. In Rhodesia and South Africa fossil skulls have been found which also show huge brow ridges and massive jaws, and at Ngandong in Java several other skulls showing the same exaggerated features have come to light. It still remains doubtful what part, if any, was played in the evolutionary history of modern man by the populations represented by these remains. Perhaps they should provisionally be regarded as local variants of the specialized Neanderthals—if so, it must be supposed that this type spread very widely over the world in a comparatively short space of geological time. But, in any case, it seems unlikely that they were directly ancestral to the local races of man which now inhabit the regions where they lived (though some authorities have assumed them to be so). Like Neanderthal man in Europe, these aberrant types almost certainly became extinct without leaving direct descendants.

We now come to the problem of 'pre-Mousterian man', that is to say, the nature of the populations which immediately preceded the Early Mousterians. Here we are faced with the difficulty that their fossil record is still too meagre to allow firm conclusions. A fairly complete though partly defective skull found at Steinheim in Germany dates from the last interglacial period, or perhaps even earlier from the second interglacial period. It closely resembles the Early Mousterians and shows pronounced brow ridges of the skull, but the forehead region is quite well developed and the occipital region is full and rounded as in modern man. The cranial capacity is estimated to be about 1,100 cc.; this comes well within the range of variation of modern *H. sapiens* but is considerably below the mean value (1,350 cc.). Two fragments of skulls found at Fontéchevade in France, also reckoned to be of pre-Mousterian date, show a much closer resemblance to modern *H. sapiens*, for the brow ridges are only moderate in size. Finally, a most important discovery was made of portions of a skull at Swanscombe in Kent in 1935—important because their antiquity may be assigned with considerable assurance to the second interglacial period which, by methods of relative dating, can hardly have been less than 100,000 years ago and may well have been more. Unfortunately, only three of the main bones forming the roof and back of the skull (the two parietals and the occipital) were found, and although they are excellently preserved they do not tell us very much. But careful anatomical and statistical studies have shown that, while in some respects they are certainly unusual, as far as can be ascertained they do not exceed the range of variation of *H. sapiens* in their dimensions, shape, and individual structural features. They are unusual in the thickness of the skull wall and in the width of the occipital region. There is also definite evidence that the air sinuses of the face region were extensively developed, and this suggests the probability that the facial skeleton was rather massive, and that the frontal region may have had large brow ridges like the Steinheim skull. Until further remains of the contemporary population are discovered, all we can say, therefore, is that *on the evidence of the three skull bones available* Swanscombe man was probably similar to the Steinheim man, and at least very closely akin to modern *H. sapiens*. The brain, incidentally, was quite large (the cranial capacity has been estimated at about 1,320 cc.), and the impressions on the inner surface of the skull bones also show that it was richly convoluted.

From this condensed statement of the fossil record of the Hominidae during the Pleistocene, it is evident that a closely graded morphological series links *Australopithecus* through *H. erectus* with our own species *H. sapiens*. Although this record extends over about three million years, evolution appears to have proceeded with unusual rapidity during that time, particularly in the rate of expansion of the brain. Probably this was the result of the intensive action of selective forces operating on small hunting communities widespread over the Old World and in vigorous competition with each other for the means of subsistence. The earlier phases of evolution preceding *Australopithecus* are still not certainly known, and it is here that there still exists a serious gap in the palaeontological history of the Hominidae. From the data of comparative anatomy which we have presented, supplemented by the evidence of the palaeontological record, there can be no reasonable doubt that the Hominidae were initially derived, probably in the Miocene period (or perhaps even earlier) from primitive ape-like creatures which, in spite of the fact that they were far more generalized in their morphology than the anthropoid apes of to-day, are properly referable to the same taxonomic group, the Pongidae. But just when and where the hominid sequence of evolution diverged and became segregated from the Pongidae will only be determined with the accumulation of more fossil material.

Perhaps one of the most remarkable considerations which has emerged from our outline of the Primate series is this—that the inferences regarding their evolutionary history which were primarily based on the indirect evidence of the comparative anatomy of living types have been so substantially confirmed by the direct evidence of the fossil record. Among the Primates of to-day, the series tree-shrew—lemur—tarsier—monkey—ape—man suggests progressive levels of organization in an actual evolutionary sequence. And that such a sequence did occur is demonstrated by the fossil series beginning with the early plesiadapids and extending through the Palaeocene and Eocene prosimians, and through the cercopithecoid and pongid Primates of the Oligocene, Miocene, and Pliocene, to the hominids of the Pleistocene.

Thus the foundations of evolutionary development which finally culminated in our own species, *H. sapiens*, were laid when the first little tree-shrew-like creatures advanced beyond the level of the lowly insectivores which lived during the Cretaceous period and embarked on an arboreal career without the restrictions and limitations imposed by specializations for a terrestrial mode of life. It was evidently the opportunities provided by life in the trees that were responsible for the subsequent development in progressive stages of the distinctive characters whereby the Primates are contrasted with other mammalian orders—the enhancement of visual acuity, the retrogression of the olfactory sense, the early and rapid expansion of the brain, the preservation of pentadactyl limbs with a more efficient functional prehensility, the retention (in the central lines of evolution) of a relatively unelaborated type of dentition, and the avoidance of many of the structural specializations of other parts of the body which are commonly found in ground-living mammals.

Finally, it was the modification of the limbs, trunk and skull in adaptation to erect bipedalism, rapidly followed by an accelerated expansion of the brain, which led to the emergence of man himself.

Bibliography

The literature on the relationships and evolution of the various groups of Primates has come to be so extensive that it would be by no means practicable to attempt here a complete list of all the relevant books and scientific papers. We have been content, therefore, to give a selected list of works to which the reader may find it useful to refer, but a fuller bibliography will be found in those more comprehensive treatises which have been marked with an asterisk.

1. *ABEL,O. (1931). *Die Stellung des Menschen im Rahmen der Wirbelthier e.*Jena.
2. ARAMBOURG, C. (1955). A recent discovery in human palaeontology: *Atlanthropus* of Ternifine (Algeria). *Amer. J. Phys. Anth.*, **13**, 191.
3. BARNETT, S. A. (Editor) (1958). *A Century of Darwin.* London.
4. BEATTIE, J. (1927). The anatomy of the common marmoset. *Proc. Zool. Soc.*, 593.
5. DE BEER, G. (1937). *The Development of the Vertebrate Skull.* Clarendon Press.
6. —— (1958). *Embryos and Ancestors.* 3rd edition. Clarendon Press.
7. BLACK, D. (1930). On the endocranial cast of the adolescent *Sinanthropus* skull. *Proc. Roy. Soc. London*, B, **112**, 263.
8. *BOULE, M. and VALLOIS, H. V. (1957). *Fossil Men.* London.
9. BOYDEN, A. A. (1953). Fifty years of systematic serology. *Systematic Serology.*, **2**, 19.
10. BRODMANN, K. (1909). *Vergleichende Lokalisationlehre der Grosshirnrinde.* Leipzig.
11. BRONOWSKI, J. and LONG, W. M. (1952). Statistics of discrimination in anthropology. *Amer. J. Phys. Anthr.*, **10**, 385.
12. *BUETTNER-JANUSCH, J. (1963). An Introduction to the Primates, in *Evolutionary and Genetic Biology of the Primates.* Vol. 1. Academic Press.
13. CAMPBELL, C. B. G. (1966). Taxonomic status of tree-shrews. *Science*, **153**, 436.
14. CARPENTER, C. R. (1940). A field study of the behaviour and social relations of the gibbon. *Comp. Psych. Mon.*, **16**, 1.
15. CHACKO, L. W. (1954). The lateral geniculate body in the Lemuroidea, *Tarsius* and the New World Monkeys. *J. Anat. Soc. India.*, **3**, 23, 61, 75.
16. CHOW, M. (1964). A lemuroid Primate from the Eocene of Lantian. *Vertebrata Palasiatica*, **8**, 257.

17. CHU, E. H. Y. and BENDER, M. A. (1961). Chromosome cytology and evolution in Primates. *Science*, **133**, 1399.

18. CLARK, J. D. (1961). Sites yielding hominid remains in Bed I Olduvai gorge. *Nature*, **189**, 903.

19. CONOLLY, C. J. (1950). *External Morphology of the Primate Brain.* Illinois.

20. DAY, M. H. (1969). Femoral fragment of a robust australopithecine from the Olduvai Gorge, Tanzania. *Nature*, **221**, 230.

21. *DOBZHANSKY, T. (1941). *Genetics and the Origin of Species.* New York.

22. DORAN, A. H. G. (1879). Morphology of the mammalian ossicula auditus. *Trans. Linn. Soc.*, **1**, 371.

23. EDINGER, T. (1929). Die fossilen Gehirne. *Erg. Anat. Entwichl.*, **83**, 105.

24. ELFTMAN, H. and MANTOS, J. (1935). The evolution of the human foot. *J. Anat.*, **70**, 56.

25. ELLIOT SMITH, G. (1902). *Catalogue of the Physiological Series of Comparative Anatomy in the Museum of the Royal College of Surgeons.* Vol. 2.

26. —— (1903). On the morphology of the brain in the Mammalia. *Trans. Linn. Soc.*, **8**, 319.

27. *—— (1927). *Essays on the Evolution of Man.* 2nd edition. Oxford University Press.

28. EVERNDEN, J. F., CURTIS, G. H. and KISTLER, R. (1957). Potassium-argon dating of Pleistocene volcanics. *Quaternaria*, **4**, 1.

29. ———— J. F. (1958). Dating of Tertiary and Pleistocene rocks by the potassium-argon method. *Proc. Geol. Soc. London*, No. 1565, 17.

30. FORD, E. B. (1938). The genetic basis of adaptation, in *Evolution*, Edited by G. de Beer, London.

31. *—— (1957). *Mendelism and Evolution.* 6th edition. London.

32. GAZIN, C. L. (1958). A review of the Eocene Primates of N. America. *Smithsonian Misc. Coll.*, **136**, 1.

33. GLASSTONE, S. (1938). A comparative study of the development *in vivo* and *in vitro* of rat and rabbit molars. *Proc. Roy. Soc.*, B, **126**, 315.

34. GLICKSTEIN, M. (1967). Laminar structure of the dorsal geniculate nucleus in the tree-shrew (*Tupaia glis*). *Journ. Comp. Neurol.*, **131**, 93.

35. GOODMAN, M. (1963). Man's place in the phylogeny of the Primates as reflected in serum proteins, in *Classification and Human Evolution.* Ed. by S. L. Washburn. Methuen and Co., London.

36. —— (1966). Phyletic position of the tree-shrews. *Science*, **153**, 1550.

37. —— (1967). Deciphering primate phylogeny from macro-molecular specificities. *Amer. Journ. Phys. Anthr.*, **26**, 255.

38. *GRASSÉ, P. (1955). *Traité de Zoologie*, vol. 17. Mammifères. Paris.

39. *GREGORY, W. K. (1910). The Orders of Mammals. *Bull. Amer. Mus. Nat. Hist.*, **27**, 1.

40. *GREGORY, W. K. (1916). Studies on the evolution of the Primates. *Mus. Nat. Hist.*, **35,** 239.
41. ——— (1922). *The Origin and Evolution of the Human Dentition.* Baltimore.
42. *——— (1951). *Evolution Emerging.* New York.
43. ——— and HELLMAN, M. (1926). The dentition of *Dryopithecus* and the origin of man. *Anthr. Papers of Amer. Mus. Nat. Hist.*, **28,** 1.
44. ——— ——— and LEWIS, G. E. (1938). Fossil anthropoids of the Yale-Cambridge Indian Expedition of 1935. *Carnegie Inst. of Washington Publ.*, No. 495.
45. GRÖSSER, O. (1933). Human and comparative placentation. *Lancet,* **224,** 999, 1053.
46. HAFLEIGH, A. S. and WILLIAMS, C. A. (1966). Reactions to human serum albumin. *Science,* **151,** 1530.
47. HARTMAN, C. G. and STRAUS, W. L. (1953). *The Anatomy of the Rhesus Monkey.* London.
48. HAUG, H. (1958). *Quantitative Untersuchungen an der Sehrinde.* Stuttgart.
49. *HEBERER, G. (1956). Die Fossilgeschichte der Hominoidea. *Primatologia,* **1,** 379 Basel.
50. HILL, J. P. (1932). The developmental history of the Primates. *Phil. Trans. Roy. Soc.*, B, **221,** 45.
51. ——— (1965). On the placentation of *Tupaia*. *Journ. Zool.*, **146,** 278.
52. *HILL, W. C. O. (1953). *Primates* Vol. I; (1955) Vol. II; (1957) Vol. III; (1960) Vol. IV; (1962) Vol. V; (1966) Vol. VI. Edinburgh University Press.
53. HÖFER, H. (1957). Uber das Spitzhörnchen. *Natur. u. Volk.*, **87,** 145.
54. HOOIJER, D. A. (1960). Quaternary gibbons from the Malay archipelago. *Zool. Verhandl. Rijksmuseum, Leiden,* No. 46.
55. HOPKINS, G. H. E. (1949). The host association of the lice of mammals. *Proc. Zool. Soc.*, **119,** 387.
56. *HOWELL, F. C. (1957). The evolutionary significance of variation and varieties of 'Neanderthal' man. *Quart. Rev. Biol.*, **32,** 330.
57. ——— (1958). Upper Pleistocene men of the Southwest Asian Mousterian. *Neanderthal Centenary,* Utrecht, 185.
58. *HRDLICKA, A. (1930). The skeletal remains of early man. *Smithsonian Misc. Coll.*, **83,** 1.
59. HÜRZELER, J. (1948). Zur Stammgeschichte der Necrolemuriden. *Schweiz. Palaeont. Abhandl.*, **66,** 1.
60. ——— (1960). Der Mensch in Raum und Zeit mit besonderer Berücksichtigung des *Oreopithecus*—Problems. *Veroff. aus d. Nat. Mus. Basel,* No. 1.
61. *HUXLEY, J. S. (1942). *Evolution, the Modern Synthesis.* London.
62. JEPSEN, G. L. (1934). A revision of the American Apatemyidae and the description of a new genus *Sinclairella. Proc. Amer. Phil. Soc.*, **74,** 287.
63. *——— MAYR, E. and SIMPSON, G. G. (1949). *Genetics, Palaeontology and Evolution.* Princeton University Press.

64. JOHNSON, M. L. (1968). Application of blood protein electrophoretic studies to problems in mammalian taxonomy. *Systematic Zoology*, **17**, 23.
65. VAN KAMPEN, P. N. (1924). Die Tympanalgegend des Säugethier-eschädels. *Morph. Jahrb.*, **34**, 321.
66. *KEITH, A. (1923). Man's posture: its evolution and disorders. *Brit. Med. Journ.*, **1**, 451, 499, 545, 624, and 669.
67. KLAATSCH, H. (1892). Zur Morphologie der Mesenterialbildungen am Darmkanal der Wirbelthiere. *Morph. Jahrb.*, **18**, 385, 609.
68. V. KOENIGSWALD, G. H. R., GENTNER, W. and LIPPOLT, H. J. (1961). Age of basalt flow at Olduvai, East Africa. *Nature*, **192**, 720.
69. KOLMER, W. (1927). Geruchsorgan. *Handb. d. Anat. d. Menschen*, **3**, 192.
70. —— (1930). Zur Kenntniss des Auges der Primaten. *Zeits. f. Anat. v. Entwickl.*, **93**, 679.
71. KULP, J. L. (1961). Geologic time scale. *Science*, **133**, 1105.
72. LEAKEY, L. S. B. (1959). A new fossil skull from Olduvai. *Nature*, **184**, 491.
73. —— (1960). Recent discoveries at Olduvai Gorge. *Nature*, **188**, 1050.
74. —— EVERNDEN, J. F. and CURTIS, G. H. (1961). Age of Bed I, Olduvai Gorge, Tanganyika. *Nature*, **191**, 478.
75. LE GROS CLARK, W. E. (1926). On the anatomy of the pen-tailed tree-shrew (*Ptilocercus lowii*) *Proc. Zool. Soc.*, 1179.
76. —— (1936). The problem of the claw in Primates. *Proc. Zool. Soc.*, 1.
77. —— (1950). Hominid characters of the Australopithecine dentition. *J. Roy. Anth. Inst.*, **80**, 37.
78. *—— (1955). *The Fossil Evidence for Human Evolution*. Chicago University Press.
79. —— (1959). The crucial evidence for human evolution. *Proc. Amer. Phil. Soc.*, **103**, 159.
80. *—— (1960). *History of the Primates*. 7th edition. London.
81. —— and LEAKEY, L. S. B. (1951). The Miocene Hominoidea of East Africa. Fossil mammals of Africa, No. 1. Brit. Mus. Nat. Hist. London.
82. —— and THOMAS, D. P. (1951). Associated jaws and limb bones of *Limnopithecus macinnesi*. *Fossil mammals of Africa*, No. 3. Brit. Mus. Nat. Hist. London.
83. LEWIS, G. E. (1934). Preliminary notice of new man-like apes from India. *Am. J. Sc.*, **27**, 161.
84. LIBBY, W. F. (1961). Radiocarbon dating. *Science*, **133**, 621.
85. LYON, M. W. (1913). Tree-shrews—an account of the mammalian family Tupaiidae. *Proc. U.S. Nat. Hist. Mus.*, **45**, 1.
86. MARTIN, R. D. (1966). Tree-shrews: Unique reproductive mechanism of systematic importance. *Science*, **152**, 1402.
87. *MAYR, E. (1942). *Systematics and the Origin of Species*. New York.

88. McCown, T. D. and Keith, A. (1939). *The Stone Age of Mount Carmel*. Clarendon Press.

89. Meister, W. and Davis, D. D. (1956). Placentation of the pygmy tree-shrew, *Tupaia minor*. *Fieldiana : Zoology*, **35,** 73.

90. Mitchell, P. C. (1905). On the intestinal tract of mammals. *Trans. Zool. Soc.*, **17,** 437.

91. —— (1916). Further observations on the intestinal tract of mammals. *Proc. Zool. Soc.*, **1,** 437.

92. Morant, G. M. (1927). Studies of Palaeolithic man. II A biometric study of Neanderthal skulls and their relationships to modern racial types. *Ann. Eugenics*, **2,** 318.

93. More, G. W. and Goodman, M. (1968). A set theoretical approach to immunotaxonomy. *Bull. Math. Biophysics*, **30,** 279.

94. Morton, D. J. (1922). Evolution of the human foot. *Amer. J. Phys. Anthr.*, **5,** 305.

95. —— (1926). Evolution of man's erect posture. *J. Morph. Physiol*. **43,** 147.

96. —— (1927). Human origin. *Amer. J. Phys. Anthr.* **10,** 173.

97. —— and Fuller, D. D. (1952). *Human Locomotion and Body, Form*. Baltimore.

98. Mossman, H. W. (1937). Comparative morphogenesis of the foetal membranes and accessory uterine structures. *Contr. Embryol. Carnegie Inst.*, **26,** 129.

99. *Movius, H. L. (1960). Radiocarbon dates and upper palaeolithic archaeology in central and western Europe. *Current Anthropology*, **1,** 355.

100. Napier, J. R. (1960). Studies of hands of living Primates. *Proc. Zool. Soc.*, **134,** 647.

101. —— (1961). Prehensility and opposability in the hands of Primates. *Symp. Zool. Soc. Lond.*, No. 5, 115.

102. ——(1964). The evolution of bipedal walking in the hominids *Arch. de Biologie*, **75,** 673.

103. —— (1967). Evolutionary aspects of primate locomotion. *Amer. Journ. Phys. Anthr.*, **27,** 333.

104. —— and Davis, P. R. (1959). The forelimb skeleton and associated remains of *Proconsul africanus*. *Fossil Mammals of Africa*, No. 16. Brit. Mus. Nat. Hist. London.

105. *—— and Napier, P. H. (1967). *A Handbook of Living Primates*. Academic Press.

106. Nuttall, G. H. F. (1904). *Blood Immunity and Blood Relationships*. Cambridge University Press.

107. Oakley, K. P. (1954). The dating of the Australopithecinae of Africa. *Amer. J. Phys. Anth.*, **12,** 9.

108. *—— (1956). *Man the Tool-maker*. 3rd edition. London.

109. —— (1956). Dating fossil men. *Mem. and Proc. Manchester Lit. and Phil. Soc.*, **98,** 1.

110. —— (1959). Tools makyth man. *Smithsonian Rep. for 1958*. 431.

111. OAKLEY, K. D. (1969). *Frameworks for Dating Fossil Man*, 3rd ed. Weidenfeld and Nicolson, London.

112. OWEN, R. (1859). *On the Classification and Geographical Distribution of the Mammalia*. Appendix B. *On the Orang, Chimpanzee and Gorilla*. London.

113. *PARKES, A. S. (1956). Editor of *Marshall's Physiology of Reproduction*. 3rd edition. Vols. 1 and 2. London.

114. PATTERSON, B. and HOWELLS, W. W. (1967). Hominid humeral fragment from Early Pleistocene of north-western Kenya. *Science*, **156**, 64.

115. PAULLI, S. (1899). Über die Pneumaticität des Schädels bei den Säugethieren. *Morph. Jahrb.*, **28**, 483.

116. *PIVETEAU, J. (1957). *Traité de Paléontologie*, Vol. 7. *Primates*. Paris.

117. POLYAK, S. (1957). *The Vertebrate Visual System*. Chicago University Press.

118. REMANE, A. (1927). Studien über die Phylogenie des menschlichen Eckzahnes. *Zeits. f. Anat. u. Anthr.*, **42**, 311.

119. *——— (1956). Paläontologie und Evolution der Primaten. *Primatologia*, **1**, 267. Basel.

120. REYNOLDS, E. (1931). Evolution of the human pelvis. *Papers Peabody Museum*, **11**, 255.

121. ROBINSON, J. T. (1956). The dentition of the Australopithecinae. *Mem. Transvaal Mus. Pretoria.*, No. 9, 1.

122. *ROMER, A. S. (1941). *Man and the Vertebrates*. 3rd edition. Chicago University Press.

123. *——— (1945). *Vertebrate Palaeontology*. 2nd edition. Chicago University Press.

124. RUGE, G. (1902). Die äusseren Formverhältnisse der Leber bei der Primaten. *Morph. Jahrb.*, **29**, 450; **30**, 42; **35**, 75; **36**, 93.

125. RUSSELL, D. E. (1959). Le crâne de *Plesiadapis*. *Bull. de la Soc. Geol. de France*, **1**, 312.

126. SABAN, R. (1956 and 1957). Les Affinités du genre Tupaia, d'après les caractères morphologiques de la tête osseuse. *Ann. de Paléont.*, 42 and 43.

127. SCHULTZ, A. H. (1930). The skeleton of the trunk and limbs of higher Primates. *Human Biol.*, **2**, 203.

128. *——— (1936). Characters common to higher Primates and characters specific for man. *Quart. Rev. Biol.*, **11**, 259, 425.

129. ——— (1944). Age changes and variability in gibbons. *Amer. J. Phys. Anthr.*, **2**, 1.

130. ——— (1956). Postembryonic age changes. *Primatologia*, **1**, 887, Basel.

131. ——— (1957). Past and present views on man's specializations. *Irish J. Med. Sc.*, August, 341.

132. ——— (1960). Einige Beobachtungen und Masse am Skelett von *Oreopithecus*. *Zeits. Morph. Anthr.*, **50**, 136.

133. ——— (1962). Schädelkapazität mannlicher Gorillas und ihr Hochstwert. *Anthr. Anz.*, **25**, 197.

134. SIMONS. E. L. (1959). An anthropoid frontal bone from the Fayum Oligocene of Egypt. *Amer. Mus. Novit.*, No. 1976.
135. *—— (1961). New fossil Primates: A review of the past decade. *Amer. Scient.*, **48**, 179.
136. —— (1960). *Apidium* and *Oreopithecus. Nature,* **186**, 824.
137. —— (1961). Notes on Eocene tarsioids and a revision of some Necrolemurinae. *Bull. Brit. Mus. Nat. Hist.*, Geology, Vol. 5, No. 3.
138. —— (1963). A critical reappraisal of Tertiary Primates, in *Evolutionary and Genetic Biology of Primates.* Ed. by J. Buettner-Janusch. Academic Press.
139. —— (1967). Fossil Primates and the evolution of some primate locomotor systems. *Amer. Journ. Phys. Anthr.*, **26**, 241.
140. —— (1967). The earliest apes. *Scientific American,* **217**, 28.
141. —— and PILBEAM, D. R. (1965). Preliminary revision of the Dryopithecinae (Pongidae, Anthropoidea). *Folia Primatologia*, **3**, 81.
142. —— and RUSSELL, D. E. (1960). Notes on the cranial anatomy of *Necrolemur. Breviora.* Museum of Comp. Zool. No. 127.
143. SIMPSON G. G. (1935). The Tiffany fauna, Upper Palaeocene. II Structure and relationships of *Plesiadapis. Amer. Mus. Novit.*, No. 816.
144. —— (1936). Studies of the earliest mammalian dentitions. *Dental Cosmos,* August and September.
145. —— (1937). The beginning of the Age of Mammals. *Biol. Rev.*, **12**, 1.
146. *—— (1940). Studies on the earliest Primates. *Bull. Amer. Mus. Nat. Hist.*, **77**, 185.
147. *—— (1945). The principles of classification and classification of mammals. *Bull. Amer. Mus. Nat. Hist.*, **85**, 1.
148. *—— (1950). *The Meaning of Evolution.* Oxford University Press.
149. —— (1955). The Phenacolemuridae: New family of the early Primates. *Bull. Amer. Mus. Nat. Hist.*, **105**, 417.
150. *—— (1961). *Principles of Animal Taxonomy.* Oxford University Press.
151. —— (1962). Primate taxonomy. *Ann. N.Y. Acad. Sc.*, **102**, 497.
152. SPATZ, W. B. (1966). Zur Ontogenese der Bulla Tympanica von *Tupaia glis. Folia Primatologia,* **4**, 26.
153. —— (1968). Die Bedeutung der Augen für die sagittale Gestaltung des Schädels von *Tarsius. Folia Primatologia,* **9**, 22.
154. SPRANKEL, H. (1961). Über Verhaltensweisen und Zucht von *Tupaia glis. Zeits. f. wissen. Zool.*, 165.
155. STEPHAN, H. (1958). Vergleichend-anatomische Untersuchungen an Insectivorengehirnen. III. Hirn-Körpergewichtsbeziehungen. *Morph. Jahrb.*, **99**, 853.
156. STIRTON, R. A. (1951). Ceboid monkeys from the Miocene of Colombia. *Univ. Calif. Publ., Bull. Dep. Geol. Sc.*, **28**, 215.

157. STRAUS, W. L. (1936). The thoracic and abdominal viscera of the Primates. *Proc. Amer. Phil. Soc.*, **76,** 1.

158. SZALAY, F. S. (1968). The beginnings of Primates. *Evolution,* **22,** 19.

159. TIGGES, J. (1963). Untersuchungen über den Farbenzinn von *Tupaia glis. Zeitschr. f. Morph. Anthr.,* **53,** 109.

160. TOBIAS, P. V. (1967) *Olduvai Gorge. The Cranium of Australopithecus (Zinjanthropus) boisei.* Camb. Univ. Press.

161. VALLOIS, H. V. (1949). The Fontéchevade fossil man. *Amer. J. Phys. Anth.,* **7,** 339.

162. VAN VALEN, L. (1965). Tree-shrews, Primates and fossils. *Evolution.* **19,** 137.

163. *WATSON, D. M. S. (1951). *Palaeontology and Modern Biology.* Yale University Press.

164. *WEBER, M. (1928). *Die Säugethiere.* Jena.

165. WEIDENREICH, F. (1937). The dentition of *Sinanthropus pekinensis. Palaeont. Sinica.,* Series D, Peking, No. 101.

166. —— (1943). The skull of *Sinanthropus pekinensis. Palaeont. Sinica,* Peking, **10,** 1.

167. WEIGELT, J. (1933). Neue Primaten aus der Mitteleozänen Braunkohle des Geisetals. *Nov. Act. Leop. Carol.,* **1,** 97 and 321.

168. WEINER, J. S. (1958). The pattern of evolutionary development of the genus *Homo. S. Afr. J. Med. Sc.,* **23,** 11.

169. *WISLOCKI, G. B. (1929). On the placentation of the Primates, with a consideration of the phylogeny of the placenta. *Contr. Embryol., Carnegie Inst.,* **20,** 51.

170. WISLOCKI, G. B. (1936). The external genitalia of the simian Primates. *Human Biology,* **8,** 309.

171. WOO, J. (1956). Human fossils found in China and their significance in human evolution. *Scientia Sinica,* **5,** 389.

172. —— and CHOW, M. (1957). *Hoanghonius stehlini. Vertebrata Palasiatica,* 1, 267.

173. —— —— (1957). New materials of the earliest Primate known in China. *Vertebrata Palasiatica,* **1,** 267.

174. *WOOD-JONES, F. (1916). *Arboreal Man.* London.

175. WOOLLARD, H. H. (1925). The Anatomy of *Tarsius Spectrum. Proc. Zool. Soc.,* 1071.

176. —— (1927). The retina of Primates. *Proc. Zool. Soc.,* 1.

177. —— (1942). The cortical lamination of *Tarsius. J. Anat.,* **60,** 86.

178. ZAPFE, H. (1952). Die *Pliopithecus*—Funde aus der Spaltenfüllung von Neudorf an der March. *Sond. Verhandl. Geol. Bund.,* No. 1.

179. ZEUNER, F. E. (1940). The age of Neanderthal man. *Occasional papers of the Inst. of Arch.,* No. 3.

180. *—— (1952). *Dating the Past.* 3rd edition. London.

Index

Figures in italic indicate illustrations. n denotes a footnote reference

Aberrancy, 10, 38, 169, 321
  in dentition, 76, 98
  of *Daubentonia*, 68, 73, 92, 99,
    101, 136, 189, 189n, 247, 248,
    269-70, 321, 330
  of lemurs, 93, 139-40, 144, 147,
    187, 324, 328
  of *Tarsius*, 147
Absarokius, 106, 107
Acetabulum, 180, 216, 217, 224
Acromion process, 175
Adapidae, 69, 89, 326, 327, 336
  dentition, 96, 109, 325, 326
  skull, 136, 137, 140, *142*, 142, 326
Adapinae, 69, 327
  dentition, 86, 96, 106, 327
Adapis, 69, 86, 106
  brain, 230, *239*, 248, *253*, 253
  dentition, 95, 96, 97, 327
  limbs, 194
  skull, 140, 141, 142, 143
Adapis magnus, 69; *parisiensis*, 69,
    136, *142*
Aegyptopithecus, 53-4, 120, 339
Aelopithecus, 53, 120
Affinities, assessment of, 4, 11, 13,
    15, 16-17, 76, 316-17
  of Ceboidea, 340
  of Cercopithecoidea, 339-40, 341
  of Hominidae, 345-60
  of lemurs, 99-101, 108-10, 143,
    324-31
  of Pongidae, 339-45
  of Tarsioidea, 143, 151, 331-8
  of Tupaioidea, 71, 72-3, 236,
    240-1, 317-23
Agger nasi, 271
Alimentary tract, Evidence of the,
    289-95
Alisphenoid, 130, 131, *131*, 132,
    137, 148, 152, 154, 157, 160,

    161, 332, 333, 338, 344
Allantois, 38, 308, 309, 310, 311,
    312, 313, 331, 338
Allometry, 155-6, 162
Alouatta, 62, 112, 156, 202, 205,
    282
Amnion, 308, *308*, 310, 312, 331,
    345
Amphipithecus, 53n, 341
  dentition, *87*, 116-17, 123, 340
Amphitherium, 77
Anagale, 72
Anaptomorphidae, 65
Anaptomorphinae, 106
Anaptomorphus, 109n
Anathana, dentition, 88, *90*, 325
  skull, 135
Anathana wroughtoni, 90
Ancestry, 2, 4, 8, 9, 10, 20, 21, 24,
    25, 26, 28-30, 32, 38, 40, 45,
    46, 47, 139-40, 259, 271, 294,
    316, 324, 325, 326, 332, 334,
    340, 343, 345
Ansa coli, 294, 329, 331
Anthropoid apes, *see* Pongidae
Anthropoidea, 43, 45
  brain, 230, 231, *244*, 244-5, 246,
    247, 250, 252, 254-64, 276,
    278, 344
  brain-case, 47, 272, 344
  caecum, 295
  colon, 293, *293*, 293-4, 345
  dentition, 11, 84, 108, 110, 113,
    114-21, 344
  ear, 47, 282, 283, 344
  evolutionary tendencies, 344-5
    344
  eye, 47, 273, 274-6, 278, 280,
  general survey, 47-63
  intestines, 345
  limbs, 344

Anthropoidea (*cont.*)
 liver, 298
 manus, 47
 nails, 48, 344
 nose, 47, 267-8, 270-2, 344-5
 placenta, 311, 314, 345
 pollex, 48, 344
 reproductive organs, 301, 302,
  303, 304, 305-6, 345
 skull, 148, 149, 154, 155, 157, 344
 tongue, 289, 345
 vision, 275-6, 278
Anthropomorpha, 57n
Antihelix, *281*, 282
Antitragus, *281*, 282
Aotus, 61
 eye, 61, 275
 limbs, 202
Apatemyidae, 73, 101n
 dentition, 73, 91-2, 100, *100*, 109,
  330
Aphanolemur, 70
Apidium, 56, 60
 dentition, 56, 113, 114, 340
Appendix, 295, 345
Arboreal mode of life, and fossilization, 336-7, 339
 and special senses, 265, 284
 structural adaptations, 42, 49-50,
  126, 171, 200-1, 213, 214, 227,
  233, 252, 265, 281, 319, 322-3,
  347, 360
Archaeolemur, 70, 329
 brain, 248
 dentition, 99, 329, 330
 skull, 70, 141, 144, *144*, 329
Archaeolemur majori, 144
Archaeolemuridae, 70
Archipallium, 233, 234
Arctocebus, 69, 184, 188
Arteria promontorii, 135, 138, 139,
  148
Artery, stapedial, 135, 138, 139,
  151, 157
Association areas, 228, 239, 241,
  243, 245-6, 256, 260, 280
Ateles, 62
 brachiation, 200, 213
 dentition, 111, 112
 limbs, 200, 204, 205, 213
 nose, 268
 reproductive organs, 302, 304,
  306

Ateles (*cont.*)
 tail, 62
Ateles ater, *63*
Atlanthropus, 351
Auditory, aperture, *129*, 129, *139*,
  140, 282
 bulla, 129, 134, 147
 meatus, 148, 149, 160, 167, 336,
  344
 tube, *133*, 134
Australian aboriginal, *44*, *168*, 225
Australopithecinae, 48, 166, 348-50
Australopithecus, 48
 bipedalism, 48, 167, 167n, 224,
  347-8
 brain, 262-3, *346*, 347, 348
 dentition, 16, 27, 48, *115*, *117*,
  122, 125, 348, 349
 femur, 226
 ilium, 224
 mandible, 117, 350
 os innominatum, *216*
 pelvis, 27, 48, 224, *225*, 348
 skull, 156, 164, *165*, 166-7, *167*,
  167n, 168-9, 348, 349, 350, *354*
 species of, 349-50
 taxonomic status, 27-9, 48, 262,
  348, 349-50, 359-60
Aurignacian culture, 353
Aye-aye, *see Daubentonia*

Baboon, 58, *59*, 113, 343
 dentition, 113, 154
 digits, 206
 ear, 282
 fossil, 60
 hallux, 206
 limbs, 201, 202
 pes, 206
 quadrupedalism, 58-9, 200
 skull, *154*, 154, 155, 156, 158
 tail, 59
Baculum, 300, 302, 304, 331, 345
Basicranial axis, of Hominidae, 166
 of lemurs, 127, *127*, *136*
 of monkeys, 156, 344
 of Primates, *127*, 127
 of Tarsioidea, 148, 149
Bilophodonty, 35, *87*, 98, 112, 113,
  114, 121, 125, 339, 340
Biogenetic Law, 22
Biometrics, 15-16, 167
Bipedalism, erect, 14

Bipedalism (*cont.*)
functional adaptations, 208, 215, 216-17, 223-6, 347-8, 360
in chimpanzee, 51
in gibbon, 51, 215, 223
in Hominidae, 14, 21, 48, 51, 167, 167n, 208, 215, 216, 223-6, 347-8, 360
Body weight, 170, 171, 179, 187, 205, 215-16, 217, 221-3, 224
Boyden, A. A., 19
Brachial index, 178, 209, 212, 212n, 213, 215
Brachiation
in monkeys, 200-1, 213
in Pongidae, 49-50, 51, 52, 54, 200-1, 208, 209-14, 215, 223, 224, 226, 323, 343, 347-8
structural adaptations, 49-50, 200-1, 203, 208, 209-14, 221-6, 347-8
Brachyteles, 202
Brain, cell density, 260, 261-2, 280
cranial capacity, 260-1, 262-3, 347, 356, 358, 359
elaboration of, 5, 42, 227, 264, 277, 360
evidence of the, 227-64
evolutionary history of, 5, 232, 263-4
expansion of, 42, 126, 227, 240, 259, 360
of *Adapis*, 230, *239*, 248, *253*, 253
of Anthropoidea, 230, 231, *244*, 244-5, 246, 247, 250, 252, 254-64, 276, 278, 344
of *Archaeolemur*, 248
of *Australopithecus*, *346*, 347, 348
of Cebinae, 256
of Ceboidea, 6-7, 9-10, 254, 256, *257*, 257-8, 344
of *Cebus*, 250, *257*
of *Centetes*, 232-5, *234*, *238*
of Cercopithecoidea, 6-7, 15, 256, *257*, 257-8, 260, 264, 344
of *Cercopithecus*, *246*
of chimpanzee, 260, 261, *346*
of *Daubentonia*, 69, *247*, 247-8, 329
of *Dryopithecus* (*Proconsul*), 264
of *Echinosorex*, *237*, *239*, 240
of gibbon, 51, 259, 264, 346

Brain (*cont.*)
of gorilla, 260, 261, *263*, 345, *346*, 347
of *Homo*, 48, 263, 342-3, 347
of *Homo erectus*, 48, 261-2, 262n, 263, 347, 349, 351
of *Homo sapiens*, 7, 15, 22, 260, *261*, 261-3, *263*, 345, 346-7, 359-60
of lemurs, 135, 236, 242-9, *244*, *245*, *246*, 249, 253, 258, 326-7, 331
of *Libypithecus*, 264
of *Macaca*, *246*, *257*
of mammals, 232-5
of marmosets, 9, 254-6, *255*, *256*, 258
of *Megaladapis*, 248-9
of *Mesopithecus*, 264
of *Microcebus*, 236, *237*, 237, *238*, *239*, 239, 240, *242*, 242-5, 250, 251, 252
of Neanderthal man, 356
of *Necrolemur*, *253*, 253, 333
of *Perodicticus*, 247
of *Plesiadapis*, 241
of Pongidae, 6-7, 49, 231, 262-3, 346, *346*, 347
of Primates, 6-7, 42, 126, 227-32, 264, 272, 323, 360
of *Progalago*, 248, 328
of Ptilocercinae, 235
of *Ptilocercus*, 235, 241
of *Saimiri*, 61, 254
of Tarsioidea, 249-54, 330, 331, 332, 333, 338
of *Tarsius*, 239, 240, *244*, 249-54, *249*, *250*, *251*, *253*, 256, 331, 332, 333, 338
of *Tetonius*, 253, 333
of *Tupaia*, 235-7, *236*, *237*, *238*, *239*, 239-41, 242, 243, 320
of Tupaiinae, 235, *244*
of Tupaioidea, 71, 230, 235-41, 243, 317, 318, 319, 320
size of, 227, 239-40, 241, 259-63, 345-6, *346*, 347
weight of, 227, 236-7, 248, 254, 259-60
Brain-case, of Adapidae, *142*, 142
of Anthropoidea, 47, 272, 344
of *Archaeolemur*, 144
of *Australopithecus*, *165*, 166, 167

Brain-case(*cont.*)
  of Ceboidea, 152, 154
  of *Dryopithecus*, 163
  of gibbons, 159-60
  of gorilla, 160, *165*, 164-7, *354*, 355
  of Hominidae, 164
  of *Homo erectus*, 168, 351
  of *Homo sapiens*, 158
  of Insectivora, 145
  of lemurs, 135, 141, 142, 143, 144, 145
  of *Megaladapis*, 144
  of monkeys, 155, 156, 157, 272
  of *Necrolemur*, 149, *150*, 150
  of *Notharctus*, 143
  of orang-utan, 162
  of *Oreopithecus*, 163
  of Primates, 126, 127
  of *Pronycticebus*, *142*, 143
  of *Tarsius*, 148, *150*, 272, 333
  of *Tetonius*, 149, 150
  of Tupaioidea, 128, 134, 135, 145
Bramapithecus, 56
Breeding, 306
Brodmann, K., 237, 243
Bronowski, J., 16

Caecum, *291*, 293, 294-5
Caenopithecus,
  dentition, 106, 109, 336
  skull, 149, 336
  taxonomic status, 110
Calcaneus, 181, *182*, 194, 195, 198, 199, 200, 223, 331, 332, 338
Callicebus, 61
  caecum, *296*
  dentition, 113, 327
  reproductive organs, 304
Callimico, 61
Callitrichidae, 61, 110, 201, 206, 314, 344
Callitrix
  brain, 9, 254-6, *255*, *256*
  manus, 203
  pelvis, *192*
  pes, *203*, 205-6
  reproductive organs, 304, 306
  skull, *152*, 152, 154
  tongue, *287*
Callitrix jacchus, *152*
Canine teeth, eutherian, 76, 80

Canine teeth (*cont.*)
  of Hominidae, 122, 123-5, 344, 347
  of *Homo sapiens*, 6
  of lemurs, 92, 93, 94, 98, 109, 141, 288, 325, 328, 330
  of monkeys, 112, 113, 344
  of Plesiadapidae, 89-90, 330
  of Pongidae, 114-15, 122, 123-4, 125, 347, 355
  of Primates, 84, 86
  of Tarsioidea, 101, 102, 105, 107, 109, 337
  of Tupaioidea, 88, 89, *90*, 325
Capuchin monkey, *see Cebus*
Carpenter, C. R., 215
Carpodaptes, 107
Carpolestes, 107, 107n, 335
Carpolestes nigridens, *107*
Carpolestidae, 65, 321
  dentition, 65, 85, 106, 107, 335
  taxonomic status, 106-8, 335
Carpus, *174*
  bones of, *177*, 178
  of Hominoidea, 209
  of *Lemur*, *189*, *190*
  of lemurs, 189-91, 204, 325, 326, 331
  of *Macaca*, *189*
  of monkeys, 204, 344
  of *Tarsius*, *189*, 197-8, 204, 338
  of Tupaioidea, 178
Catarrhine monkeys, *see* Cercopithecoidea
Cebidae, 61-3
  dentition, 111-12
  nails, 203-4
  tail, 61
Cebinae, 202, 256
Ceboidea, 43, 57, 61-3, 70, 327-8
  appendix, 295
  brain, 6-7, 9-10, 254, 256, *257*, 257-8, 344
  caecum, 295
  dentition, 57, 110-13, *111*, *112*, 327, 335, 340
  ear, 282, 283
  eye, 275, 278
  limbs, 200-7
  nails, 203-4, 206
  nose, 57, 268, *268*
  origin, 340
  os ectotympanicum, 138, *138*,7 15

Ceboidea (*cont.*)
  pelvis, 204-5
  phylogenetic relationships, 340
  placenta, 313
  pollex, 204
  reproductive organs, 301, 302,
    304, 306, 307
  skull, 57, 143, 152-4, 157-8, 344
  tongue, 289
  vibrissae, 284
Cebupithecia, 63, 340
Cebus, 61, *62*
  brain, 250, *257*
  dentition, 111, 112
  ear, *281*
  limbs, 201, 205
  nose, *268*
  pes, *222*
  reproductive organs, 304, 306
  skull, *139*, 152, *153*, 154, 156
Cebus fatuellus, *139*, *153*
Centetes, 232-5, *234*, *238*
Cephalization coefficient, 254, 259-
  61
Cercocebus, 57, 275
Cercopithecidae, 31, 57, 159
Cercopithecinae, 59, 154, 306-7
Cercopithecoidea, 33, 43, *44*, 57-61,
  70
  affinities of, 339-40, 341, 343, 360
  brain, 6-7, 15, 256, *257*, 257-8,
    260, 264, 344
  caecum, 295
  cheek pouches, 57, 291, 341
  dentition, 11, 33, 57, *87*, 109n,
    110, *111*, 113-14, 154, 339, 341
  ear, 282, 283
  eye, 276, 278, *279*, 280
  ischial callosities, 33, 57, 205, 341
  limbs, 200-7, 213, 215
  nails, 202-4
  nose, 57, 268, *268*
  origin, 339-40
  os ectotympanicum, *138*, 138-9,
    157
  pelvis, 205
  phylogenetic relationships, 33,
    338-40
  placenta, 313
  pollex, 204, 209
  quadrupedalism, 208, 213
  reproductive organs, 301, 302,
    304, 306-7

Cercopithecoidea (*cont.*)
  skull, 57, 148, 154-8, 159
  tongue, 289
  vibrissae, 284-5
Cercopithecus, 58
  brain, *246*
  dentition, 58
  eye, *279*
  limbs, *201*, 201
  pes, *222*
  reproductive organs, *306*
  skeleton, *201*
Cerebellum, elaboration of, 232,
  344
  lobes: anterior, 232, 235; floc-
    cular, *236*, 240, 245, 256, 259,
    264; lateral, 232, 245, 252,
    259; middle, 232, 235;
    posterior, 232, 235
  of *Centetes*, *234*, 235
  of lemurs, 245, 248
  of *Microcebus*, 240, 245
  of monkeys, 259, 344
  of Primates, 232
  of *Tarsius*, 240, 252, 253, 338
  of *Tupaia*, 240
  subarcuate fossa, 259, 264
Chalmers Mitchell, P., 289
Cheek pouches, 57, 59, 291, 341
Cheirogaleinae, 139n, 314
Cheirogaleus, 68
  caecum, 294, 295
  colon, 294
  ear, *281*, 282
  eye, 273
  reproductive organs, 302
Chimpanzee, *44*, 49, *51*, 51, 343
  bipedal posture, 51
  brachial index, 209
  brain, 260, 261, *346*
  carpus, 209
  dentition, 116, *118*
  femur, 218
  forelimb, 209, 211, *214*
  hallux, 221
  intermembral index, 212
  os innominatum, *216*
  pelvis, 225
  pes, *207*, *221*, 221, 223
  pollex, 209
  quadrupedalism, 51
  reproductive organs, 302, 307
  skeleton, *210*

Chimpanzee (*cont.*)
  skull, 160-1, *160, 168*
Chin, 159, 163
Chiromyoidea, 100
Chorion, *308*, 308, 310, 311, 312, 338, 345
Chorionic villi, 308, 309, 310
Chromosome cytology, 19-20
Chumashius, 335
Cingulum, 80, 84, 96, 102, 327
  external, *79*, 81, 82, *85*, 89
  internal, 81, *85*, 85, 96, 106, 113, 120
Classification, 18
  of Anthropoidea, 316
  of Hominidae, 19, 49, 345
  of *Necrolemur*, 74, 149-51
  of Pongidae, 19, 49, 345
  of Primates, 18, 40-7, 73-5, 316-17
  of Tarsioidea, 65, 144, 312, 316-17, 332-4, 337
  of Tupaioidea, 43n, 71, 74-5, 145-6, 183, 317, 321-2
  old schemes of, 57n
  purpose of, 33-4
  Simpson's, 43
Clavicle, 42, 170, 171, 175, 175n, *176*, 184, 197, 201, 208
Claws, 6, 42, 171, *171*, 284
  function of, 172
  of *Daubentonia*, 66, 68, 173, 188-9, 192, 248, 329
  of lemurs, 189, 191-2, 331
  of marmosets, 61, 173, 203-4, 206, 207, 344
  of *Tarsius*, 198-9, 338
  of Tupaioidea, 71, 173, 175, 179, 183
  structure of, *172*, 172-3
Clitoris, *303*, 303-5, 331
Colic loop, *see* Ansa coli
Colobinae, 59, 114, 339
  liver, 297
  skull, 154, 157
  stomach, 58, 59, 291, 297
Colobus, 59, 60, 204, 268
Colon, 290, 293-4, 325, 326, 329, 331, 338, 345
Commissure, anterior, 233, *234*, 237, 251, 254-5, 338
  fornix, 234, *234*, 243, 255
  neopallian, 234

Comparative anatomy, 3-20
  features: adaptive, 10, 12, 13-14, 109, 325; advanced, 4, 5, 7, 22, 316-17, 322; general, 4, 7, 9; isolated, 11, 13; non-adaptive, 13-14, 14n; of common inheritance, 25-6, 317; of independent acquisition, 25-6, 27, 123-4, 317-18, 331; primitive, 4, 5, 6, 7, 9, 316, 317, 319, 320, 321-2, 329, 331; specialised, 4, 5, 7, 14, 22, 123, 319
Cones, 273, 275
Connecting stalk, *308*, 309, 311, 313
Convergence, 16-17, 19, 29, 62, 197, 213, 318, 319, 321-2, 334
Cope-Osborn Theory, 82
Coronoid process, *129*, 130, 141, 148
Corpora quadrigemina, *234*, 235, 240, 241, 245, 251, 252
Corpus callosum, *229*, 229, 230, 231, *236*
  of *Centetes*, 233-4, *234*
  of *Daubentonia*, 248
  of marmosets, 254-5
  of *Microcebus*, 243
  of Primates, 234
  of Tarsius, *250*, 251
  of *Tupaia*, *236*, 237, 239
Cortex, auditory, 230, *231*, 318
  cerebral: evolutionary history of, 5, 6-7; of *Adapis*, *239*, 248; of cat, *229*; of Ceboidea, *246*, 256, 257-8, *257*, 344; of *Centetes*, 232-4; of Cercopithecoidea, *246*, 256, *257*, 257-8, 344; of *Echinosorex*, *239*; of *Homo sapiens*, 261-2; of lemurs, 245-7, *246*, 326-7, 329, 331; of marmosets, 254-6, *255*, *256*; of *Microcebus*, *237*, 237, *239*, 242-3; of Primates, 42, 227-32; of *Tarsius*, 249, *251*, 254, 331, 332, 334, 338; of *Tupaia*, *236*, *237*, 237, *239*, 320
  convolutional pattern, 227, 229, 230, 245, 247, 249, 256-7, *258*, 259-60, 262n, 264, 329, 344, 359
  differentiation of, 227-9, *231*, 239, 338, 344

Cortex (*cont.*)
  elaboration of, 227, 239, 240, 265
  expansion of, 227, 232, 237-8,
    248, 280, 344
  functions of, 228-9
  limbic, *231*
  lobes of: frontal, *231*, 231, *237*,
    238, 239, 243, 245, 246, 247,
    248, 250, 254, 256; occipital,
    127, 230, *231*, *237*, 237-8, 239,
    242, 243, 249, 250, 251, 253,
    254, 258, 332, 338; parietal,
    231, *237*, 239, 243, 245, 246,
    247, 256, 259, 260; piriform,
    233, *234*, *236*, *237*, 237, *238*,
    239, 240, 242, 249, 254, 255;
    temporal, 230-1, *231*, *237*,
    238-9, 242, 243, 248, 249, 250,
    252, 253, 255, 258, 326-7, 331
  motor, 228, *237*, 243, 245, 246,
    252, 257
  olfactory, 227, *231*, 233, 245,
    249, 253, 254, 333, 344
  somatic, 228, *231*, *237*, 243, 246,
    256, 257
  visual: expansion of, 127, 278,
    319, 326-7; lamination of, 240,
    241, 250, 256; of Anthro-
    poidea, 260, 277, 280, 344; of
    lemurs, 243-4, *246*; of marmo-
    set, 256; of Primates, 127, 227,
    228, 230, *231*, 265, 272, 278;
    of Tarsioidea, 250, 252, 253-4,
    332, 334, 338; of Tupaioidea,
    239, 240, 317, 318, 319
Craniometry, 163-7
Cretaceous period, 38
Crista obliqua, *98*
Cusp, 5, 42, 76n, 78, 79, 80
  definition, 81n
  evolution, 5-6, 77-80, *85*, 85-6,
    87
  pattern: basic, 79, 101-2, 108;
    of Anthropoidea, 11, 116, 117,
    343; of Cercopithecoidea, 11,
    341; of lemurs, 96, *98*, 327;
    of Tarsioidea, 109-10, 335; of
    Tupaioidea, 317-18
Cuspules, 81, 81n, 104
Cuvier, 30
Cynocephalus, 93
Cynomorpha, 57n
Cynopithecus, 59

AOM 2 B

Cytoarchitectonic charts, 228, 229,
  237, 243, 249, *251*, *256*, *257*

Darwin, Charles, 1, 10, 32n
Darwin's tubercle, 282
Dasycercus, 140
Daubentonia, 66, 68
  aberrancy, 68, 73, 92, 99, 101,
    136, 189, 189n, 247, 248, 269-
    70, 321, 330
  brain, 69, *247*, 247-8, 329
  caecum, 294
  claws, 66, 68, 173, 188-9, 192,
    248, 329
  colon, 294, 329
  dentition, 68, 73, 92, 99-101,
    100, 135, 248, 329, 330
  digits, 66, 69, 188, 193, 329
  evolution of, 329-30
  hallux, 173, 192, 193, 329
  nose, 269-70
  premaxilla, 136, 330
  reproductive organs, 301, 302-3
  skull, *100*, 136, 137, 248, 329, 330
  taxonomic status, 69, 99-101
  tongue, 287-8
Daubentonia robustus, 100
Daubentonioidea, 43, 67, 68
Dental comb, 69, 83, 88, *90*, 92-3,
  94, 318, 325-6, 329, 330
Dental formula, determination of,
  128
  of *Anaptomorphus*, 109n
  of *Archaeolemur*, 99
  of Ceboidea, 110, 340
  of Cercopithecoidea, 109n, 110,
    113
  of *Daubentonia*, 99
  of Indriidae, 95
  of lemurs, 95
  of marmosets, 110
  of *Necrolemur*, 105
  of *Parapithecus*, 117
  of Pongidae, 114
  of *Tarsius*, 101, 109
  of *Teilhardina*, 102
  of *Tetonius*, 106
  of Tupaioidea, 88
  primitive eutherian, 80
Dentition, bilophodont, 35, *87*, *98*,
  112, 113, 114, 121, 125, 339,
  340
  dilambdodont, 89

Dentition (*cont.*)
eutherian, 79-83, *79*
evidence of the, 5, 76-125
evolution of, 56, 83-6, 98, 99-101,
108-10, 113, 123-5, 226, 327-8
heterodont, 76, 77
homodont, 76-7
mammalian, 76-80
of *Absarokius*, 106, 107
of Adapidae, 96, 109, 325, 326
of Adapinae, 86, 96, 106, 327
of *Adapis*, 95, 96, 97, 327
of *Aegyptopithecus*, 54
of *Alouatta*, 112
of *Amphipithecus*, *87*, 116-17,
123, 340, 341
of *Amphitherium*, 77, 77-8
of *Anaptomorphus*, 109n
of *Anathana*, 88, 90, 325
of Anthropoidea, 11, 84, 108,
110, 113, 114-21, 344
of Apatemyidae, 73, 91-2, 100,
100, 109, 330
of *Apidium*, 56, 113, 114, 340
of *Archaeolemur*, 98-9, 329, 330
of *Ateles*, 111, 112 '
of *Australopithecus*, 16, 27, 48,
115, *117*, 122, 125, 348, 349,
350
of *Caenopithecus*, 106, 109, 336
of *Callicebus*, 113, 327
of Carpolestidae, 65, 85, 106,
107, 335
of Cebidae, 111-12
of Ceboidea, 57, 110-13, *111*,
112, 327, 335, 340
of *Cebus*, *111*, *112*
of Cercopithecoidea, 11, 33, 57,
87, 109n, 110, *111*, 113-14,
154, 339, 341
of chimpanzee, 116, *118*
of *Daubentonia*, 68, 73, 92, 99-
101, *100*, 135, 248, 329, 330
of *Dinopithecus*, 114
of *Dryopithecus*, *85*, *87*, 120-1,
342, 343
of *Galago*, 93, 95, 96, 328
of gibbons, 116, *118*, 340, 341
of gorilla, *115*, 116, *119*, 345, 355
of *Hemiacodon*, 65, 102
of Hominidae, 84, 113, 121-3,
124, 164, 341-2, 344, 345-6
of *Homo*, 11, 116, 122

Dentition (*cont.*)
of *Homo erectus*, 48, 122, 123,
124, 168, 350, 351
of *Homo sapiens*, 6, 84, *85*, *115*,
117, *119*, 122-3, 124, 158, 345
of Indriidae, 68, 84, 95, 99
of *Lemur*, 84, *94*, 95, 96
of lemurs, 69, 84, 85, 88, 92-101,
135-6, 141, 143, 288, 325, 326,
327, 328, 329, 330
of *Lepilemur*, 92
of *Loris*, 95, 96
of *Macaca*, *111*, *112*
of marmosets, 61, 111-12
of *Megaladapis*, 94, 98
of *Mesopithecus*, 114
of *Microchoerus*, 104-5, 335, 336,
338
of *Moeripithecus*, 113
of *Nannopithex*, 102, 108
of *Necrolemur*, 74, 104, 105, *105*,
106, 151, 335, 336, 337, 338
of Necrolemurinae, 108, 123
of Notharctinae, 86, 96, 97, 327
of *Notharctus*, 93, *94*, 95, 96, *98*,
327
of *Nycticebus*, *85*, 96
of *Oligopithecus*, 54
of *Omomys*, 102, *104*
of orang-utan, 116, *117*, *119*
of *Oreopithecus*, 56, 348
of Paramomyinae, 102, 106
of *Parapapio*, 114
of *Parapithecus*, *87*, *105*, 114,
117-18, *120*, 120, 123, 340, 341
of *Pelycodus*, 89, 95, 96, 97, 327
of *Periconodon*, 106, 335
of *Perodicticus*, 95
of Plesiadapidae, 73, 86-92, 99-
100, 109, 319, 326, 330
of *Plesiadapis*, *91*, 91, *92*
of *Pliopithecus*, 54, 120, 341
of *Pondaungia*, 116, 123
of Pongidae, 49, 84, 114-21, 122,
123, 124, 125, *154*, 162-3,
341-3, 346, 347
of *Presbytis*, 118
of Primates, 5, 42, 83-6, *87*, 360
of *Progalago*, 93, 97, 328
of *Pronycticebus*, 69, 95, 96, 97,
110, 327
of *Propliopithecus*, 53, 117-18,
120, 123, 341

Dentition (*cont.*)
of *Pseudoloris*, *85*, 102, 108, 333, 338
of *Ptilocercus*, 88, 89, *90*, 91, 317-18
of *Ramapithecus*, 56, 121, 125, 342
of *Saimiri*, 106
of *Simopithecus*, 114
of *Stehlinella*, 100, *100*
of Tarsioidea, 84, 101-10, 143, 149, 151, 331, 332, 333, 335, 337-8
of *Tarsius*, *85*, *87*, 101-2, *103*, *104*, *105*, 105, 108, 333, 338
of *Teilhardina*, 102
of *Tetonius*, *104*, 106, 107, 107n, 335, 337
of *Tupaia*, 72, *85*
of Tupaioidea, 86-92, 317-18, 325
of *Uintanius*, 106, 107
pantothere, 77, 77
therapsid, 77
tribosphenic, 83, 87, 89, 106, 331, 335, 338
Deuterocone, 95, 116, 117
Diastema, 113, 115, 115n, 121, 123, 124
Digestive system, evidence of the, 286-98
Digital formula, 187, 188, 205-6, 331, 332, 333, 344
Digital pads, 42, 171, 173-5, 197, 199, 284
Digits, manual, 276; of *Daubentonia*, 69, 188, 329; of lemurs, 187-8, *188*, *189*, *190*, 331; of *Tarsius*, 197, 331; of Tupaioidea, 178-9
mobility of, 42, 170, 171
elongation of the fourth, 187-8, 191-2, 198, 205-6, 329, 331
pedal, of baboon, 206; of *Daubentonia*, 193, 329; of *Callitrix*, 205-6, 344; of Hominoidea, 219, 220, 221; of lemurs, 69, 187-8, *188*, *189*, *190*, 191-5, 331; of monkeys, 205-6, 344; of *Tarsius*, 198-9, 332; of Tupaioidea, 181-2, 331
pentadactyly, 5, 42, 170, 360
Dinopithecus, 60, 114, 158

Docophthirus, 319
Dolichopithecus, 60, 157, 340
Doran, A. H. G., 283
Dryopithecinae, 343-4
Dryopithecus, (*Proconsul*), 54, 339, 342-3, 346
dentition, *85*, 87, 120-1, 342, 343
limbs, 213, *218*, 218, 223, 343
skull, 163, 343
Dryopithecus africanus, 213, *214*
Dryopithecus fontani, 213
Dryopithecus major, 163
Dwarf lemur, *see Cheirogaleus*

Ear, evidence of the, 280-3
external, 47, 66, *267*, 280-2, *281*, 331, 338, 344
mobility of, 280, 281, 282, 332
ossicles of the middle, 128, 134, 283, 318
size of, 280-1
Echelle des êtres, 4, 322
Echinosorex, *239*, 240
Ectoparasites, 319, 326
Ectoturbinal, 269, *270*, 271
Elliot Smith, 276
Elphidotarsius, 107
Embryology, 5, 20-3, 307-14
Endoturbinals, 269, *270*, 270, 271, 331, 338
Entoconid, *79*, 82, 86, *87*, 89, *98*, 102
Entotympanic, 134
Eocene period, 36
Eppelsheim femur, 218
Erinaceus, 132n, 237, 251
Erythrocebus, 58, 206
Ethmoid bone, 131, 132, 132n, 137, 137n, 148, 157, 270, 330, 338, 344
Ethmoturbinal, *269*, 269, *271*
Euoticus, 69
Eutheria, 39
Evidence of the alimentary tract, 289-95
brain, 227-64
dentition, 5, 76-125
digestive system, 286-98
ear, 280-3
eye, 272-80
limbs, 170-226
liver, 295-7

Evidence (*cont.*)
olfactory sense, 266-72
placenta, 307-14
reproductive system, 299-314
skull, 126-69
special senses, 265-85
tactile sensation, 283-5
tongue, 286-9
Evolution, convergence, 16-17, 19,
29, 62, 197, 213, 318, 319,
321-2, 334
environment, 7-9, 10, 11, 315,
322-3, 336
irreversibility, 7-9, 28, 29, 232
of *Homo*, 9, 24, 36, 49, 262-3,
346, 352
of *Homo sapiens*, 1, 2, 22, 24, 28,
29, 31, 49, 169, 345-60
of mammals, 37-40
of Primates, 1-36, 38-40, 322-
323
opportunism of, 10, 11
parallelism, 16-17, 29, 70, 258,
315, 316, 327, 329
populations as units of, 33
rates, 27-8, 345
regression, 6, 347
retrogression, 7, 9, 124-5, 173,
175n, 252, 280, 281, 358
selection, 10, 252, 316
unsuccessful lines of, 1-2, 315,
337
Evolutionary change, direction of,
5, 173-4
differentiation, 1, 2, 38, 39, 144,
171, 193, 315, 324, 345-6
divergence, 2, 22-3, 24-5, 45-7,
164, 262-3, 298, 315, 322, 324,
327, 328
gradational series, 28, *44*, 45, *46*,
47, 322-3, 360
history, 145-7, 360
possibilities and probabilities,
14, 316
process and the Primates, 1-36
radiations: of lemurs, 143-7, 324-
31; of monkeys, 338-45; of
Pongidae, 341-3; of Primates,
74, 286, 315-60; of Tarsioidea,
108-9, 334-8; of Tupaioidea,
317-19
reversal, 8-9, 98, 99, 316, 327,
330

Evolutionary change (*cont.*)
sequence, 1, 3
hominid, 9, 22-3, 24, 25, 26,
28, 29, 31, 49, 159, 163, 207-
8, 217, 226, 262, 298, 345,
348, 360
linear, 45-7
lemuroid, 98, 145-7, 326-30
of monkeys, 159, 258, 338-45
pongid, 22-3, 25, 26, 31, 53n,
159, 163, 207-8, 217, 226,
262, 298, 341-3, 345, 348, 360
tarsioid, 333-4
simplification, 7, 124, 347
specialization, 7, 8, 14, 64, 107-8,
139-40, 187-8, 200-1, 316, 321,
322, 324, 325, 326, 327, 328,
335, 343, 346-7, 360
trends, 1, 2, 3, 23, 26, 42; of
Hominidae, 45, *46*, 47, 343-4,
345-52, 360; of lemurs, 98,
329, 330-1; of Pongidae, 49,
159, 344-5; of Primates, 41-2,
45, *46*, 47, 48-9, 75, 86, 315,
316, 322-3; of Tarsioidea, 253,
335, 336-7
Eye, evidence of the, 272-80
of Anthropoidea, 47, 273, 274-6,
278, 280, 344
of Lorisinae, 69
of marmosets, 275
of monkeys, 275, 276-7, 278
of Primates, 272, 275
of *Tarsius*, 64, 250, *274*, 274-5,
277, 278, 331, 333, 338
of Tupaioidea, 71, 273-4, 318

Falcula, *see* Claws
Family, 30, 41
Fayum desert, 53-4, 120, 339
Femur, *174*, *180*, 180-1
Eppelsheim, 218
of *Australopithecus*, 224
of Callitrichidae, 205
of chimpanzee, *218*
of *Dryopithecus*, *218*
of gibbon, 217, 218
of Hominidae, 218-19
of lemurs, 191
of monkeys, 205
of *Nannopithex*, 199
of *Necrolemur*, *199*, 199
of *Notharctus*, *186*, 191

Femur (*cont.*)
  of *Plesiadapis*, *183*
  of Pongidae, 217-18
  of *Ptilocercus*, *180*, 180-1, 183
  of *Tarsius*, 198, *199*
Fibula, 170, 171, *174*
  of lemurs, 191
  of monkeys, 205
  of *Nannopithex*, 199
  of *Notharctus*, *186*, 191
  of *Ptilocercus*, *174*, *180*, 181
  of *Tarsius*, 198, *199*
Fluorine, 36
Foetalization, 21
Foramen,
  entocarotid, 134-5, 138, *139*, 145,
    151, 335
  hypoglossal, *133*, 134, *139*
  infraorbital, 128, *130*
  jugular, *139*
  lacerum, 139, 330
  magnum, 127, *127*, *131*, 134,
    139; of Hominidae, *165*; of
    lemurs, 135, *136*, *139*, 142,
    144; of monkeys, 156, 344; of
    Primates, *127*, 127, 161; of
    Tarsioidea, 148, 149, 152, 332,
    333, 338
  mental, 130, 167
  ovale, 130, 132, 134
  postglenoid, *139*
  rotundum, 132
  stylomastoid, *130*, *133*, 134, 139
Forelimb, 170
  lengthening of, 50, 200, 344
  of lemurs, 184, 186
  of monkeys, 57, 200-1
  of Pongidae, 50, 209, 213-14, *214*
  of *Tarsius*, 197, 338
  of Tupaioidea, 175, 178
  use of, 126, 170, 171, 200-1
Fossa, 148, 149, 154
Fossils, arboreal life and fossilisa-
    tion, 336-7, 339
  classification of, 73-4
  relative age of, 23-4, 34-6
  types of, 47n
Fovea, 273, *274*, 274, 275, 278
Frontal bone, 128, *129*, *130*, *131*,
    131, *133*, 142, 148, *152*, 152,
    156, 157, 159, 161, 344

Galaginae, 69, 144, 328, 331

Galaginae (*cont.*)
  hallux, 194
  limbs, 69, 187, 197, 331
Galago, 69, 184
  brain, 254, 256, 328
  caecum, 294
  colon, 294
  dentition, 93, 95, 96, 328
  ear, 281, 282
  gestation, 314
  nose, 266
  placenta, 311
  reproductive organs, 302, 304
  skull, 136, *136*, 149, 328
  tongue, *287*
Galago garnetti, 93, *136*
Galeopithecus, 93, 240
Genitalia
  external: female, 303-4, *303*, 324,
    331, 338; male, 299-303, *300*,
    301, 331, 338, 345
  internal, 304-7, *305*, *306*
Genus, 40-1
Geochronology, 34
Geological antiquity, estimation of,
    34-6
Gestation, 42, 313-14
Gibbon, 49, *50*, 50-1, 53, 300n, 301,
    307, 342, 343
  brachiation, 50, 209
  brain, 51, 259, 264, 346
  carpus, 209
  dentition, 116, *118*, 340, 341
  femur, 217
  intermembral index, 212
  ischial callosities, 51
  liver, 298
  manus, *188*
  mode of progression, 51, 215, 223
  pelvis, 216
  pes, 50, *188*, 223
  pollex, 50, 209
  skull, 156, *159*, 159-60, 162
Gigantopithecus, 56-7
Glenoid fossa, 130, *133*, 134, 148,
    167, 175
Gorgopithecus, 60
Gorilla, 49, 52-3, *55*, 345-6
  brachial index, 209
  brain, 260, *261*, *263*, 345, 347, 355
  carpus, 209
  dentition, 116, *118*, 345, 355
  forest g., 53, 211, *221*, 221

Gorilla (*cont.*)
  hallux, 221
  limbs, 52, 53, 55, 345-6
  liver, 297-8
  manus, *211*
  mountain g., 53, *212*, *221*, 221
  pes, *212*, *220*, *221*, 221, 223
  pollex, 52, 209
  posture, 52, 55
  quadrupedalism, 52
  reproductive organs, 302
  skull, 160-1, *165*, *167*, *354*, 354-5
  tarsus, *195*
Guenon monkey, *see Cercopithecus*
Guereza monkeys, *see Colobus*
Gyrus, 229, 247, 250
  dentate, *234*, 234, *236*, 237, 243, *250*, 251, 255

Hadropithecus, 70
Haeckel, E., 15, 22
Hafleigh, A. S. and Williams, C. A., 321
Halbaffen, 66
Hallux, 42
  hypertrophy of, 192-5, 221
  of baboon, 206
  of *Daubentonia*, 193, 329
  of chimpanzee, 221
  of Galaginae, 194
  of gorilla, 221
  of *Homo sapiens*, 219-21
  of lemurs, 66, 192-5
  of marmosets, 61, 193, 206, 207
  of monkeys, 206, 207
  of *Notharctus*, 194
  of Pongidae, 219, 220
  of *Tarsius*, 198-9
  of Tupaioidea, 181, 318
  opposability of, 181-2, 192-3, 207, 220-1
  specialization of, 192, 193-5, 207, 220
Hapale, *271*
Hapale, *see Callitrix*
Hapalidae, *see* Callitrichidae
Haplorhini, 332
Heidelberg man, 123, 351
Helix, 281, *281*
Hemiacodon, 65
  dentition, 65, 102
  limbs, *195*, 200, 332
Heterodonty, 76, 77

Hill, J. P., 307, 310, 311-12
Hindlimb, 170
  function of, 171
  of lemurs, 68, 69, *184*, 191, 332
  of monkeys, 200-1, 344
  of Pongidae, 215, 344
  of Primates, 170
  of *Tarsius*, 64, 197, 198-9, 200, 331, 332, 338
  of Tupaioidea, 175, 179
Hippocampal formation, 234
Hippocampus, *234*, 234, *236*, 237, 243, 250, 251, 255
Hoanghonius, 65, 66, 334n
Höfer, H., 325
Hominid, 31, 262, 346
Hominidae, 2, 9, 31, 48-9, 345-7
  affinities of, 345-60
  and Pongidae, 19, 49, 121-3, 207 8, 211-12, 217, 220n, 262-3, 298, 345, *346*, 347, 360
  appendix, 295
  bipedalism, 14, 21, 48, 51, 167, 167n, 208, 215, 216, 223-6, 347-8, 360
  classification, 19, 345
  dentition, 84, 113, 121-3, 124, 164, 347,
  emergence of, 345-52
  evolutionary sequence, 9, 22-3, 24, 25, 26, 28, 29, 31, 49, 159, 163, 207-8, 217, 226, 262, 298, 341-3, 348, 360
  fossil record, 348-9, 353-9
  limbs, 21, 27n, 208, 211-13, 216, 217-18, 220n, 223-4, 345-52
  phylogenetic relationships, 123-4, 159, 212-15, 298, 349
  placenta, 313
  skull, 138, 158-9, 166-7, 168, 354-5
Hominoidea, 25, 27, 33, 43, 49, 345
  carpus, 209
  limbs, 201, 207-26
  liver, 296
  reproductive organs, 301
  skull, 138, 158-69
Homo, 48
  brain, 48, 263, 346-7
  dentition, 11, 116, 122
  emergence of, 352-60
  evolution of, 9, 24, 36, 49, 262-3, 346, 352

Homo (cont.)
  intermembral index, 212-13
  limbs, 207-8, 215, 224-6
  pelvis, 216-17
  pes, *222*, 224
  skull, 169
Homodonty, 76-7
Homo erectus
  brain, 48, *346*, 347, 349, 351
  braincase, 168, 351
  classification, 24
  dentition, 48, 122, 123, 124, 168,
    350, 351
  evolution, 28, 29, 31, 350-2, 359
  jaws, 12-13, 48, 350, 351
  limb skeleton, 12, 348-9, 351
  skull, 12-13, 156, *168*, 168-9, *354*
  supra-orbital ridges, 9, 22
Homo sapiens, 48
  antiquity of, 352-3
  bipedalism, 223-6, 360
  brachial index, 212
  brain, 7, 15, 22, 260, *261*, 261-3,
    263, 345, *346*, 346-7, 358
  caecum, 295
  dentition, 6, 84, *85*, *115*, *117*,
    119, 122-3, 124, 158, 345
  emergence of, 352-60
  evolution, 1, 2, 22, 24, 28, 29, 31,
    49, 169, 345-60
  eye, *297*
  hallux, 219-21
  intermembral index, 212, 212n
  limbs, 211-12, 345-6
  liver, *296*, 296-7
  mandible, *117*
  os innominatum, *216*
  pes, 193, *195*, *207*, *219*, 220, *221*,
    221-3, *222*
  pollex, 212
  skeleton, 352
  skull, 12, 158, 161n, 169, 354,
    355, 358
  tarsus, *195*
Homunculus, 62, 340
Hopkins, G. H. E., 319
Howler monkeys, *see Alouatta*
Human, 31, 351
Humerus, *174*, *176*, 350
  of *Arctocebus*, 184
  of *Cebus*, 201
  of *Homo sapiens*, 211
  of lemurs, 184, *186*. 186

Humerus (*cont.*)
  of monkeys, 201, 212-13
  of *Notharctus*, *186*, 186
  of *Papio*, 201
  of *Plesiadapis*, 183
  of Pongidae, 208-9, 213
  of *Ptilocercus*, *174*, 175, *176*, 177,
    183
  of *Tarsius*, 197
Huxley, T. H., 15, 47n, 322, 357-8
Hylobates, see gibbon
Hylobatidae, 49
Hylobatinae, 49, 116-17, 159, 298,
  340
Hypocone, 79, *85*, 85, 95, 96, *98*,
  102, 112, 327
Hypoconid, 79, 82, 86, *87*, 89, *98*,
  102
Hypoconulid, 79, 82, 86, *87*, 89, *98*,
  102, 112, 113, 121, 122
Hypothalamus, 235, *238*, 240

Ilium, *174*, *176*
  of Hominidae, 216-17, 224
  of lemurs, 191
  of monkeys, 204-5, 205n
  of Pongidae, 205n, 215-16, 217
  of *Tarsius*, 198
  of Tupaioidea, 179-80
Incisor teeth, 76
  eutherian, 80
  of *Daubentonia*, 68, 99, 135, 288,
    330
  of Hominidae, 122, 355
  of lemurs, 92, 93, 98, 99, 135,
    141, 288, 325, 326, 328, 329,
    330
  of monkeys, 112, 113, 344
  of Plesiadapidae, 89, 90, 91, 330
  of Pongidae, 114, 120, 162-3,
    342, 344
  of Primates, 83-4, 86
  of Tarsioidea, 101, 102, 108, 109,
    337
  of Tupaioidea, 72, 88, *90*, 128,
    318, 325
Indraloris, 71, 324, 328
Indri, 66, 294
Indriidae, 67, 68
  dentition, 68, 84, 95, 99
  limbs, 68
  reproductive organs, 302
Infraorders, 43

Insectivora, 39, 132, 137, 140, 145,
 183-4, 240-1, 317
Intermembral index, 212, 212n
Ischial callosities, 51, 57, 205, 341
Ischial tuberosities, 191, 205, 216,
 217, 224
Ischium, *174*, *176*, 179-80, 216, 217

Jaws, 126, 130, 155, 156, 158, 159,
 160, 163, 166, 168, 270, 272,
 339, 340, 345, 346, 349, 350,
 351, 355, 356, 358
Jepsen, G. L., 91
Jurassic period, 37

Karyotypes, 19-20
Keith, Sir Arthur, 52

Lacrimal bone, *130*, 131, 132, 136,
 137, 143, 148, 154, 319
 foramen, 131, 136, 143, 148, 154
Lagothrix, 302
Laminae, *see*, Cortex, visual, and
 lateral geniculate nuclei
Large intestine, 290, *293*, 293-5
Lateral geniculate nuclei, 240, *244*,
 244, 277-80
 cell laminae, 278, *279*, 320
 lamination, 240, 244-5, 259, 277-
 9, 318, 320
 of Anthropoidea, 245, 259, 278,
 344
 of cat, 278
 of Cercopithecoidea, 278
 of *Cercopithecus*, *279*
 of *Homo sapiens*, *279*
 of lemurs, 244-5, 278, *279*, 320,
 331
 of Primates, 244-5, 278, 320
 of *Tarsius*, 252, 278, 332, 338
 of *Tupaia*, 240, 241, 278, 318, 320
Lemur, 66, 184
 brain, *245*, *246*
 caecum, 295
 carpus, *189*, 189-91
 dentition, 84, *94*, 95, 96
 eye, 273, *279*
 manus, 188-9, *189*, *190*
 nose, *271*
 pelvis, 191, *192*
 pes, *190*
 skull, *139*
 tail, 67, 325

Lemur (*cont.*)
 tarsus, *195*
 tongue, *188*
Lemur catta, *67*, 139, 245; *nigrifons*,
 246; *varius*, 94
Lemuridae, 67, 302
Lemuriformes, 43, 67, 324
 evolution of, 146-7, 324-9
 fossil, 69-70, 339
 nose, 270, 324
 os ectotympanicum, 137-8, *138*,
 139-40, 143, 145, 324, 330
 phylogenetic relationships, 108-9
 137-40, 143-4, 324-5, 327, 329,
 330
 reproductive organs, 302, 304,
 331
 skull, 130, *131*, 132, 135, 136,
 146-7, 148, 151, 318, 325, 326,
 330
 vibrissae, 284, 285
Lemuriform lemurs, 66, 67n, 146-
 7, 328-30
Lemuroidea, 43, 67, 87, 95
Lemurs, *44*, 45, 66-71
 aberrancy of, 93, 139-40, 144,
 147, 187, 324, 328
 affinities of, 99-101, 108-10, 143,
 324-31
 alimentary tract, 292, *293*, 331
 brain, 135, 236, 242-9, *244*, *245*,
 246, 249, 253, 258, 326-7, 331
 caecum, 294-5
 colon, *293*, 294, 325, 326, 331
 dentition, 69, 84, 85, 88, 92-101,
 135-6, 141, 143, 288, 325, 326,
 327, 328, 329, 330
 digits, 69, 187-8, *188*, *189*, *190*,
 191-5, 331
 dwarf, l., 68
 ear, 66, 282, 283, 331
 evolution of, 66, 137-8, 145-7,
 326-30
 eye, 273-4, 277, 331
 gestation, 314
 hallux, 66, 192-3
 lemuriform l., 66, 67n, 146-7,
 328-30
 limbs, 184-95, 325, 326, 331
 liver, 297
 lorisiform l., 66, 69, 146-7, 188,
 328-9
 Maadagscar l., 66, 94, 144, 328-30

Lemurs (*cont.*)
  Malagasy l., 67, 319
  manus, 66, 184, 187-91, 331
  mouse l., 68
  nails, 66, 188, 191-2
  nose, 65, 266, *266*, 268, 269-70,
    272, 324, 331
  parasites of, 319, 326
  pes, 66, 182, 184, 187, *188*, 191-5,
    207, 331
  placenta, 310-11, 331
  pollex, 66, 187-8, 189, 191
  Recent, 66
  reproductive organs, 300, 301,
    302, 307, 324, 331
  skeleton, *185*
  skull, *127*, 127, 135-47, *139*, 318,
    324, 330
  snout, 66, 69, 135, 143
  tarsus, *195*, 331
  tongue, 287-8, *288*, 326, 331
Lepilemur, 92
Leptictidae, 71, 140
Libypithecus, 60, 157-8, 264
Limbs, evidence of the, 170-226
  of Anthropoidea, 344
  of chimpanzee, *214*
  of Galaginae, 69, 187, 197, 331
  of gorilla, 52, 53, 55
  of Hominidae, 21, 27n, 208, 211-
    13, 216, 217-18, 220n, 223-4,
    345-52
  of Hominoidea, 201, 207-26
  of Indriidae, 68
  of lemurs, 184-95, 325, 326, 331
  of Lorisinae, 69
  of monkeys, 57, 200-7, 208, 209,
    342, 344
  of *Notharctus*, 185-7, *186*, *190*,
    191, 194, 327
  of Pongidae, 21, 49, 207-26,
    345-8
  of Primates, 170-5
  of Tarsioidea, 197-200, 331, 332,
    338
  of Tupaioidea, 175-84, 318
  specialization, 170, 331, 347-8
  structure of mammalian, 170-1
Limnopithecus, see Pliopithecus
Liver, evidence of the, 295-8
Long, W. M., 16
Loris, 69
  caecum, 295

Loris (*cont.*)
  colon, 294
  dentition, 95, 96
  eye, 273
  gestation, 314
  reproductive organs, 301
  skull, 138
Lorisidae, 69
Lorisiformes, 43, 67, 69, 324
  ear, 281
  evolution of, 146-7, 328-9
  nose, 270, 324
  os ectotympanicum, 137, *138*,
    145, 324, 330
  pes, 194-5
  phylogenetic relationships, 324-5
  reproductive organs, 301, 304,
    331
  skull, *131*, 132, 135, 136, 137,
    138, 139, 140, 146-7, 148, 324,
    330, 332
  vascularization, 195
  vibrissae, 284
Lorisinae, 69
  pes, 194, 207
  tail, 69, 322
Lushius, 65
Lytta, 288, *289*

Macaca, 57, 58, 60, 340
  brain, *246*, *257*
  caecum, *296*
  carpus, *189*
  dentition, *111*, *112*
  ear, 282
  eye, 278
  manus, *188*
  nose, *268*
  pes, *188*, 207
  pollex, 204
  skull, 154, 156
  tail, 57
  tarsus, *195*
Macaca mulatta, 58
Macaque monkeys, *see Macaca*
Macrotarsius, 66, 108, 335
Macula lutea, 273, 276
  of Anthropoidea, 273, 344
  of monkeys, 275
  of *Tarsius*, 250, 274, 275, 332,
    334, 338
Madagascar lemurs, 66, 94, 144,
  328-30

Magdalenian culture, 352-3
Malagasy, lemurs, 67, 319
Man, 31, 351
Mandible,
  of *Amphitherium*, 77
  of *Australopithecus*, *177*, 350
  of *Caenopithecus*, 149
  of Hominidae, 167
  of *Homo sapiens*, *117*
  of lemurs, 140-1
  of monkeys, 157, 341
  of orang-utan, *117*
  of Pongidae, 162
  of *Ptilocercus*, *129*, 130
  of *Tarsius*, 148, 341
  of Tupaioidea, 130
Mandrill, 58-9, 154
Mandrillus, *see* Mandrill
Manus, 47-8
  of gibbon, *188*
  of gorilla, *211*
  of *lemur*, 187, *189*, *190*
  of lemurs, 66, 184, 187-91, 331
  of *Macaca*, *188*, *189*
  of marmoset, 202-4, *203*
  of monkeys, 202-4
  of *Notharctus*, *190*, 191
  of *Nycticebus*, 187, *188*
  of orang-utan, *211*
  of *Ptilocercus*, *177*, 178-9, *179*,
    189
  of *Tarsius*, *188*, *189*, 197-8, 331
  of Tupaioidea, 178
Marmoset, 61
  brain, 254-6, *255*, *256*, 258
  claws, 61, 173, 203-4, 206, 207,
    344
  dentition, 61, 111-12
  ear, 283
  eye, 275
  gestation, 314
  hallux, 61, 193, 206, 207
  limbs, 201, 202, 206, 344
  manus, 202-4, *203*
  nose, 270, *271*
  pelvis, *192*
  pes, *203*, 206, 207
  reproductive organs, 301, 304
  skull, 151, *152*
  tongue, 189
  vibrissae, 284
Marsupialia, 37, 38, 39, 140
Masseter, 126, 129, 161

Maxilla, 128, *129*, 129, 130, 132,
  133, 134, 149, 350
Maxilloturbinal, 269, *269*, 270, 271,
  271
Medulla, 235
Megaladapis, 70, 328-9
  brain, 248-9
  dentition, 94, 98
  skull, 144-5, *145*
Megaladapis grandidieri, 145
Mesopithecus, 60, 339-40
  brain, 264
  dentition, 114, 339-40
  skull, 60, 157, *158*
Mesopithecus pentelicus, 158
Mesopropithecus, 70
Mesostyle, 81, *85*, 88, *98*
Mesozoic epoch, 37, 39
Metacarpus, *174*, *177*, 178
Metacone, *79*, 81, *85*, 95, *98*
Metaconid, *79*, 82, 86, *87*, 89, 96,
  98, 102
Metaconule, *79*, 81, *85*, *98*, 104
Metastyle, 81, *85*, 88, 89
Metatarsi-fulcrumation, 193-4, 206
Metatarsus, *174*, 181, *182*, 192, 193,
  198, 206, 219, 220, *220*
Metatheria, 38, 39
Microcebus, 68, *68*
  brain, 236, *237*, 237, *238*, *239*,
    239, 240, *242*, 242-5, 250, 251,
    252
  colon, 294
  ear, 282
  skull, 137n, 146, *146*
Microcebus murinus, 68, 242
Microchoerinae, 65n
Microchoerus, 65
  dentition, 104-5, 335, 338
  skull, 148
Midbrain, of lemurs, 245
  of mammals, 235
  of *Tarsius*, 252
  of *Tupaia*, 240
Miocene period, 36
Moeripithecus, 113
Molarization, 84, 85, 88, 89, 95,
  101, 109, 327, 330
Molar teeth, 76, 77
  eutherian, 80-3
  of Cercopithecoidea, 11, 340
  of Hominidae, 11, 122
  of lemurs, 96-7, 109, 327

Molar teeth (*cont.*)
  of monkeys, 110, 111, 112, *112*, 113, 327, 339, 341
  of Pongidae, 114, 116, 117, *118*, *119*, 121, 122, 123, 124, 339, 341, 342, 343
  of Primates, 5-6, 33, 42, *85*, 85-6
  of Tarsioidea, 102, 104-5, 106, 108, 109, 331, 335, 336, 338
  of Tupaioidea, 89, 91
  quadritubercular, 85, *85*, 86, 96, 102, 104, 108, 112, 327, 335, 338, 344
  tritubercular, 80, 81-2, *85*, 96, 102, 108, 112, 327
Monkeys, 57-63
  affinities of, 338-45
  brachiation, 200-1, 213
  dentition, 110-14
  eye, 275, 276-7, 344
  limbs, 57, 200-7, 208, 209, 342, 344
  pes, 206-7, 219-21, *220*, *221*, 223
  posture, 57, 208
  quadrupedalism, 57, 200, 205n, 206, 208, 213, 215
  skull, 152-8, 341, 344
  tails, 57
Monotremata, 37n, 38, 39
Morphology, 4, 6, 11, 12, 18, 25-6, 28, 29
  change, 7-8, 29, 216-17, 329
  equivalence, 16
  pattern, 8-9, 12, 16, 38, 166-7, 319, 324
  sequence, 107
  series, 169, 335, 359-60
  total morphological pattern, 12, 13, 40-1, 71, 145-6, 224, 317, 318, 354-5
Morton, D. J., 193
Mouse lemur, 68
Mousterian man, 353-6, *354*, 357, 358, 359
Multitubercular Theory, 5, 82
Multituberculata, 37, 39, 77, 78, 82, 107
Multivariate analysis, 15, 16, 167
Myology, of *Notharctus*, 186
  of Pongidae, 211, 217, 221
  of Tupaioidea, 183-4, 318

Nails, 6, 42, 284
  evolution of, 173-5, 189, 193

Nails (*cont.*)
  function, 173
  of Anthropoidea, 48, 344
  of Ceboidea, 203-4, 206
  of Cercopithecoidea, 202-4
  of *Daubentonia*, 193, 329
  of lemurs, 66, 188, 191-2
  of marmosets, 61, 207, 344
  of *Notharctus*, 189
  of Primates, 171, 172, 173-5
  of *Tarsius*, 174, 198
  structure of, 173
Nannopithex, 65, 102, 108, 149, 199-200, 332
Napier, J. R., 204, 226
Nasal cavity, 67, 128, 131, 132, 134, 254, 268, *269*, 269, *270*, 270, *271*, 271, 272, 318, 324, 331, 332, 333, 338, 344
Nasalis, 59, 60
Nasoturbinal, 269, *269*, *270*, 271, *271*
Neanderthal man, 31, 48, 169
  brain, 356
  cranial capacity, 356
  generalised type, *355*, 356
  skull, *354-5*, 356-7
  specialised type, *355*, 356, 357, 358
Necrolemur, 65
  brain, *253*, 253, 333
  calcaneus, *195*, 199
  classification of, 74, 149-51
  dentition, 74, 104, 105, *105*, 106, 151, 335, 336, 337, 338
  limbs, 151, 199, *199*, 332
  nose, 272
  os ectotympanicum, 140, 149-50, 325, 338
  skull, 140, 149-52, *150*, 335-6
Necrolemur antiquus, 105
Necrolemuridae, 74
Necrolemurinae, 65, 104, 106, 108, 123, 335
Neocranium, *see* Brain-case
Neopallium, of *Centetes*, 233, *234*, 234-5, *238*
  of lemurs, 242-3, 245
  of marmosets, 254-5
  of *Microcebus*, *237*, 239, 242
  of monkeys, 259
  of Primates, 233, 235
  of *Tarsius*, 252

Neopallium (cont.)
  of Tupaia, 237, 238, 239-40, 241
Neosaimiri, 63, 340
Neoteny, 21
New World monkeys, see Ceboidea
Nose, see nasal cavity, nostrils,
  rhinarium
Nostrils, 57, 265-6, 268, 268
Notharctinae, 69, 70, 143, 327
  dentition, 86, 96, 97, 327
Notharctus, 70
  dentition, 93, 94, 95, 96, 98, 327
  limbs, 185-7, 186, 190, 191, 194
  manus, 187, 189, 190, 191
  nails, 189
  pes, 190, 194, 222
  skull, 142, 143
Notharctus osborni, 94, 142, 186
Nuchal crest, 129, 129, 142, 144,
  160-1, 166, 167
Nuttall, G. H. F., 18
Nycticebus, 69, 184
  dentition, 85, 96
  eye, 273
  gestation, 314
  limbs, 187, 188, 191
  reproductive organs, 302
  skull, 138, 140

Occipital bones, 128, 129, 130, 132,
  133, 152
  condyles, 129, 129, 130, 133, 134,
    144, 160, 161, 165, 166, 167,
    168, 355
  protuberance, 129, 129, 165, 166,
    355
Occiput, 148, 161
Occlusion, 78, 78, 81, 113, 116
Old World monkeys, see Cercopithecoidea
Olduvai remains, 350, 351-2
Olfactory bulb, 233, 234, 235, 241
  cortex, 227, 231, 233, 245, 249,
    253, 254, 333, 344
  mechanism, reduction of, 237,
    245, 253, 265, 268, 272, 317
  of Adapis, 248
  of Daubentonia, 247, 270
  of lemurs, 245
  of marmosets, 254
  of Megaladapis, 248-9
  of Microcebus, 242
  of Necrolemur, 253, 272

Olfactory bulb (cont.)
  of Tetonius, 272
  of Tarsius, 249, 253, 254, 272
  of Tupaia, 237, 241, 317
  sense, 227, 266-72, 323, 360
  tract, 234, 234, 249
  tubercle, 233, 234, 236, 237, 254
Oligocene period, 36
Oligopithecus, 54, 120, 123, 339
Omomyidae, 65
Omomyinae, 102, 335, 336
Omomys, 102, 104
Ontogeny, 20
Optic axis, 275-6
  foramen, 131
  nerve fibres, 278, 320
Orang-utan, 49
  brachiation, of, 51, 209
  brain, 346
  dentition, 116, 117, 119, 343
  limbs, 210, 211, 212, 223
  reproductive organs, 301n, 302
  skull, 117, 161, 162
Orbital aperture, 126, 132, 136,
    145, 157, 160, 276, 344, 345
  cavities, 126, 130, 143, 145, 147,
    151, 152, 163, 254, 272, 331,
    333, 336, 338
Orbitosphenoid, 130, 131, 131,
    132
Order, 1, 40, 41
Oreopithecidae, 56
Oreopithecus
  dentition, 56, 114, 121, 348
  limbs, 56, 121, 213, 215, 348
  skull, 56, 121, 163
Orthogenesis, 1, 8, 29, 123
Orthoselection, 8
Orycteropus, 269
Os ectotympanicum,
  of Ceboidea, 138, 138, 157
  of Cercopithecoidea, 138, 138-9,
    157, 344
  of Lemuriformes, 137-8, 138,
    139-40, 143, 145, 324, 330
  of Lorisiformes, 137, 138, 145,
    324, 330
  of Necrolemur, 140, 149-50, 325
  of Primates, 138, 140
  of primitive mammals, 137-8,
    138, 151
  of Tarsioidea, 138, 138, 148, 149-
    51, 152, 338

Os ectotympanicum (*cont.*)
   of Tupaioidea, 134, 138, 324
   phylogeny, 138-40, 149-51
Os innominatum, *176*, 179-80, *183*,
   216, 216
Osseous mosaic, 128, 130, 132, 138
   145
   of *Daubentonia*, 137
   of lemurs, 130-1, 137
   of Tupaioidea, 130, 132, 318
Ossicles of middle ear, 128, 134,
   283, 318
Owen, Sir Richard, 52

Paedomorphosis, 22
Paidopithex, 218
Palaeocene period, 36
Palaeontology, 18, 23-30, 315-16,
   319, 332
Palaeopropithecus, 70, 329
Palaeosimia, 56, 343
Palaeozoic epoch, 37
Palatine bone, *130*, 131, 132, 134,
   137, 319, 329, 330
Palmar pads, *see* Pressure pads
Pan, *see* chimpanzee
Pan paniscus, 52
Pantotheria, 37, 39, 77, 77, 78
Papillae, carpal, 178, *179*, 197, 202,
   284
   conical, 285, 286, *288*
   fungiform, 287, *287*
   vallate, 287, *287*, 288
Papio, *see* Baboon
Paracone, *79*, 81, *85*, 95, *98*
Paraconid, 79, 82, 86, *87*, 89, 96,
   97, *98*, 102, 104, 108, 109, 112,
   117
Paraconule, 81, *85*, 102, 104
Parallelism, 16-17, 29, 70, 258, 315,
   316, 327, 329
Paramomyidae, 72, 73, 75n
Paramomyinae, 102, 104n, 106
Paranthropus, 48, 166, 349-50
Parapapio, 60, 114
Parapithecidae, 60, 114, 341
Parapithecus, 53, 60
   dentition, *87*, *105*, 114, 117-18,
   120, 120, 123, 340
   taxonomic status, 117-18, 341
Parastyle, 81, *85*, 88, 89, *98*
Parietal bones, 128, *129*, *130*, 132,
   152, 154, 159, 161, 351

Patella, *174*, 181
Pelvis, 170, 179
   of *Australopithecus*, 27, 48, 224,
   225, 348
   of gibbons, *192*, 216
   of Hominidae, 27n, 216-17, 224,
   225
   of lemurs, 191, *192*
   of marmosets, *192*
   of monkeys, 204-5
   of *Nannopithex*, 199
   of Pongidae, 215-17, *225*
   of *Tarsius*, *192*, 198
   of Tupaioidea, 179-80
Pelycodus, 70
   dentition, 89, 95, 96, 97, 327
Pelycodus frugivorus, 96 ; *ralstoni*,
   96 ; *trigonodus*, 96, 97
Penis, 300, 301-2, 331
Pentadactyly, 5, 42, 170, 360
Perissodactyla, 175, 258
Periconodon, 106, 335
Perodicticus, 69
   brain, 247
   caecum, 294
   colon, 294
   dentition, 95
   digits, 188
   reproductive organ, 301
Periconodon, 106, 335
Permian period, 37
Pes, 181-2
   axis of, 206-7, *207*, 219-23, *220*
   bones of, 182
   mechanism of Primate, 193-4
   of *Adapis*, 195
   of *Cebus*, *222*
   of *Cercopithecus*, *222*
   of chimpanzee, *207*, 221, *221*, 223
   of gibbon, *188*, 223
   of gorilla, *212*, *220*, *221*, 221, 223
   of Hominidae, 223-4
   of *Homo*, *222*, 224
   of *Homo sapiens*, 193, *195*, *207*,
      219, 220, *221*, 221-3, *222*
   of *Lemur*, *190*
   of lemurs, 66, 182, 184, 187, *188*,
      191-5, *207*, 331
   of *Macaca*, *188*, *207*
   of marmoset, *203*, 206, 207
   of monkeys, 206-7, 219-21, *220*,
      221, 223,
   of *Notharctus*, *190*, 194, *222*
   of orang-utan, *212*, 223

Pes (*cont.*)
of Pongidae, 219-23
of *Ptilocercus*, 181-2, *182*
of *Tarsius*, *188*, *195*, 198-9, 332, 338
Petrous bone, 128, 134, 137, 138, 141, 148, 167
Phenacolemur, 74, 104n
Phenacolemuridae, 74, 104n
Photoreceptors, 272, 273, 274, 275
Phylogenetic relationships, assessing, 13, 16, 20, 23, 28-9, 30, 232, 307
  of Anthropoidea, 151, 332-4, 338-40
  of *Daubentonia*, 92, 99-100, *100*
  of Hominidae, 123-4, 159, 212-15, 298
  of lemurs, 108-9, 137-40, 143-4, 324-5, 327, 329, 330
  of monkeys, 33, 203-4, 307, 338-40
  of Plesiadapidae, 73, 86-8, 182-3, 319, 326
  of Pongidae, 116, 159, 212-15, 298, 338-40
  of Primates, 3, 4, 10, 11, 23, 27-8, 40-1, 43, 86, 140, 193, 232, 286, 315-16
  of Tarsioidea, 108-9, 253, 316-17, 332-4, 340
  of Tupaioidea, 86-8, 182-3
  sequence, 28-9, 40
Phylogeny, 3, 9, 30, 32, 33, 140, 143, 286
Picrodontidae, 75n
Pithecia, 62, 63
Pithecometra thesis, 15
Placenta, evidence of the, 307-14
  deciduate, 310, 311, 345
  endotheliochorial, 309, 312, 320
  epitheliochorial, 309, 310, 311, 320, 331
  haemochorial, 310, 311, 312, 313, 320, 334, 338, 345
  non-deciduate, 310, 331
  of Anthropoidea, 311, 312-13, 345
  of lemurs, 310-11, 320, 331
  of monkeys, 313
  of Pongidae, 313
  of *Tarsius*, 311-12, 313, 334, 338
  of *Tupaia*, 312, 320
  syndesmochorial, 309

Placentalia, 37, 38
Platyrrhine monkeys, *see* Ceboidea
Plesiadapidae, 72, 73, 241, 326, 330, 360
  dentition, 73, 86-92, 99-100, 109, 319, 326, 330
  phylogenetic relationship, 73, 86 8, 182-3, 319, 326
  taxonomic status, 87-8
Plesiadapis, 73
  brain, 241
  dentition, *91*, 91, 92
  limbs, *183*, 187
  skull, *141*, 141
Plesiadapis anceps, 91, *91*; *gidleyi*, 92
Plesianthropus, 48, 349-50
Plica fimbriata, 289
Pliocene period, 36
Pliopithecus (*Limnopithecus*), 54, 120, 339, 341, 342,
Pollex, 42
  atrophy, 32, 50, 200-1, 209, 211
  of *Ateles*, 62, 200, 204
  of Anthropoidea, 48, 344
  of Ceboidea, 204
  of Cercopithecoidea, 204, 209
  of *Colobus*, 60, 204
  of gorilla, 52, 209
  of *Homo sapiens*, 44, 212
  of lemurs, 66, 187-8, 189, 191
  of *Macaca*, 204
  of Pongidae, 50, 204, 209
  of *Tarsius*, 197, 198
  of Tupaioidea, 178-9, 181
  opposability of, 178, 187-8, 198, 204, 212
  opposition of the, 178-9, 204
Polyak, Dr S., 274
Pondaungia, 53n, 116, 123
Pongidae, 30-2, 49-57
  affinities of, 339-40
  alimentary tract, 293
  and Hominidae, 19, 49, 121-3, 207-8, 211-12, 217, 220n, 262-3, 298, 345, *346*, 347, 360
  brachiation, 49-50, 51, 52, 54, 200-1, 208, 209-14, 215, 223, 224, 226, 323, 343, 347-8
  brain, 6-7, 49, 231, 262-3, 347, 360
  caecum, 295
  classification of, 19, 49, 345

Pongidae (*cont.*)
  dentition, 49, 84, 114-21, 122,
    123, 124, 125, *154*, 162-3, 341-
    3, 346, 347
  digits, 219
  ear, 282
  evolution of, 22-3, 25, 26, 27, 31,
    49, 53n, 159, 163, 207-8, 217,
    338-45, 348, 360
  eye, 274
  femur, 217-18
  hallux, 219, 220
  ilium, 205n, 215-16, 217
  limbs, 21, 49, 207-26, 346, 347-8
  pes, 219-23
  phylogenetic relationships, 116,
    159, 212-15, 298, 338-40
  placenta, 313
  pollex, 50, 204, 209
  posture, 49, 208, 215, 223
  reproductive organs, 304-5, 306,
    307
  skull, 49, 155, 166-7, *354*, 354-5
  tongue, 289
Ponginae, 49
Pongo, 49
Pons, 235
Post-natal life, 42, 161, 212
Posture, erect, 14, 21, 22, *55*, 158,
    164, 166, 167n, 205, 208, 215,
    216-17, 223-26, 351
  of lemurs, 68
  of monkeys, 208
  of Pongidae, 49, 208, 215, 223
  orthograde, 49, 126, 127, 205,
    208, 215, 223, 348, 351
  pronograde, 57, 142, 208
  semi-erect, 49, 215
Potassium-argon method, 35, 350
Potto, 136, 191
Precipitin reaction, 18-19, 340
Prehensile functions, 63, 170, 171,
    175, 187, 192, 194, 197, 198,
    200-1, 206, 207, 215, 219, 360
Premaxilla, of *Daubentonia*, 136,
    138
  of lemurs, 136, *141*, 142, 325,
    329, 330
  of monkeys, 156
  of Pongidae, 161
  of Tupaioidea, 128, *129, 133*, 134
Premolar teeth, eutherian, 76, 80
  of Hominidae, 122, 350

Premolar teeth (*cont.*)
  of lemurs, 95-6, 325, 326, 327,
    330
  of monkeys, 112, 113, 340, 344
  of Plesiadapidae, 89, *90*, 91, *91*,
    92
  of Pongidae, 116, 117, 121, 124,
    125, 344
  of Primates, 84-5
  of Tarsioidea, 101-2, 104-6, *105*,
    107, 107, 108, 109, 333, 335,
    336, 337
  of Tupaioidea, 88, *90*, 91
  sectorial, 113, 116, 123, 124
Presbytis, 59, 61, *118 153* 213
Pressure pads, 42, 265, 276
  palmar, 178, *179*, 182, 187, 197,
    202
  plantar, 182, 193, 206
Primates, ancestral type, 2, 4, 9, 10,
    20, 21, 24, 25, 26, 28-30, 32,
    38, 40, 45, 46, 47, 139-40, 259,
    271, 294, 316, 324, 325, 326,
    345
  brain, 6-7, 42, 126, 227-32, 264,
    272, 323, 360
  classification, 18, 40-7, 73-5,
    316-17
  dentition, 5, 42, 83-6, *87*, 360
  emergence of, 323
  evolutionary radiations of, 74,
    286, 315-60 : trends, 41-2, 45,
    46, 47, 48-9, 75, 86, 315, 316,
    322-3
  evolution of, 1-36, 38-40, 322-3
  in space and time, 37-75
  order of, 38, 39, 41-2, 43
  pes, 181-2, 193-5
  phylogeny, 3, 4, 10, 11, 23, 27-8,
    40-1, 43, 86, 140, 193, 232,
    286, 315-16
  skull, 126-8
  vision, 265, 272-80, 323
Proboscis monkey, *see Nasalis*
Progalago, 70, 339
  brain, 248, 328
  dentition, 93, 97, 328
  skull, 137, 140, 144, 324, 328
Prognathism, 121, 164
Progression, mode of, 51, 52, 59,
    171, 193-4, 195, 197, 200-1,
    205, 208, 215, 343
Pronothodectes, 89

Pronycticebus, 68, 142
  dentition, 69, 95, 96, 97, 110, 327
  skull, 69, 97, 142
Pronycticebus gaudryi, 142
Propithecus, 66, 184, *185*, 191
Propliopithecus, 53, 342
  dentition, 53, 117-18, 120, 123, 341
Prosimii, 43, 316
Protoadapis, 70
Protocone, 79, 81, *85*, *98*, 327
Protoconid, 79, 82, 86, 87, 89, *98*, 102
Protoconule, 79, 81, *98*
Prototheria, 38, 39
Pseudohypocone, 86, 96, *98*, 327
Pseudoloris, 65, 336
  dentition, *85*, 102, 108, 333, 338
  nose, 272
  skull, 149, 275, 333
Ptilocercinae, 71, 89, 281
Ptilocercus, 72
  alimentary tract, *290*, *291*, *293*, 293
  brain, 235, 241
  dentition, 88, 89, *90*, 91, 317-18
  ear, 281
  eye, 273
  hallux, 181
  limbs, *174*, 175-84, *176*, *179*, *180*, *183*
  liver, 297
  manus, *177*, 178-9, *179*, 189
  nose, 269, *269*
  pes, 181-2, *182*
  reproductive organs, 299-300, *300*, *301*, 301, *303*, 303-4, *305*, 305
  skeleton, *174*
  skull, 128-35, *129*, *130*, *133*, 141, *146*
  tail, 72
  tongue, *289*
Pubis, *174*, *176*, 179-80, 191, 198, 204
Pulvinar, 252, 258
Purgatoriinae, 75n

Quadrupedalism, oblique, 52, 215
  of baboons, 58-9, *59*, 200
  of monkeys, 57, 200, 205n, 206, 208, 213, 215
  of Pongidae, 214, 215, 216, 223

Quadrupedalism (*cont.*)
  secondary, 51
  semi-erect, 215
  structural adaptations of, 171, 208, 213, 223-4
  terrestrial, 59, 171

Radioactive carbon, 34, 353
Radiometric method of dating, 36
Radius, 170, 171, *174*, *176*, 178
  of lemurs, 184, *186*
  of monkeys, 202
  of *Notharctus*, *186*, 187
  of Pongidae, 209
  of *Ptilocercus*, *176*, 177, 178
  of *Tarsius*, 197
Ramapithecus, 56, 342
  dentition, 56, 121, 125, 342
Ramus, 130, 141, 161
Recent types, 10
Reproductive organs, female, 303-7; male, 299-303, 345
  system, evidence of the, 299-314
Rete mirabile, 195
Retina, 272, 273, 320
  of lemurs 273-4 331
  of monkeys, 275, 276-7, 344
  of *Tarsius*, 250, *274*, 274-5, 332, 334, 338
  of Tupaioidea, *266*, 266, 268, 318,
  types of : cone, 273; rod, 274, 275, 331
Rhinarium, 47, *266*, 266, *267*, 267, 268
  of Anthropoidea, *266*, 267-8, 344
  of lemurs, 66, *266*, 266, 331
  of monkeys, *268*, 268
  of *Tarsius*, *266*, 267, *267*, 268, 332, 334, 338
Rhinopithecus, 59
Rhodesian man, 346
Rods, 273

Saban, R., 134n, 146
Sacrum, *174*, 179, 217
Sagittal crest, 134, 142, 143, 144, 157-8, 160-1, 166, 349
Saimiri, 61, 63, 106, 292, 294, 304
Saki monkeys, 62
Scapula, *174*, 175, *176*, 201
  of *Homo sapiens*, 211
  of lemurs, 184
  of monkeys, 201

Scapula (*cont.*)
of Pongidae, 208
of *Ptilocercus*, 175, *176*
of *Tarsius*, 197, 201
Secondary
retrogression, 6-7, 124, 173,
175n, 188, 252, 280, 281, 358
simplification, 7, 104, 124, 140,
147, 203, 275, 324, 327, 347
Serology, 18-19, 321, 340
Sexual cycles, 306-7
Shoulder
girdle, 170, 175
joint, 170, 171, 208
Simian shelf, *116*, 163
Simons, E. L., 114
Simopithecus, 60, 114
Simpson, G. G., 3, 10, 17, 27, 28,
37n, 43, 83, 89, 104n, 321
Sivapithecus, 56
Skull, evidence of the, 126-69
mastoid region of, *130*, 134, 140,
166
measurement of, 155-6, 163-9
of Adapidae, 136, 137, 140, *142*,
142, 326
of *Adapis*, 140, 141, 143
of *Aegyptopithecus*, 53-4, 163
of *Alouatta*, 62, 156
of Anthropoidea, 148, 149, 154,
155, 157, 344
of *Archaeolemur*, 70, 141, 144,
144, 329
of *Australopithecus*, 156, 164,
165, 166-7, *167*, 167n, 168-9,
348, 349, 359
of baboon, *154*, 154, 155, 156,
158
of *Caenopithecus*, 149, 336
of Ceboidea, 57, 143, 152-4, 157-
8, 344
of *Cebus*, *139*, 152, *153*, 154, 156
of Cercopithecoidea, 57, 148,
154-8, 159
of Colobinae, 154, 157
of *Daubentonia*, *100*, 136, 137,
248, 329, 330
of *Galago*, 136, *136*, 149, 328
of gibbon, 156, *159*, 159-60, 162
of gorilla, 160-1, *165*, *167*, *354*
of Hominidae, 138, 158-9, 166-7,
168, 348-9, 354-5
of Hominoidea, 138, 158-69

Skull (*cont.*)
of *Homo erectus*, 12-13, 156, *168*,
168-9, *354*
of *Homo sapiens*, 12, 158, 161n,
169, *355*
of lemurs, 67, 127, *127*, 130, *131*,
135-47, *139*, 318, 324, 330
of marmoset, 151, *152*
of *Megaladapis*, 144-5, *145*
of *Mesopithecus*, 60, 157, *158*
of *Microcebus*, 137n, 146, *146*
of monkeys, 152-8, 340, 344
of Mousterian man, 353, *354-5*,
356
of Neanderthal man, 354-5, *355*,
356
of *Necrolemur*, 140, 149-52, *150*
of orang-utan, *117*, 161, *162*
of *Oreopithecus*, 56, 121, 163
of Pongidae, 49, 155, 166-7,
354-5
of *Presbytis*, 153
of Primates, 126-8
of *Progalago*, 137, 140, 144, 324,
328
of *Pronycticebus*, 69, 97, 142
of *Pseudoloris*, 149, 275, 333
of *Ptilocercus*, 128-35, *129*, 130,
133, 141, *146*
of *Stehlinella*, 100, *100*
of Tarsioidea, 143, 147-52, 331,
333, 335-6, 338
of *Tarsius*, *131*, 135, *139*, *147*,
147-9, *150*, 151, 152, 157,
331-2, 333, 338
of *Tetonius*, 107, 149, *150*, 275
of Tupaioidea, 128-35, *139*, 146,
318, 319
size of, 152-6, 162-9
Steinheim, *354*, 358, 359
Swanscombe, 351, 358-9
Small intestine, *291*, 291-2, *292*
Smell, sense of, 42, 126, 265, 266-
72
Smilodectes, 70, 142
Snout, 42
of lemurs, 66, 69, 135, 143
of monkeys, 155, 157, 158, 344
of Tarsioidea, 135, 148, 149, *150*,
268, 332, 333, 336
of Tupaioidea, 128
recession of s. region, 126, 268,
270, 332, 336, 338, 344

Special senses, evidence of the, 265-85
Species, 40
Sphenoidal fissure, *130*, 131
Spider monkeys, *see Ateles*
Squamosal, 128, *129*, *152*, 154, *165*, 167
Squirrel monkeys, *see Saimiri*
Stapes, 135, 283
Stehlinella, 100
Steinheim skull, *354*, 358, 359
Sternum, *174*, 175
Stomach, 59, *290*, 290-1
Straus, W. L., 298
Styles, 81, 104
Stylids, 83
Subfamilies, 41
Sublingua, *287*, 287, *288*, 288, 289, 318, 331, 332, 338
Suborders, 41, 43, 316
Sugrivapithecus, 56
Sulcus, 229, 229n, 245, 247, 258
   arcuate, *246*, *257*, 257
   axial, 230, 331
   calcarine, *236*, 239, 241, 242, 247, 251, *255*, 255, 258
   centralis, *246*, 246, 247, 248, 256-7, *257*, *261*, 344
   inferior temporal, *257*
   intraparietal, 245, *246*, 248, *257*, 257
   lateral, 230, 242, *246*, 251, *255*, 255, *257*, 258
   limiting, 230, 247, 258, *261*, 331, 344
   lunatus, *246*, *257*, 257, 344
   parallel, *246*, *255*, 255, *257*, 258
   post-Sylvian, *247*, 247
   pseudosylvian, 229, *229*, 247, *247*, 251
   rectus, 245, *246*, *257*
   retrocalcarine, 241, 242, 247, 255
   rhinal, 233, *234*, *236*, *238*, 239, 241, 242, 247, 254
   simian, 7, 257, 260
   supra-sylvian, *247*, 247
   sylvian, *255*, 257
Superfamily, 43
Supinator crest, 177, 197, 201
Supracondylar ridge, 117, 184, 186, 197
Supra-orbital
   height index, 167

Supra-orbital *(cont.)*
   ridges, 9, 22, *165*, 169, 355, 356, 358, 359
   torus, 160, 161, 166
Supratragus, 282
Swanscombe skull, 351, 358-9
Sylvian fissure, *see* Sulcus lateral fossa, 239, 251
Symmetrodonta, 37, 77, 78
Symphalangus, 49n, *118*, *159*
Symphysis menti, 130, 141, 148-9, 157, 163, 167, 336

Tactile organs, 173, 227, 283-5
Tail, 20-1, 49, 54, 57, 59, 61, 62, 67, 325
Talonid, *79*, 82, 86, *87*, 96, 102, 104, 108, 112, 116, 117, 121
Talus, 181, *182*, *194*, 223
Tarsifulcrumation, 193-4, 199, 206
Tarsiiformes, 63, 144, 147, 150, 317, 333, 337-8
Tarsioidea, 43, 45, 63-6
   affinities of, 143, 151, 331-8
   brain, 249-54, 330, 331, 332, 333, 338
   classification, 65, 144, 312, 316-7, 332-4, 337
   dentition, 84, 101-10, 143, 149, 151, 331, 332, 333, 353, 337-8
   evolution, 108-9, 151-2, 334-8
   limbs, 197-200, 331, 332, 338
   phylogeny, 108-9, 253, 316-17, 332-4, 340
   skull, 143, 147-52, 331, 333, 335-6, 338
   specialization of, 63, 333-4, 336-7
Tarsipes, 175
Tarsius, *44*, 63-4, *64*, 331
   aberrancy of, 147
   alimentary tract, 292, *293*, 293, 294, 338
   brain, 239, 240, *244*, 249-54, *249*, *250*, *251*, *253*, 256, 331, 332, 333, 338
   caecum, 295
   carpus, *189*, 197-8, 204, 338
   dentition, *85*, *87*, 101-2, *103*, *104*, *105*, 105, 108, 333, 338
   ear, 64, *267*, *281*, 282, 283, 332, 338
   eyes, 64, 250, *274*, 274-5, 277, 278, 331, 333, 338

Tarsius (cont.)
 face, *267*
 limbs, 64, *192, 195, 196,* 197-200,
  199, 331, 338
 liver, 297
 manus, *188, 189,* 197-8, 331
 nails, 174, 198
 nose, *266, 267,* 267-8, 271, *272,*
  272, 332, 334, 338
 pes, *188, 195,* 198-9, 331
 placenta, 311-12, 313, 334, 338
 pollex, 198
 reproductive organs, 302, 303,
  304, 305, 306, 307, 320, 334,
  338
 skeleton *196*
 skull, *131,* 135, *139, 147,* 147-9,
  150, 151, 152, 157, 331-2, 333,
  338
 tail, 64
 taxonomic status, 63-4, 311-12,
  316-17, 332-3
 tongue, *287,* 288-9, 332, 338
 vibrissae, *267,* 284
Tarsius spectrum, 64, *103, 104, 139,*
  147, 250, 251
Tarsus *174,* 181-2, 194, *195,* 198,
 200, 206-7, 223, 331, 344
Tasmanian wolf, 12, 17
Taxonomy and phylogeny, 32-3,
 40, 316-17, 340
 relevance, 11-12, 13, 26, 28, 329
 status, of *Australopithecus,* 27-9,
  48, 262, 347, 348-50
 of *Caenopithecus,* 110
 of *Carpolestidae,* 106-7, 335
 of *Daubentonia,* 69, 99-101
 of lemurs, 66
 of *Parapithecus,* 117-18, 341
 of Plesiadapidae, 87-8
 of *Tarsius,* 63-4, 311-12, 316-17,
  332-3
 of Tupaioidea, 182-3, 319, 321-2
Tee-tee monkeys, *see Callicebus*
Teilhardina, 65, 332
 dentition, 102
 limbs, *195, 199,* 332
Telanthropus, 349-50
Temporal bone, 128, *129,* 129, 134,
 144-5, 161
Tenrec, *see Centetes*
Terminology, 30-4
Tertiary epoch, 35, 35n, 36

Testes, 300-1
Tetonius, 63, 107n
 brain, 253, 333
 dentition, *104,* 106, 107, 107n,
  335, 337
 nose, 272
 skull, 107, 149, *150,* 275
Thalamus, 234-5, *238,* 240, 241,
 243-4, 251, 258
Theory of multituberculy, 82
Therapsida, 37, 39, 77
Thylacinus, 12
Tibia, 170, 171, *174, 180,* 181, 191,
 198, 205
Tibio-fibula, 171, 198, *199,* 200,
 332, 338
Tongue, evidence of the, 286-9
 functions of, 286
 of lemurs, 286, 326, 331
 of *Tarsius, 287,* 288-9, 332, 338
Tools, use of, 351, 356-7
Tragus, *181,* 182
Tree-shrews, *see* Tupaioidea
Triconodonta, 37, 77
Trigone, *79,* 81
Trigonid, *79,* 82, 86, *87,* 89, 96,
 104, 108, 112, 117
Trituberculy, 5-6, 81-2, 112
Tupaia, 71, 321
 alimentary tract, 293, *293*
 brain, 235-7, *236, 237, 238,* 239,
  239-41, 242, 243, 320
 dentition, 72, *85*
 ear, 281, *281,* 283, 318
 eye, 273-4
 pes, 181, 182
 placenta, 312, 320
 reproductive organs, 301, 307
 skull, 135, 319
 tail, 72, 325
Tupaia glis, 72, 318; *minor,* 72, 236,
  238; *javanica,* 237
Tupaiinae, 71, 134, 135, 138, 288
Tupaioidea, 43, 45, 46, 71-5, *72,* 89
 affinities of, 71, 74-5, 236, 240-1,
  317-23
 brain, 71, 230, 235-41, 243, 317,
  318, 319, 320
 caecum, 294
 classification of, 43n, 71, 74-5,
  145-6, 183, 317, 321-2
 claws, 71, 173, 175, 179, 183
 dentition, 86-92, 317-18, 325

Tupaioidea (*cont.*)
  eye, 71, 273-4, 317, 318
  gestation, 314
  hallux, 181, 318
  limbs, 175-84, 318
  liver, 297
  myology of, 183-4, 318
  nose, 266, *266*, 268, *269*, 269,
    270, 318
  os ectotympanicum, 134, 138,
    324
  pes, 181, *182*
  phylogeny, 86-8, 182-3
  placenta, 312, 318, 320
  pollex, 178-9, 181
  reproductive organs, 299-300,
    303, 305, 307
  skull, 128-35, *139*, 146, 318, 319
  taxonomic status, 182-3, 319,
    321-2
  tongue, 287, 288, *289*, 318
  vibrissae, 284
Turbinal system, 268-72, *269*, *270*,
  271, 318, 345
Tympanic bulla, 137, *138*, 138, 139,
  140, 141, 144, 148, 149, 150,
  152, 324, 330, 336, 338, 344
  cavity, 128, 129, 135, 137, *138*,
    138, 145, 148, 157, 283
  ring, *see* os ectotympanicum

Uintanius, 106, 107
Ulna, 170, 171, *174*, *176*, 177, 178,
  184, *186*, 187, 197, 202, 209
Uranium-lead, 34-5
Urogale, 88, 135
Uterus, 305-6, 338

Vertebrae, 130, *174*
Vibrissae, 178, *179*, 283-5
Vision, 126, 227
  binocular, 42
  colour, 277, 279, 318
  neural apparatus of, 227, 265, 272
  of man, 276-7, 280, 360
  of Primates, 265, 272-80, 323,
    360
  photopic, 273
  scotopic, 273
  stereoscopic, 276, 277
  twilight, 273
Von Baer, K. E., 20, 22

Waisting of molars, 113
Watson, D. M. S., 18

Zinjanthropus, 349, 350
Zygomatic arch, 126, 129, *129*,
  130, 132, 134, 141, 144, 161
  bone, 126, 129, *130*, *133*, 136,
    148, *152*, 154, 157, 161, 167, 336

A Note on the Author

SIR WILFRID E. LE GROS CLARK is Professor of Anatomy at Oxford University and a Fellow of the Royal Society, the Royal College of Surgeons, and the Royal Anthropological Institute. A recipient of the Viking Fund Medal of the Wenner-Gren Foundation for Anthropological Research, Professor Clark has also written *Early Forerunners of Man* (1934), *Morphological Aspects of the Hypothalamus* (1936), *The Tissues of the Body* (1939), *History of the Primates* (1949), and *The Fossil Evidence for Human Evolution* (1955).